Handbook of Enzyme Biotechnology

Handbook of Enzyme Biotechnology

Edited by Caden Hamilton

Syrawood Publishing House
New York

Published by Syrawood Publishing House,
750 Third Avenue, 9th Floor,
New York, NY 10017, USA
www.syrawoodpublishinghouse.com

Handbook of Enzyme Biotechnology
Edited by Caden Hamilton

International Standard Book Number: 978-1-64740-388-1 (Hardback)

Cataloging-in-publication Data

Handbook of enzyme biotechnology / edited by Caden Hamilton.
p. cm.
Includes bibliographical references and index.
ISBN 978-1-64740-388-1
1. Enzymes. 2. Enzymes--Biotechnology. 3. Enzymology. I. Hamilton, Caden.
QP601 .E59 2023
572.7--dc23

TABLE OF CONTENTS

PREFACE

Enzymes are proteins that act as biocatalysts and help increase the rate of biochemical reactions in living organisms. These are important for biological processes such as digestion and liver function. An excess or lack of a certain enzyme can lead to health problems. Enzymes are also used in multiple biotechnological industrial processes after being extracted from cells. They are used extensively in industries such as food processing industry, pharmaceutical industry, dairy industry, biofuel industry, and brewing industry. Proteases, amylases and lipases are commonly used enzymes in biological washing powders and various cleaning products. In food processing industry, amylase enzyme is used in the production of sweetening agents. Enzymes also play a fundamental role in analytical devices and assays that have forensic, clinical and environmental applications. This book includes some of the vital pieces of works being conducted across the world, on various topics related to enzyme biotechnology. It is a valuable compilation of topics, ranging from the basic to the most complex advancements in this field. The book will serve as a valuable source of reference for graduate and postgraduate students.

This book is the end result of constructive efforts and intensive research done by experts in this field. The aim of this book is to enlighten the readers with recent information in this area of research. The information provided in this profound book would serve as a valuable reference to students and researchers in this field.

At the end, I would like to thank all the authors for devoting their precious time and providing their valuable contribution to this book. I would also like to express my gratitude to my fellow colleagues who encouraged me throughout the process.

Editor

1

Preparation and Applications of Amylose Supramolecules by Means of Phosphorylase-Catalyzed Enzymatic Polymerization

Jun-ichi Kadokawa

Graduate School of Science and Engineering, Kagoshima University, 1-21-40 Korimoto, Kagoshima 890-0065, Japan; E-Mail: kadokawa@eng.kagoshima-u.ac.jp

---

Abstract: This paper reviews preparation and applications of amylose supramolecules by means of phosphorylase-catalyzed enzymatic polymerization. When the enzymatic polymerization of α-D-glucose 1-phosphate (G-1-P) as a monomer was carried out in the presence of poly(tetrahydrofuran) (PTHF) of a hydrophobic polyether as a guest polymer, the supramolecule, *i.e.*, an amylose-PTHF inclusion complex, was formed in the process of polymerization. Because the representation of propagation in the polymerization is similar to the way that vines of plants grow twining around rods, this polymerization method for the preparation of amylose-polymer inclusion complexes was proposed to be named "vine-twining polymerization". Various hydrophobic polyethers, polyesters, poly(ester-ether), and polycarbonates were also employed as the guest polymer in the vine-twining polymerization to produce the corresponding inclusion complexes. To obtain the inclusion complex from a strongly hydrophobic guest polymer, the parallel enzymatic polymerization system was developed as an advanced extension of the vine-twining polymerization. In addition, it was found that amylose selectively includes one side of the guest polymer from a mixture of two resemblant guest polymers, as well as a specific range in molecular weights of the guest PTHF. Amylose also exhibited selective inclusion behavior toward stereoisomers of poly(lactide)s. Moreover, the preparation of hydrogels through the formation of inclusion complexes of amylose in vine-twining polymerization was achieved.

Keywords: phosphorylase; enzymatic polymerization; amylose supramolecule; inclusion complex; vine-twining polymerization

---

1. Introduction

Amylose is a linear natural polysaccharide with helical conformation, which is composed of glucose residues linked through α-(1→4)-glycosidic linkages [1]. This is one component of starch and acts as an energy resource in nature with the other component of starch, *i.e.*, amylopectin, which has a branched structure composed of α-(1→4)-glucans inter-linked α-(1→6)-glycosidic linkages (Figure 1) [2]. Besides having such an important role in nature, amylose is known as a high performance material because it acts as a host molecule and forms amylose supramolecules by inclusion of various guest molecules (inclusion complexes) owing to its helical conformation [3]. However, a pure amylose is currently not available from starch because the complete separation of natural amylose from amylopectin is difficult. The only method used for production of the pure amylose with the desired average molecular weight is a phosphorylase-catalyzed enzymatic polymerization [4].

Figure 1. Structures of amylose and amylopectin.

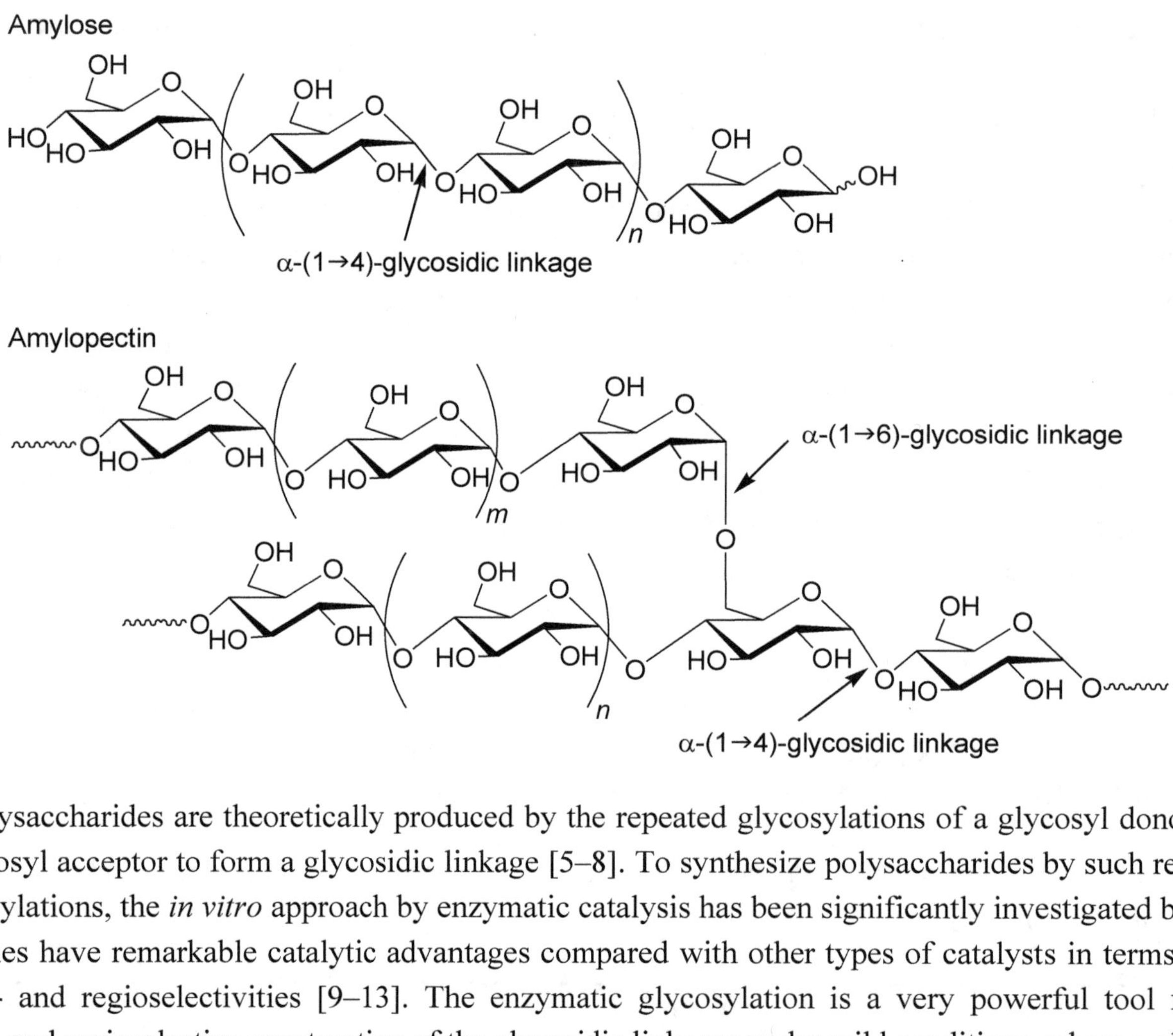

Polysaccharides are theoretically produced by the repeated glycosylations of a glycosyl donor with a glycosyl acceptor to form a glycosidic linkage [5–8]. To synthesize polysaccharides by such repeated glycosylations, the *in vitro* approach by enzymatic catalysis has been significantly investigated because enzymes have remarkable catalytic advantages compared with other types of catalysts in terms of the stereo- and regioselectivities [9–13]. The enzymatic glycosylation is a very powerful tool for the stereo- and regioselective construction of the glycosidic linkages under mild conditions, where a glycosyl donor and a glycosyl acceptor can be employed in their unprotected forms, leading to the direct formation of an unprotected glycoside in aqueous media [14,15]. Thus, repetition of the enzymatic glycosylations, *i.e.*, enzymatic polymerization, forms polysaccharides with well-defined structure.

Phosphorylase is an enzyme that catalyzes the reversible phosphorolytic reaction of α-(1→4)-glucans, such as amylose and glycogen, at a non-reducing end in the presence of inorganic phosphate to produce α-D-glucose 1-phosphate (G-1-P) and those with one smaller degree of polymerization (DP) [16]. Because the reaction by phosphorylase catalysis exhibits reversible nature, α-(1→4)-glycosidic linkage can be constructed by the phosphorylase-catalyzed glycosylation using G-1-P as a glycosyl donor (Figure 2). As a glycosyl acceptor, which is often called a 'primer', maltooligosaccharides with DPs higher than the smallest one recognized by phosphorylase, are used. The smallest glycosyl acceptor for the phosphorylase-catalyzed glycosylation is typically maltotetraose (G_4), whereas that for the phosphorolytic cleavage is typically maltopentaose (G_5). In the glycosylation, a glucose unit is transferred from G-1-P to a non-reducing end of the primer to form α-(1→4)-glycosidic linkage. When the excess molar ratio of G-1-P to the primer is present in the reaction system, the successive glycosylations occur as a propagation of polymerization to produce the α-(1→4)-glucan chain, *i.e.*, amylose [17–22]. Because the phosphorylase-catalyzed polymerization proceeds analogously to a living polymerization, the molecular weight of the produced amylose has a narrow distribution ($M_w/M_n < 1.2$) and can be controlled by the G-1-P/primer feed molar ratios [4].

Figure 2. Phosphorylase-catalyzed enzymatic polymerization.

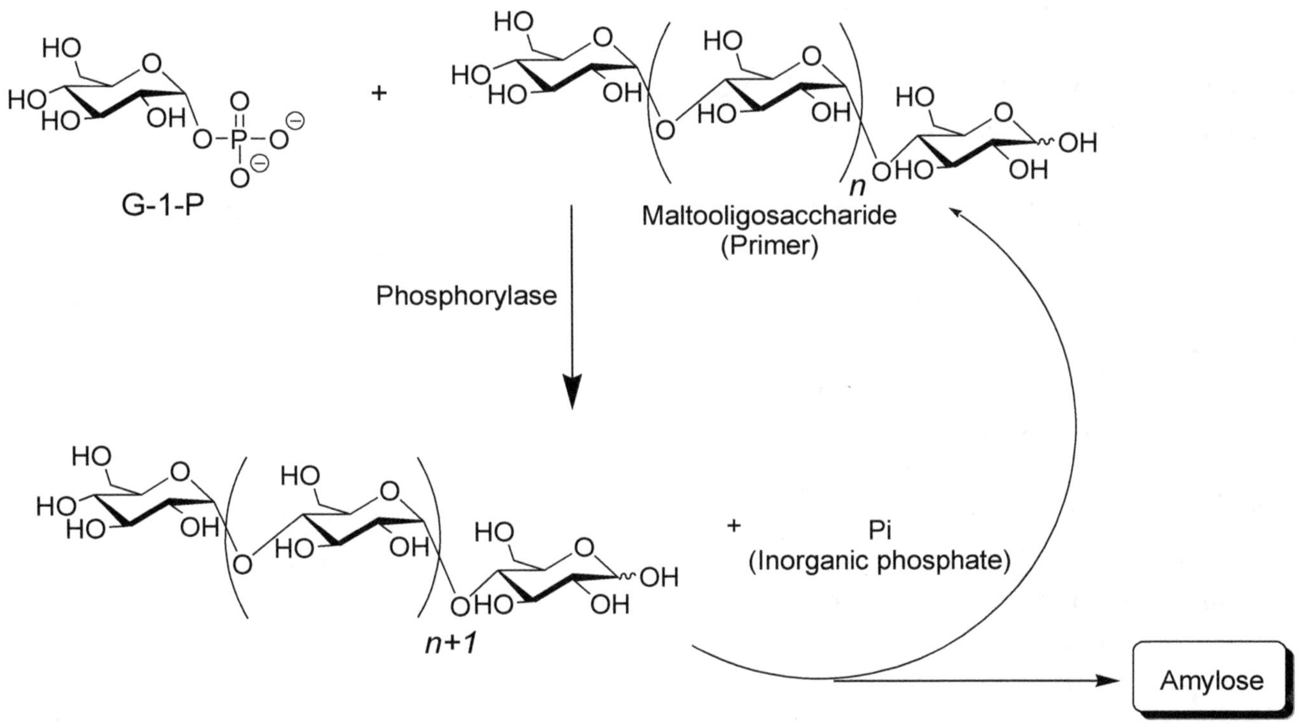

As aforementioned, amylose is the well-known host molecule that forms inclusion complexes with guest molecules having relatively lower molecular weight. The driving force for formation of inclusion complexes is hydrophobic interaction between host and guest molecules because the cavity inside the amylose helix has hydrophobic nature due to the presence of hydrophilic hydroxy groups in the glucose units on the outer side of the helix. Therefore, hydrophobicity is required as the property of guest molecules to be included by amylose. However, only limited studies had been reported regarding the formation of inclusion complexes composed of amylose and polymeric guest molecules [23–29]. Because the driving force incorporating guest molecules into the cavity of amylose is the weak hydrophobic interaction, amylose does not have sufficient ability to include the long chains of polymeric guests into its cavity.

By means of the phosphorylase-catalyzed enzymatic polymerization of G-1-P for direct construction of amylose, the author's group has developed an efficient method for the preparation of inclusion complexes of amylose with synthetic polymers. The representation of propagation in the polymerization is similar as the way that vines of plants grow twining around rods. Accordingly, it has been proposed that this polymerization method for the preparation of amylose-polymer inclusion complexes is named "vine-twining polymerization" (Figure 3) [11–13,30–32]. In this review article, the author describes the principal results, discussion, and applications of vine-twining polymerization.

Figure 3. Image of "vine-twining polymerization".

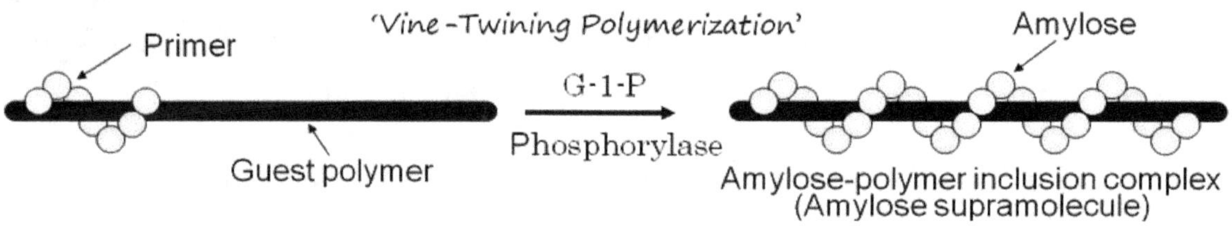

2. Preparation, Characterization, and Formation Process of Amylose-Poly(tetrahydrofuran) Inclusion Complex by Vine-Twining Polymerization

A first example of the vine-twining polymerization was achieved using poly(tetrahydrofuran) (PTHF) as a hydrophobic guest polyether (Figure 4) [33]. When the phosphorylase-catalyzed enzymatic polymerization of G-1-P from maltoheptaose (G_7) was performed in the presence of hydroxy-terminated telechelic PTHF with M_n of 4,000 dispersed in sodium citrate buffer, the product was gradually precipitated as the polymerization progressed, which was collected by filtration and characterized by powder X-ray diffraction (XRD) and ^{1}H NMR measurements.

The XRD profile of the product showed two strong diffraction peaks at 2θ = 12.4 and 19.8° corresponding to d = 0.71 and 0.45 nm, respectively (Figure 5(b)). The XRD pattern of the product was completely different from that of amylose in Figure 5(a), but similar to that of the inclusion complexes of amylose with monomeric guest molecules [34].

The ^{1}H NMR spectrum of the product in NaOD/D_2O (Figure 6) showed signals not only due to amylose but also due to PTHF in spite of washing with methanol, that was a good solvent of PTHF. The original PTHF was insoluble in an alkaline aqueous solution, and thus, no signals due to PTHF appeared in the ^{1}H NMR spectrum of the suspension of PTHF in NaOD/D_2O. These NMR results indicated that the PTHF in the product was solubilized in the alkaline solution, probably because of its inclusion in the cavity of amylose. The XRD and NMR data supported the conclusion that helical inclusion complex was obtained by the vine-twining polymerization.

Figure 4. Formation of amylase-poly(tetrahydrofuran) (PTHF) inclusion complex in phosphorylase-catalyzed polymerization in the presence of PTHF.

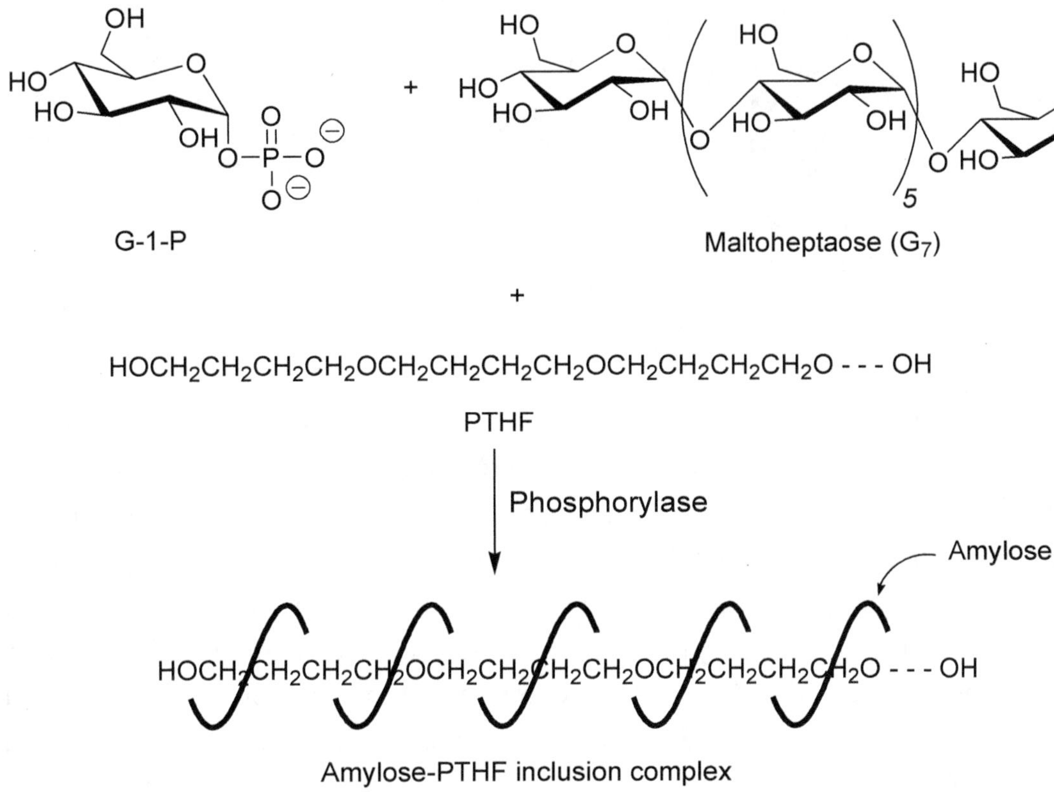

Figure 5. XRD profiles of amylose (**a**), amylose-PTHF inclusion complex (**b**), and amylose- poly(L-lactide) (PLLA) inclusion complex (**c**).

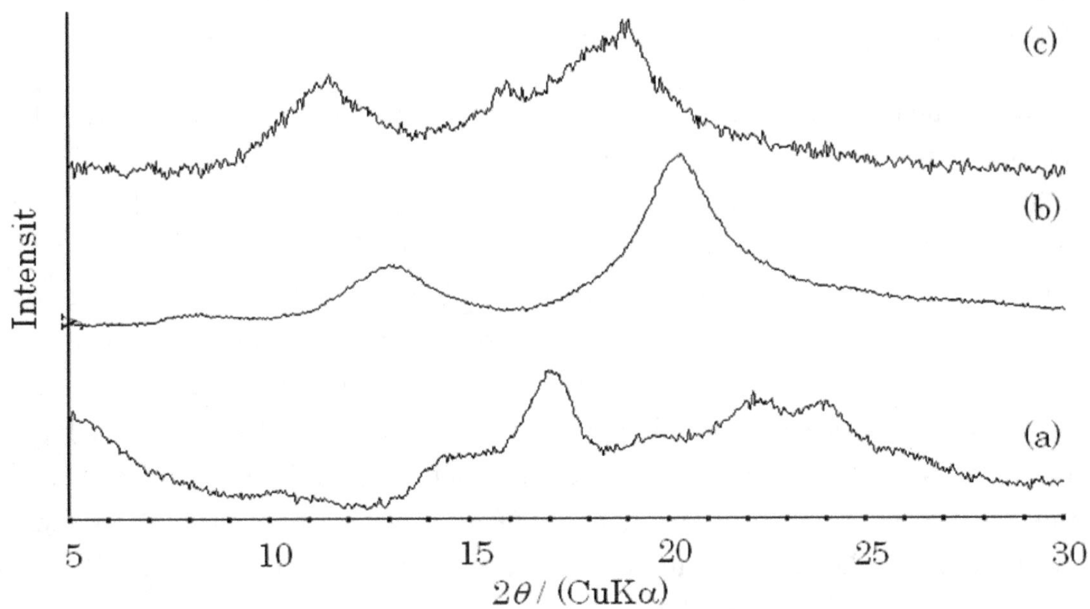

Figure 6. ^{1}H NMR spectrum of amylose-PTHF inclusion complex in $NaOD/D_2O$.

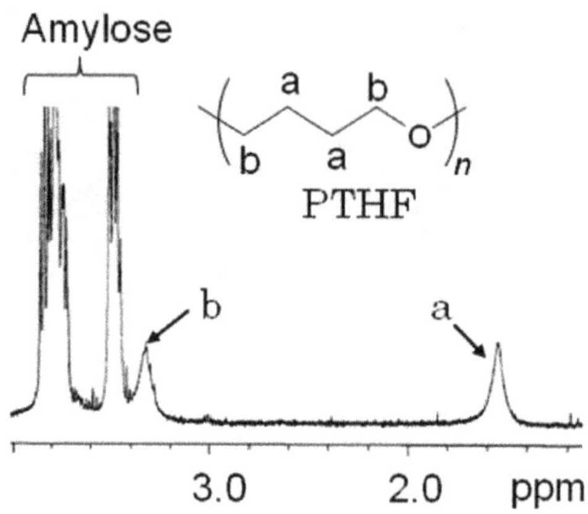

Generally, one helical turn of amylose is composed of *ca.* 6 repeating glucose units when linear molecules with small sectional area are included [35]. The repeat distance of the helix of amylose was reported as 0.795 nm, whereas the length of one unit of PTHF was calculated as *ca.* 0.60 nm. Therefore, 4.5 repeating glucose units in amylose correspond to the length of one PTHF unit. On the basis of these values, a theoretical ratio of the $C–CH_2–CH_2–C$ protons of PTHF to the anomeric H1 protons of the glucose units in the inclusion complex was calculated to be 0.89. Actually, the integrated ratio of the signal due to the $C–CH_2–CH_2–C$ protons to the anomeric H1 protons in the ^{1}H NMR spectrum of the product was in good agreement with the calculated value. This result also supported the structure of the inclusion complex.

Mixing amylose and PTHF in a buffer solvent did not induce the formation of inclusion complex. This observation strongly suggested that the inclusion complex formed during the enzymatic polymerization. Then, effects of the molecular weight and end group structure in PTHF on the formation of inclusion complexes in the vine-twining polymerization were investigated.

The formation of inclusion complexes using PTHFs with various molecular weights (1,000, 2,000, 10,000, 14,000) by the vine-twining polymerization was performed [36]. When PTHF with M_n of 1,000 or 2,000 was used as the guest polymer, inclusion complexes were obtained by the vine-twining polymerization. In contrast, PTHFs with higher M_ns such as 10,000 and 14,000 did not form inclusion complexes with amylose by the vine-twining polymerization. The PTHFs with higher M_ns were not dispersed well in buffer of the polymerization solvent, resulting in difficulty of inclusion by amylose. To obtain inclusion complexes from these PTHFs, the vine-twining polymerization was conducted in the following two-phase system. The higher molecular weight PTHF was dissolved in diethyl ether and buffer was added to the ether solution (diethyl ether:buffer = 1:5, v/v). Then, the phosphorylase-catalyzed enzymatic polymerization of G-1-P was carried out with vigorously stirring the mixture. Consequently, the XRD profiles of the products from both PTHFs with M_ns of 10,000 and 14,000 by the two-phase system indicated the formation of the inclusion complexes.

The effect of the end groups in the telechelic PTHFs on the formation of inclusion complexes in the vine-twining polymerization was also investigated [36]. The end groups employed were the hydroxy

(–OH, as an aforementioned standard PTHF), methoxy (–OCH_3), ethoxy (–OCH_2CH_3), and benzyloxy (–OCH_2Ph) groups. The products were characterized by the integrated ratios of the C–CH_2–CH_2–C signals of PTHFs to the H1 signals of amylose in the ^{1}H NMR spectra. The ratio in the product from the methoxy-terminated PTHF was close to that from the hydroxy-terminated PTHF as well as the calculated value (0.89), indicating the formation of inclusion complex. In contrast, the ratio in the product from the ethoxy-terminated PTHF was much lower than the calculated value. Moreover, no inclusion complex was formed when the benzyloxy-terminated PTHF was used as the guest polymer. These results indicated that the formation of inclusion complexes was strongly affected by bulkiness of the end groups in PTHFs.

The DPs of the enzymatically synthesized amylose by phosphorylase are generally changed by the feed molar ratios of G-1-P/G_7. The vine-twining polymerization using PTHF in lower feed molar ratio of G-1-P/G_7 such as 10 did not yield the inclusion complex because of the production of amylose with considerably low DP, which had no helical structure. On the other hand, when the polymerization system was performed in higher feed molar ratio of G-1-P/G_7, e.g., 1,000, the DP of amylose was not higher than *ca.* 75. In the present polymerization system, the enzymatic polymerization proceeds with forming the inclusion complex composed of amylose with V-type crystalline structure, which leads to precipitation of the product during the polymerization. The precipitation of the inclusion complex causes a difficulty in increasing the DP of amylose by the vine-twining polymerization even when the higher feed molar ratio of G-1-P/G_7 is employed.

3. Preparation of Inclusion Complexes by Vine-Twining Polymerization Using Other Hydrophobic Guest Polymers

3.1. Preparation of Inclusion Complexes Using Other Polyethers as Guest Polymers

To investigate the effect of alkyl chain lengths in the guest polyethers on the formation of inclusion complexes in the vine-twining polymerization, the phosphorylase-catalyzed enzymatic polymerization of G-1-P was performed in the presence of polyethers with different alkyl chain lengths from PTHF (the number of methylenes = 4), that is, poly(oxetane) (POXT, the number of methylenes = 3) and poly(ethylene glycol) (PEG, the number of methylenes = 2) [36]. The XRD pattern of the product from POXT was same as that of the amylose-PTHF inclusion complex. Furthermore, the ^{1}H NMR spectrum of the product in DMSO-d_6 showed the signals due to both amylose and POXT. These results indicated that the inclusion complex was formed by the vine-twining polymerization using POXT as the guest polyether. On the other hand, the XRD profile of the product from PEG showed the pattern for amylose, but did not show that for inclusion complex. In the ^{1}H NMR spectrum, moreover, the signals only due to amylose were observed. These analytical data indicated that amylose was produced by the enzymatic polymerization in this system, but that did not include PEG in the cavity. This was attributed to the hydrophilic nature of PEG, which caused much less hydrophobic interaction with the cavity of amylose. While the hydrophobic polyethers such as PTHF and POXT exhibited interaction with the cavity of amylose, leading to the formation of inclusion complexes by the vine-twining polymerization. The above results suggested that the hydrophobicity of guest polymers was an important factor for the formation of inclusion complexes in the vine-twining polymerization.

3.2. Preparation of Inclusion Complexes Using Polyesters as Guest Polymers

On the basis of the importance in hydrophobicity of guest polymers on the formation of inclusion complexes, hydrophobic polyesters such as hydroxy-terminated telechelic poly(ε-caprolactone) (PCL) and poly(δ-valerolactone) (PVL) were employed as the guest polymers in the vine-twining polymerization (Figure 7) [37,38]. The phosphorylase-catalyzed enzymatic polymerization of G-1-P from G_7 was performed in the presence of PCL or PVL in sodium citrate buffer. The precipitated products were characterized by the ^{1}H NMR and XRD measurements. The ^{1}H NMR spectra of the products from PCL with M_n of 1,000 and PVL with M_n of 2,000 showed signals not only due to amylose but also due to the polyesters. The XRD patterns of the products were completely different from those of amylose and were similar as those of the amylose-PTHF inclusion complex. The IR measurement was also conducted to confirm the structure of the product. The IR spectrum of the original PVL exhibited a strong absorption at 1,728 cm^{-1}, corresponding to carbonyl group of the crystalline PVL. In contrast, the carbonyl absorption of PVL in the product shifted to the region at 1,736 cm^{-1}, which was assignable to the noncrystalline PVL. This result indicated that the crystalline PVL did not present in the product because of inclusion of a PVL chain in the cavity of amylose, suppressing the formation of crystalline aggregates. This IR observation was in good agreement with the XRD data, in which the XRD profile of a sole PVL showed strong crystalline peaks, whereas no such peaks were detected in the XRD profile of the product. All the above analytical data were taken to support the structures of the inclusion complexes.

Figure 7. Preparation of inclusion complexes by vine-twining polymerization using hydrophobic guest polyesters.

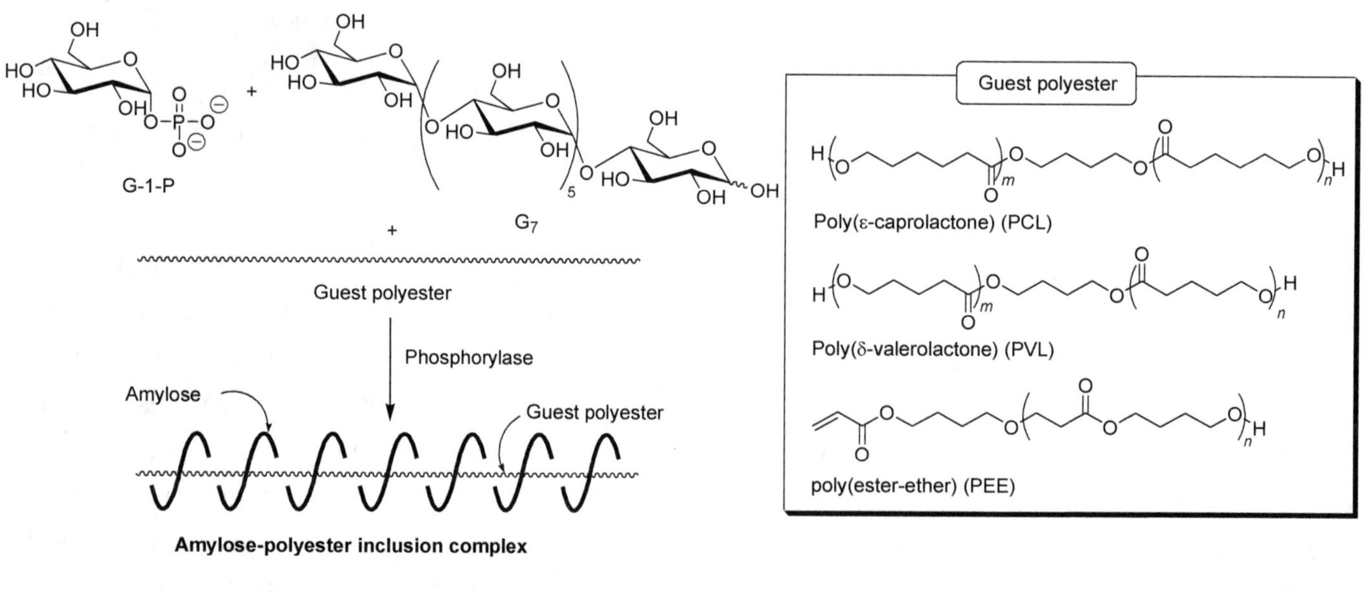

Although PCL with a higher M_n (2,000) was attempted to be used as the guest polyester in the vine-twining polymerization, the PCL was not dispersed well in sodium citrate buffer of the polymerization solvent, and accordingly the inclusion complex was not formed. To obtain the inclusion complex from such PCL, the vine-twining polymerization was conducted in a mixed solvent of citrate buffer and acetone (5:1, v/v) because the PCL was dispersed well in the mixed solvent system. The resulting product was characterized by the ^{1}H NMR and XRD measurements to be the inclusion complex. As aforementioned, the inclusion complex was formed from PVL with M_n of 2,000 by the vine-twining polymerization conducted in sodium citrate buffer. This suggested that PVL was more favorable as the guest polymer to form inclusion complex compared with PCL by the vine-twining polymerization.

As other guest polyester, a hydrophobic poly(ester-ether) ($–CH_2CH_2C(=O)OCH_2CH_2CH_2CH_2O–$), which was composed of ester and ether linkages alternatingly, was employed in the formation of an inclusion complex by the vine-twining polymerization (Figure 7) [38]. The structure of the product was confirmed by the ^{1}H NMR and XRD measurements to be the inclusion complex. When a hydrophilic poly(ester-ether) ($–CH_2CH_2C(=O)OCH_2CH_2O–$), which had a shorter methylene length, was used as the guest polymer, no inclusion complex was formed. This result also supported that the hydrophobicity of the guest polymers strongly affected the formation of inclusion complexes by the vine-twining polymerization.

Optically active polyesters, *i.e.*, poly(lactide)s (PLAs), which included three kinds of the stereoisomers, *i.e.*, poly(L-lactide) (PLLA), poly(D-lactide) (PDLA), and poly(DL-lactide) (PLDLA), were employed as the guest polyester in the vine-twining polymerization (Figure 8) [39]. First, the vine-twining polymerization using PLLA was performed by the phosphorylase-catalyzed enzymatic polymerization of G-1-P from G_7. The precipitated product was characterized by the XRD and ^{1}H NMR measurements.

Figure 8. Stereoselective inclusion of amylose by vine-twining polymerization using poly(lactide)s (PLAs).

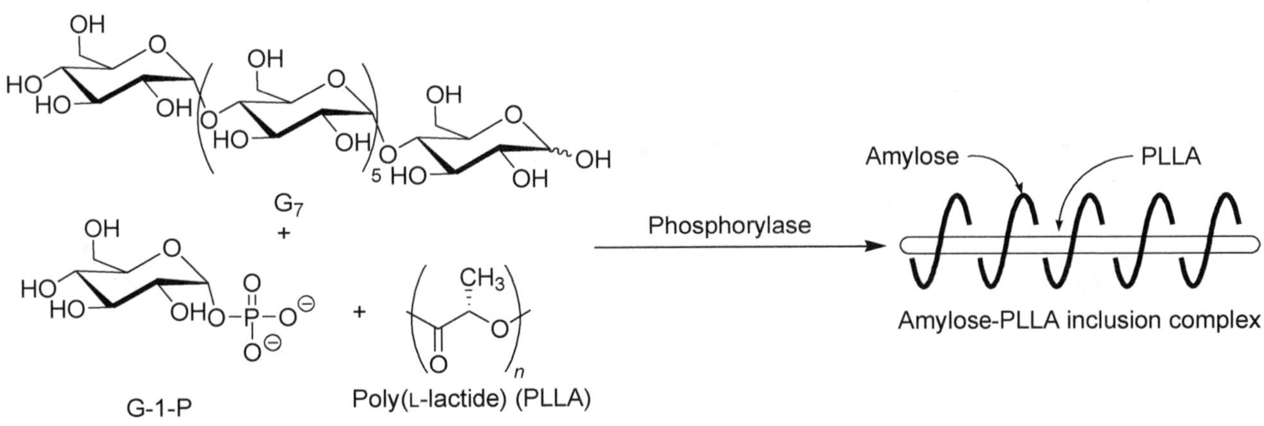

The XRD profile of the product showed two diffraction peaks at 2θ = 11°–12° and 18°–19° (Figure 5(c)), which was completely different from that of amylose, indicating that the inclusion complex structure of the product. Interestingly, the diffraction peaks of the product appeared at lower angles compared with those of the inclusion complexes of amylose with the slender guest polymers such as PTHF (Figure 5(b)) [33]. This difference in the two XRD patterns reasonably revealed that the diameter of amylose helix in the product became larger by inclusion of the bulky PLLA in the cavity of amylose. The similar XRD pattern as that of the present product was obtained when amylose included bulky alcohols [40]. The ^{1}H NMR spectrum of the product in DMSO-d_6 showed the signals not only due to amylose but also due to PLLA. These analytical data fully supported that the amylose-PLLA inclusion complex was formed in the vine-twining polymerization using PLLA as the guest polyester.

To investigate the effect of the chirality in PLAs on inclusion by amylose, the vine-twining polymerization was performed using PDLA and PDLLA. Consequently, the XRD patterns of the products showed only the diffraction peaks due to amylose and the ^{1}H NMR spectra did not show the signals due to PLAs, indicating no formation of the corresponding inclusion complexes. These results indicated that amylose perfectly recognized the chirality in PLAs on the formation of inclusion complexes in vine-twining polymerization.

3.3. Preparation of Inclusion Complexes of Amylose with Strongly Hydrophobic Polyesters in Parallel Enzymatic Polymerization System

As aforementioned, the hydrophobicity of guest polymers is the important factor for the formation of inclusion complexes in the vine-twining polymerization. In addition to no formation of inclusion complexes from hydrophilic polymers, the production of inclusion complexes had not been achieved from polymers with strong hydrophobicity due to their aggregation in aqueous buffer solvent in the vine-twining polymerization as appeared in the case using PCL with the higher M_n. As other example, the vine-twining polymerization using poly(oxepane), which was a polyether having 6 methylene units and was more hydrophobic than PTHF and POXT owing to longer alkyl chains, did not induce the formation of an inclusion complex [36].

Figure 9. Formation of inclusion complexes composed of amylose and strongly hydrophobic polyester in parallel enzymatic polymerization system.

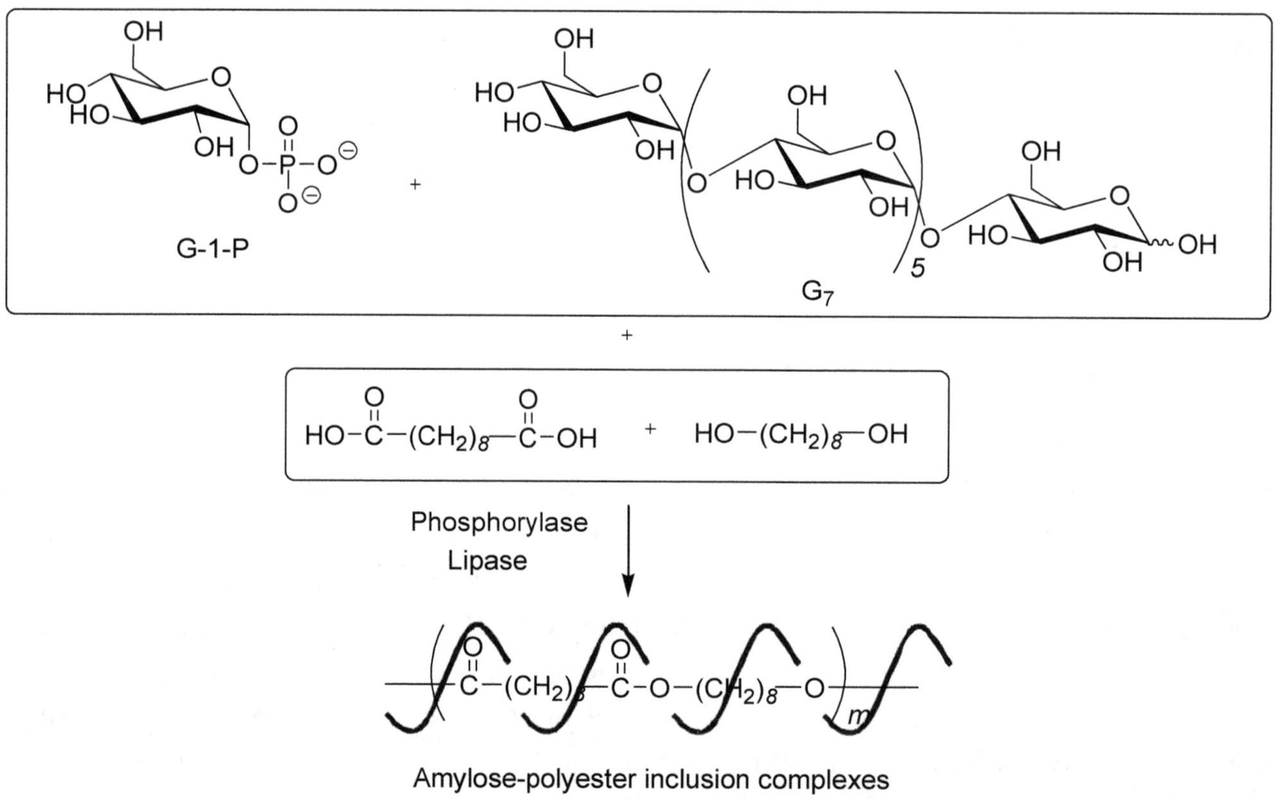

To obtain the inclusion complex from a strongly hydrophobic polyester, a parallel enzymatic polymerization system as an advanced extension of the vine-twining polymerization was investigated (Figure 9) [41]. In this system, two enzymatic polymerizations, which were the phosphorylase-catalyzed enzymatic polymerization of G-1-P from G_7, giving rise to amylose and the lipase-catalyzed polycondensation of a dicarboxylic acid and a diol, leading to an aliphatic polyester of the guest polymer [42,43], were simultaneously performed. As the monomers for the latter polymerization, the dicarboxylic acid and the diol having methylene units of 8 were used, that yielded the strongly hydrophobic polyester by the lipase-catalyzed polycondensation under aqueous conditions. The product by the parallel enzymatic polymerization system was characterized by the ^{1}H NMR and XRD measurements, which indicated the formation of the inclusion complex of amylose with the polyester.

To demonstrate that the inclusion complex could be obtained only by the parallel enzymatic polymerization system, the following two experiments were carried out. When the phosphorylase-catalyzed enzymatic polymerization was performed in the presence of the polyester having methylene units of 8, that accorded to the vine-twining polymerization manner, only amylose was produced, which did not include the polyester. As the other experiment, the lipase-catalyzed enzymatic polycondensation of the dicarboxylic acid and the diol having methylene units of 8 was carried out in the presence of amylose. Consequently, amylose included some monomers, but did not include the polyester although the enzymatic polycondensation progressed. These results of the two experiments indicated that the inclusion complex of amylose with such strongly hydrophobic polyester was formed only by the parallel enzymatic polymerization system.

3.4. Preparation of Inclusion Complexes Using Polycarbonates as Guest Polymers

Hydrophobic aliphatic polycarbonates were employed as the guest polymers in the vine-twining polymerization to give the corresponding inclusion complexes with amylose [44]. The four hydrophobic polycarbonates with different methylene chain lengths were used, which were poly(tetramethylene carbonate) (PC-4), poly(octamethylene carbonate) (PC-8), poly(decamethylene carbonate) (PC-10), and poly(dodecamethylene carbonate) (PC-12). First, PC-4 was employed as the guest polycarbonate in the vine-twining polymerization in an aqueous buffer/acetone mixed solvent. The precipitated product was characterized by means of the ^{1}H NMR, XRD, and IR measurements, which supported the structure of amylose-PC-4 inclusion complex.

The effect of the methylene chain lengths in the polycarbonates was investigated on the formation of inclusion complexes in this polymerization system using PC-4, PC-8, PC-10, and PC-12 as the guest polycarbonates. The ^{1}HNMR spectra of the products by the vine-twining polymerization suggested that less amounts of the polycarbonates having longer methylene chain lengths were included in the cavity of amylose. Such strongly hydrophobic polycarbonates were probably aggregated in the aqueous buffer/acetone mixed solvent, and accordingly separated from the cavity of amylose, causing difficulty in complex formation.

4. Selective Inclusion of Amylose in Vine-Twining Polymerization

4.1. Amylose Selectively Includes One from a Mixture of Two Resemblant Guest Polymers

On the basis of the aforementioned results in the vine-twining polymerization, it can be considered that suitable hydrophobicity for properties of the guest polymers is required in deciding whether amylose includes them or not. Taking more precise information into consideration, amylose exhibits different inclusion behavior depending on subtle changes in structures of the guest polymers. Such behavior of amylose was applied to realize the selective inclusion toward two resemblant guest polymers (Figure 10) [45,46].

Figure 10. Amylose selectively includes one of two resemblant polyethers or polyesters.

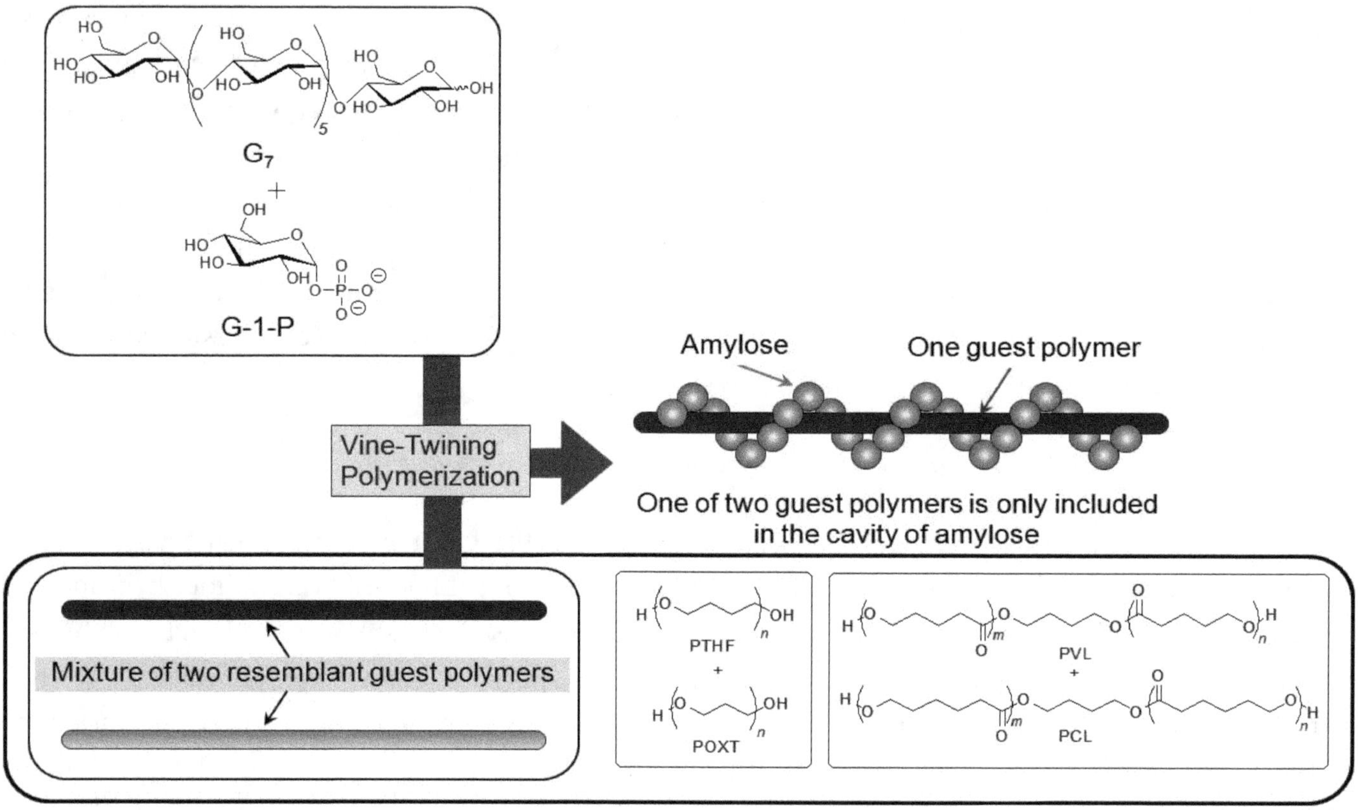

For example, the vine-twining polymerization was performed in the presence of a mixture of POXT (M_n = ~1,800) and PTHF (M_n = ~1,600) (unit molar ratio = 0.90:1.00) in sodium acetate buffer; the difference in the two polymer structures was only one methylene unit. In the ^{1}H NMR spectrum of the product (DMSO-d_6), the signals due to PTHF and amylose were prominently observed but the signals due to POXT were mostly not detected (POXT/PTHF = 0.02:1.00). This NMR data indicated that amylose almost selectively included PTHF from the mixture of the two resemblant polyethers in the vine-twining polymerization [45,46]. The slight difference in hydrophobicities of the two polyethers probably contributed to the difference in the inclusion by amylose toward them.

The selective inclusion of amylose toward two resemblant polyesters was performed by the vine-twining polymerization in the presence of a mixture of PVL (M_n = 830) and PCL (M_n = 930) (unit molar ratio = 1.00:0.92) under similar conditions as those described above using the polyethers [46]. In the ^{1}H NMR spectrum of the product (DMSO-d_6), the signals due to PVL and amylose were observed, whereas no signals due to PCL appeared, indicating that amylose selectively included PVL from the mixture of the two resemblant polyesters in the vine-twining polymerization.

4.2. Amylose Selectively Includes a Specific Range of Molecular Weights in Guest Polymers

On the basis of the above findings, the selective inclusion to a specific range in molecular weights of synthetic guest polymers by amylose was investigated in the vine-twining polymerization [47], because synthetic polymers generally have molecular weight distribution, meaning that they are reasonably considered as mixtures of analogous molecules with different numbers of the repeating units. Additionally, the numbers of repeating units contribute to exhibit different properties of the polymers with the same structure of a repeating unit. For example, PTHF with considerably low molecular weight is soluble in water although PTHF is known as a hydrophobic polymer, indicating that the hydrophobicity of PTHF is probably affected by its molecular weight. Therefore, the vine-twining polymerization in the presence of some PTHFs with different average molecular weights was examined. Three PTHFs with different M_ns and (M_w/M_n)s (PTHF-1K: M_n and M_w/M_n = 1,350 and 2.86, PTHF-3K: M_n and M_w/M_n = 3,040 and 3.13, and PTHF-6K: M_n and M_w/M_n = 6,330 and 2.45) were used as the guest polymers. To estimate M_ns and (M_w/M_n)s of PTHFs included in the cavity of amylose, PTHFs were extracted from the inclusion complexes according to the following operation. After solutions of the inclusion complexes in DMSO were stirred at 40 °C for 13 h, they were added to an excess amount of acetone. Then, the acetone soluble fractions were collected to obtain the guest PTHFs. The M_ns and (M_w/M_n)s of the extracted PTHFs were estimated by ^{1}H NMR and GPC measurements, respectively. When PTHF-3K with M_w/M_n = 3.13 was employed as the guest polymer for the vine-twining polymerization, The M_w/M_n of PTHF included in the cavity of amylose became narrower (M_w/M_n = 1.46) although its M_n (=3,590) was almost same as that of the employed one (M_n = 3,040). These results indicated that amylose selectively included a specific range in molecular weights of PTHF-3K in the vine-twining polymerization. To investigate the effect of average molecular weights of the guest PTHFs on the inclusion behavior of amylose, the vine-twining polymerization using PTHF-1K and PTHF-6K as the guest polymers was performed. Consequently, the M_ns and (M_w/M_n)s of PTHFs included in the cavity of amylose were almost the same as those using PTHF-3K as aforementioned (the included PTHF-1K: M_n and M_w/M_n = 3,120 and 1.41, the included PTHF-6K:M_n and M_w/M_n = 3,700 and 1.74). Thus, it can be concluded that, in this polymerization system, amylose selectively includes a specific range in molecular weights of PTHFs.

5. Preparation of Hydrogels through Formation of Inclusion Complexes of Amylose by Vine-Twining Polymerization

Polymeric hydrogels are three-dimensional polymer networks including a large amount of water, which have been used in a variety of applications because of their high water contents and softness [48]. They are conventionally classified as either chemical or physical gels by the types of cross-linking points. The type of cross-linking structure often determines properties of the hydrogels. Because amylose can enzymatically be prepared and hydrolyzed by the phosphorylase and amylase catalyses, respectively [17–22,49], hydrogels with cross-linking structures based on amylose have a possibility for enzymatic disruption and reproduction behaviors by two enzyme-catalyzed reactions, *i.e.*, the amylase-catalyzed hydrolysis of amylose and the formation of amylose by the phosphorylase-catalyzed polymerization.

Figure 11. Preparation of Hydrogels through Formation of Inclusion Complexes of Amylose by Vine-Twining Polymerization.

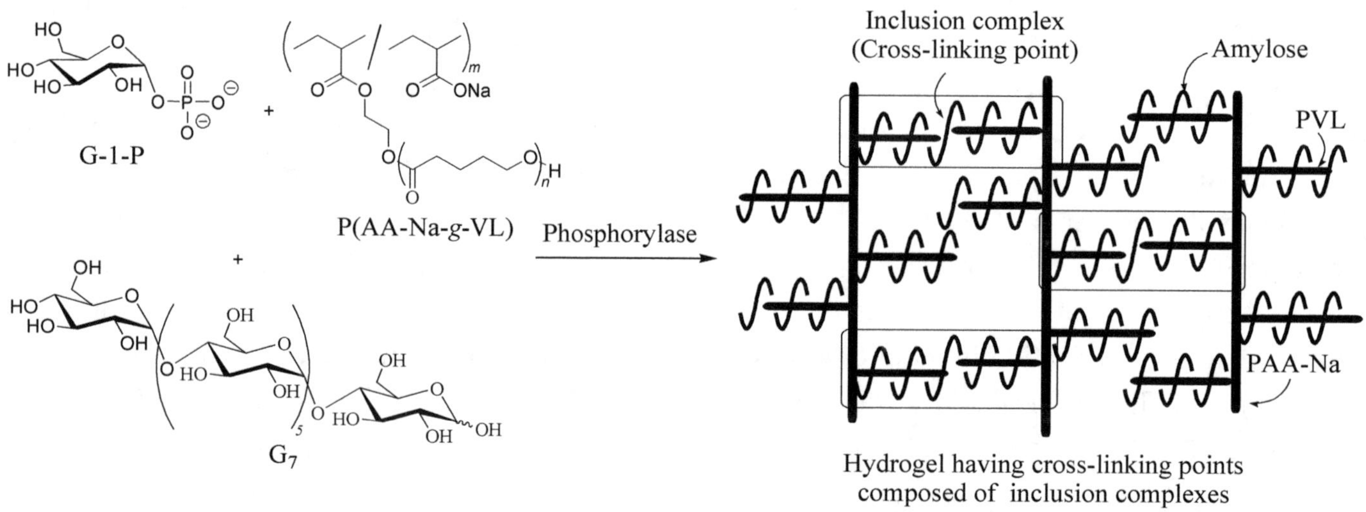

Figure 12. Cycle of enzymatic disruption and reproduction of hydrogel by two enzymatic reactions.

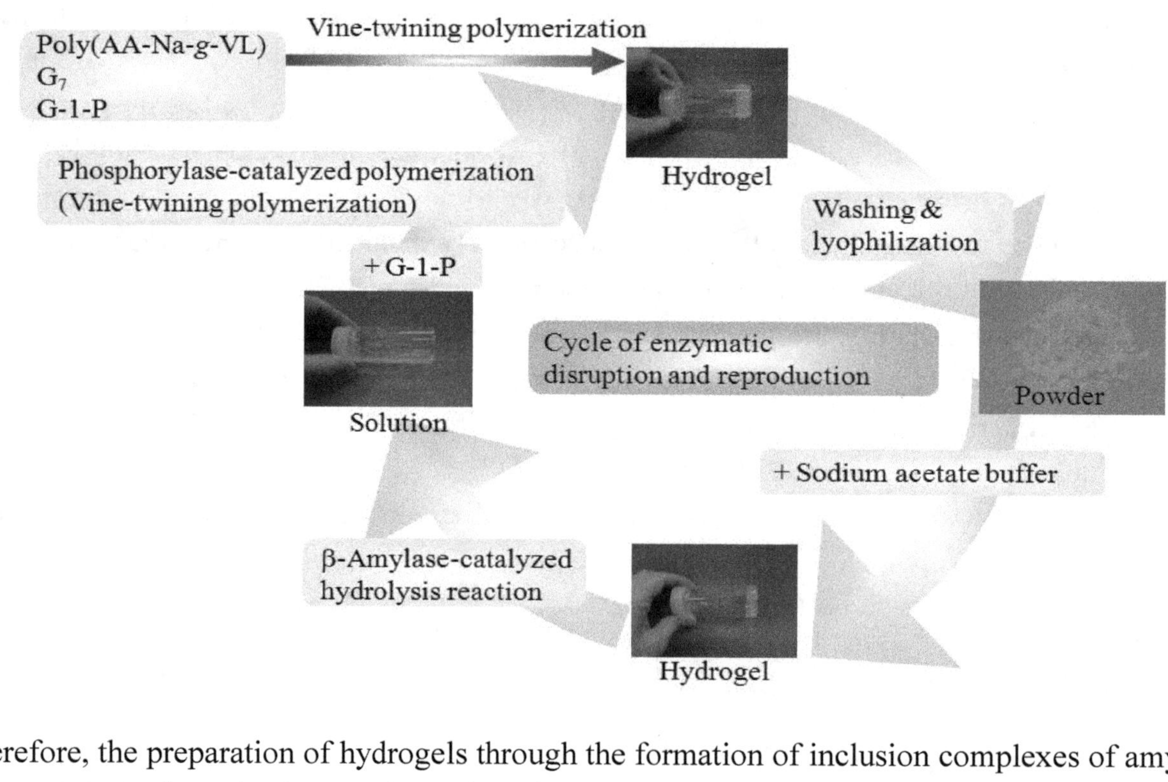

Therefore, the preparation of hydrogels through the formation of inclusion complexes of amylose in the vine-twining polymerization was investigated [50]. This was achieved by the phosphorylase-catalyzed polymerization of G-1-P from G_7 in the presence of a water-soluble copolymer having the hydrophobic PVL graft chains; poly(acrylic acid sodium salt-*graft*-δ-valerolactone) (P(AA-Na-*g*-VL)) (Figure 11). The enzymatic reaction mixture turned into a gel during the polymerization process. Because the produced amylose included the PVL graft chains in the intermolecular guest copolymers as the polymerization progressed, the formed inclusion complexes acted as the cross-linking points for the formation of the hydrogel. Furthermore, enzymatic disruption and reproduction of the hydrogels were achieved by the combination of the β-amylase-catalyzed hydrolysis of the amylose component and the formation of amylose by the phosphorylase-catalyzed polymerization (Figure 12). Therefore, it should be noted that the present hydrogel exhibited the enzymatically recyclable behavior by means of the two enzyme-catalyzed reactions.

6. Conclusions

In this review, the preparation of amylose supramolecules in the phosphorylase-catalyzed enzymatic polymerization of G-1-P, in the presence of hydrophobic polymers, such as polythers, polyesters, a poly(ester-ether), and polycarbonates according to the vine-twining polymerization manner, was overviewed. The corresponding inclusion complexes of the supramolecules were produced in the process of the polymerization. The results in the vine-twining polymerization indicated that amylose exhibited different inclusion behavior depending on subtle changes in the structures of the guest polymers. Therefore, this method was applied to the selective inclusion toward guest polymers with a subtle structural difference. Furthermore, hydrogel was produced through the formation of inclusion complexes of amylose in the vine-twining polymerization. The resulting hydrogels showed enzymatic disruption-reproduction behavior. The present method by means of the phosphorylase-catalyzed polymerization will be applied to the further production of new amylose supramolecules with well-defined structure, and the development of hierarchically controlled bio-based material in future.

Acknowledgments

The author is indebted to the co-workers, whose names are found in references from his papers, for their enthusiasistic collaborations. The author also gratefully thanks for financial supports from a Grant-in-Aid for Scientific Research from Ministry of Education, Culture, Sports, and Technology, Japan (No. 14550830, 17550118, and 19550126) and Sekisui Foundation.

References

1. Schuerch, C. Polysaccharides. In *Encyclopedia of Polymer Science and Engineering*, 2nd ed.; Mark, H.F., Bilkales, N., Overberger, C.G., Eds.; John Wiley & Sons: New York, NY, USA, 1986; Volume 13, pp. 87–162.
2. Lenz, R.W. Biodegradable polymers. *Adv. Polym. Sci.* **1993**, *107*, 1–40.
3. Putseys, J.A.; Lamberts, L.; Delcour, J.A. Amylose-inclusion complexes: Formation, identity and physico-chemical properties. *J. Cereal Sci.* **2010**, *51*, 238–247.
4. Kitamura, S. Starch, Polymers, Natural and Synthetic. In *The Polymeric Materials Encyclopedia, Synthesis, Properties and Applications*; Salamone, C., Ed.; CRC Press: New York, NY, USA, 1996; Volume 10, pp. 7915–7922.
5. Paulsen, H. Advances in selective chemical syntheses of complex oligosaccharides. *Angew. Chem. Int. Ed. Engl.* **1982**, *21*, 155–173.

6. Schmidt, R.R. New methods of the synthesis of glycosides and oligosaccharides—Are there alternative to the Koenigs-Knorr methods? *Angew. Chem. Int. Ed. Engl.* **1986**, *25*, 212–235.
7. Toshima, K.; Tatsuta, K. Recent progress in *O*-glycosylation methods and its application to natural products synthesis. *Chem. Rev.* **1993**, *93*, 1503–1531.
8. Mydock, L.K.; Demchenko, A.V. Mechanism of chemical *O*-glycosylation: From early studies to recent discoveries. *Org. Biomol. Chem.* **2010**, *8*, 497–510.
9. Kobayashi, S.; Uyama, H.; Kimura, S. Enzymatic polymerization. *Chem. Rev.* **2001**, *101*, 3793–3818.
10. Kobayashi, S. New development of polysaccharide synthesis via enzymatic polymerization. *Proc. Jpn. Acad. Ser. B* **2007**, *83*, 215–247.
11. Kobayashi, S.; Makino, A. Enzymatic polymer synthesis: An opportunity for green polymer chemistry. *Chem. Rev.* **2009**, *109*, 5288–5353.
12. Kadokawa, J.; Kobayashi, S. Polymer synthesis by enzymatic catalysis. *Curr. Opin. Chem. Biol.* **2010**, *14*, 145–153.
13. Kadokawa, J. Precision polysaccharide synthesis catalyzed by enzymes. *Chem. Rev.* **2011**, *111*, 4308–4345.
14. Shoda, S.; Izumi, R.; Fujita, M. Green process in glycotechnology. *Bull. Chem. Soc. Jpn.* **2003**, *76*, 1–13.
15. Seibel, J.; Jördening, H.-J.; Buchholz, K. Glycosylation with activated sugars using glycosyltransferases and transglycosidases, *Biocatal. Biotranform.* **2006**, *24*, 311–342.
16. Kitaoka, M.; Hayashi, K. Carbohydrate-processing phosphorolytic enzymes. *Trends Glycosci. Glycotechnol.* **2002**, *14*, 35–50.
17. Ziegast, G.; Pfannemüller, B. Phosphorolytic syntheses with di-, oligo- and multi-functional primers. *Carbohydr. Res.* **1987**, *160*, 185–204.
18. Gidley, M.J.; Bulpin, P.V. Aggregation of amylose in aqueous systems: The effect of chain length phase behavior and aggregation kinetics. *Macromolecules* **1989**, *22*, 341–346.
19. Niemann, C.; Sanger, W.; Pfannemüller, B.; Eigner, W.D.; Huber, A. Phospholytic Synthesis of Low-Molecular-Weight Amyloses with Modified Terminal Groups; Comparison of Potato Phosphorylase Andmuscle Phosphorylase b. In *ACS Symposium Series*; Comstock, M.J., Ed.; American Chemical Society: Washington, DC, USA, 1991; pp. 189–204.
20. Fujii, K.; Takata, H.; Yanase, M.; Terada, Y.; Ohdan, K.; Takaha, T.; Okada, S.; Kuriki, T. Bioengineering and application of novel glucose polymers. *Biocatal. Biotransform.* **2003**, *21*, 167–172.
21. Yanase, M.; Takaha, T.; Kuriki, T. α-Glucan phosphorylase and its use in carbohydrate engineering. *J. Food Agric.* **2006**, *86*, 1631–1635.
22. Ohdan, K.; Fujii, K.; Yanase, M.; Takaha, T.; Kuriki, T. Enzymatic synthesis of amylose. *Biocatal. Biotransform.* **2006**, *24*, 77–81.
23. Shogren, R.L.; Greene, R.V.; Wu, Y.V. Complexes of starch polysaccharides and poly(ethylene-*co*-acrylic acid) - structure and stability in solution. *J. Appl. Polym. Sci.* **1991**, *42*, 1701–1709.
24. Shogren, R.L. Complexes of starch with telechelic poly(ε-caprolactone) phosphate. *Carbohydr. Polym.* **1993**, *22*, 93–98.

25. Ikeda, M.; Furusho, Y.; Okoshi, K.; Tanahara, S.; Maeda, K.; Nishino, S.; Mori, T.; Yashima, E. A luminescent poly(phenylenevinylene)-amylose composite with supramolecular liquid crystallinity. *Angew. Chem. Int. Ed.* **2006**, *45*, 6491–6495.
26. Kida, T.; Minabe, T.; Okabe, S.; Akashi, M. Partially-methylated amyloses as effective hosts for inclusion complex formation with polymeric guests. *Chem. Commun.* **2007**, doi: 10.1039/B616231B.
27. Kida, T.; Minabe, T.; Nakano, S.; Akashi, M. Fabrication of novel multilayered thin films based on inclusion complex formation between amylose derivatives and guest polymers. *Langmuir* **2008**, *24*, 9227–9229.
28. Frampton, M.J.; Claridge, T.D.W.; Latini, G.; Brovelli, S.; Cacialli, F.; Anderson, L. Amylose-wrapped luminescent conjugated polymers. *Chem. Commun.* **2008**, doi: 10.1039/B803335H.
29. Kaneko, Y.; Kyutoku, T.; Shimomura, N.; Kadokawa, J. Formation of amylose-poly(tetrahydrofuran) inclusion complexes in ionic liquid media. *Chem. Lett.* **2011**, *40*, 31–33.
30. Kaneko, Y.; Kadokawa, J. Vine-twining polymerization: A new preparation method for well-defined supramolecules composed of amylose and synthetic polymers. *Chem. Rec.* **2005**, *5*, 36–46.
31. Kaneko, Y.; Kadokawa, J. Synthesis of nanostructured bio-related materials by hybridization of synthetic polymers with polysaccharides or saccharide residues. *J. Biomater. Sci. Polym. Ed.* **2006**, *17*, 1269–1284.
32. Kaneko, Y.; Kadokawa, J. *Modern Trends in Macromolecular Chemistry*; Lee, J.N., Ed.; Nova Science Publishers, Inc.: Hauppauge, NY, USA, 2009; Chapter 8, pp. 199–217.
33. Kadokawa, J.; Kaneko, Y.; Tagaya, H.; Chiba, K. Synthesis of an amylose-polymer inclusion complex by enzymatic polymerization of glucose 1-phosphate catalyzed by phosphorylase enzyme in the presence of polyTHF: A new method for synthesis of polymer-polymer inclusion complexes. *Chem. Commun.* **2001**, doi: 10.1039/B008180I.
34. Seneviratne, H.D.; Biliaderis, C.G. Action of α-amylases on amylose-lipid complex superstructures. *J. Cereal Sci.* **1991**, *13*, 129–143.
35. Zobel, H.F. Starch crystal transformations and their industrial importance. *Starch* **1988**, *40*, 1–7.
36. Kadokawa, J.; Kaneko, Y.; Nagase, S.; Takahashi, T.; Tagaya, H. Vine-twining polymerization: Amylose twines around polyethers to form amylose-polyether inclusion complexes. *Chem. Eur. J.* **2002**, *8*, 3321–3326.
37. Kadokawa, J.; Kaneko, Y.; Nakaya, A.; Tagaya, H. Formation of an amylose-polyester inclusion complex by means of phosphorylase-catalyzed enzymatic polymerization of α-D-glucose 1-phosphate monomer in the presence of poly(ε-caprolactone). *Macromolecules* **2001**, *34*, 6536–6528.
38. Kadokawa, J.; Nakaya, A.; Kaneko, Y.; Tagaya, H. Preparation of inclusion complexes between amylose and ester-containing polymers by means of vine-twining polymerization. *Macromol. Chem. Phys.* **2003**, *204*, 1451–1457.
39. Kaneko, Y.; Ueno, K.; Yui, T.; Nakahara, K.; Kadokawa, J. Amylose's recognition of chirality in polylactides on formation of inclusion complexes in vine-twining polymerization. *Macromol. Biosci.* **2011**, *11*, 1407–1415.
40. Yamashita, Y. Single crystals of amylose V complexes. II. Crystals with 7_1 helical configuration. *J. Polym. Sci. Part A-2* **1966**, *4*, 161–171.

41. Kaneko, Y.; Saito, Y.; Nakaya, A.; Kadokawa, J.; Tagaya, H. Preparation of inclusion complexes composed of amylose and strongly hydrophobic polyesters in parallel enzymatic polymerization system. *Macromolecules* **2008**, *41*, 5665–5670.
42. Kobayashi, S.; Uyama, H.; Suda, S.; Namekawa, S. Dehydration polymerization in aqueous medium catalyzed by lipase. *Chem. Lett.* **1997**, doi: 10.1246/cl.1997.105.
43. Suda, S.; Uyama, H.; Kobayashi, S. Dehydration polycondensation in water for synthesis of polyesters by lipase catalyst. *Proc. Jpn. Acad. Ser. B* **1999**, *75*, 201–206.
44. Kaneko, Y.; Beppu, K.; Kadokawa, J. Preparation of amylose/polycarbonate inclusion complexes by means of vine-twining polymerization. *Macromol. Chem. Phys.* **2008**, *209*, 1037–1042.
45. Kaneko, Y.; Beppu, K.; Kadokawa, J. Amylose selectively includes one from a mixture of two resemblant polyethers in vine-twining polymerization. *Biomacromolecules* **2007**, *8*, 2983–2985.
46. Kaneko, Y.; Beppu, K.; Kyutoku, T.; Kadokawa, J. Selectivity and priority on inclusion of amylose toward guest polyethers and polyesters in vine-twining polymerization. *Polym. J.* **2009**, *41*, 279–286.
47. Kaneko, Y.; Beppu, K.; Kadokawa, J. Amylose selectively includes a specific range of molecular weights in poly(tetrahydrofuran)s in vine-twining polymerization. *Polym. J.* **2009**, *41*, 792–796.
48. Lee, K.Y.; Mooney, D.J. Hydrogels for tissue engineering. *Chem. Rev.* **2001**, *101*, 1869–1879.
49. Bijttebier, A.; Goesaert, H.; Delcour, J.A. Amylase action pattern on starch polymers. *Biologia* **2008**, *63*, 989–999.
50. Kaneko, Y.; Fujisaki, K.; Kyutoku, T.; Furukawa, H.; Kadokawa, J. Preparation of enzymatically recyclable hydrogels through the formation of inclusion complexes of amylose in a vine-twining polymerization. *Chem. Asian J.* **2010**, *5*, 1627–1633.

2

In Silico Discovery of Aminoacyl-tRNA Synthetase Inhibitors

Yaxue Zhao [1,2], Qingqing Meng [1], Linquan Bai [2] and Huchen Zhou [1,2,*]

[1] School of Pharmacy, Shanghai Jiao Tong University, 800 Dongchuan Road, Shanghai 200240, China; E-Mails: yaxuezhao@sjtu.edu.cn (Y.Z.); qqmeng@sjtu.edu.cn (Q.M.)

[2] State Key Laboratory of Microbial Metabolism, Shanghai Jiao Tong University, 800 Dongchuan Road, Shanghai 200240, China; E-Mail: bailq@sjtu.edu.cn

* Author to whom correspondence should be addressed; E-Mail: hczhou@sjtu.edu.cn

---

Abstract: Aminoacyl-tRNA synthetases (aaRSs) are enzymes that catalyze the transfer of amino acids to their cognate tRNA. They play a pivotal role in protein synthesis and are essential for cell growth and survival. The aaRSs are one of the leading targets for development of antibiotic agents. In this review, we mainly focused on aaRS inhibitor discovery and development using *in silico* methods including virtual screening and structure-based drug design. These computational methods are relatively fast and cheap, and are proving to be of great benefit for the rational development of more potent aaRS inhibitors and other pharmaceutical agents that may usher in a much needed generation of new antibiotics.

Keywords: aminoacyl-tRNA synthetase; inhibitor; antibiotics; virtual screening; structure-based drug design; docking

---

1. Introduction

Aminoacyl-tRNA synthetases (aaRSs) play a central role in the process of protein synthesis. They are responsible for catalyzing the attachment of the correct amino acid to its cognate tRNA through an esterification reaction at the 3' end of tRNA. This highly specific aminoacylation reaction involves two steps:

$$\text{aa} + \text{ATP} + \text{aaRS} \rightleftharpoons \text{aaRS}\cdot\text{aa-AMP} + \text{PPi} \qquad (1)$$

$$\text{aaRS}\cdot\text{aa-AMP} + \text{tRNA} \rightleftharpoons \text{aaRS} + \text{aa-tRNA} + \text{AMP} \tag{2}$$

where aa is an amino acid. The first step is formation of an aminoacyl-adenylate (aa-AMP) activated intermediate from an amino acid and ATP. During this step, ATP and the amino acid first bind the aaRS active site, and are positioned appropriately to facilitate the α-carboxylate of the amino acid to attack the α-phosphate of ATP via in-line nucleophilic displacement. In the second step, the activated amino acid is transferred from aa-AMP to the tRNA to form the aminoacyl-tRNA (aa-tRNA) via nucleophilic attack by the 2'- or 3'-hydroxyl of the 3'-terminal adenosine of the tRNA on the α-carbonyl of the aa-AMP [1].

The aaRSs are divided into two unrelated classes (class I and class II, as shown in Table 1) based on mutually exclusive sets of sequence motifs that reflect distinct active site topologies [2]. The class I synthetase active site adopts a Rossmann-fold domain and binds ATP in an extended conformation. In contrast, class II synthetase active sites are housed on an antiparallel β-fold domain, and bind ATP in a bent conformation [3]. Synthetase enzymes can be arranged into three subclasses within each class, and subclasses group enzymes that are more closely related to each other than to other enzymes in the same class [4]. Subclasses Ic and IIc contain the synthetases for aromatic amino acids, subclasses Ib and IIb comprise the synthetases for amino acids with carboxylate side chains and their amidated derivatives, and subclasses Ia and IIa include synthetases for hydrophobic amino acids.

Table 1. Classes of aminoacyl-tRNA synthetases.

Class	Subclass	aaRS
I	Ia	MetRS ValRS LeuRS IleRS CysRS ArgRS
	Ib	GluRS GlnRS LysRS-I
	Ic	TyrRS TrpRS
II	IIa	SerRS ThrRS AlaRS GlyRS ProRS HisRS
	IIb	AspRS AsnRS LysRS-II
	IIc	PheRS

Accurate protein synthesis and hence, cell survival, requires aaRSs to discriminate between chemically similar, non-cognate amino acids by a factor of at least 10^4. This is difficult to achieve in one step, especially for aliphatic and hydrophobic amino acids that lack distinguishable molecular features. For example, the weakness of the additional van der Waals interactions of isoleucine compared with valine in the active site of IleRS was predicted to yield an error rate of up to one in five [5]. To overcome this problem, some aaRSs have a specific editing activity that hydrolyzes misactivated aa-AMPs (pre-transfer editing) and mischarged aa-tRNAs (post-transfer editing). This is known as the double sieve mechanism. The first sieve occurs during classical aminoacylation at the aaRS synthetic active site which binds cognate amino acids but cannot adequately distinguish between amino acids with highly similar (isosteric) or slightly smaller structures. The second sieve occurs at an editing active site which hydrolyzes non-cognate amino acids that are misactivated or mischarged. Synthetases with this additional editing site include IleRS, LeuRS, and ValRS from class I, and ThrRS, AlaRS, PheRS and ProRS from class II enzymes [6–8].

The aaRSs have become key targets for antibiotics. Inhibition of aaRSs depletes charged tRNAs, inhibits protein synthesis and leads to arrest of cell growth and ultimately cell death [9]. Inhibitors of aaRSs are being developed as antibacterials, antifungals and anti-parasitic drugs [10–13], and they also possess potent immunosuppressive activity [14]. Both synthetic and editing active sites are targets for inhibition. Mupirocin and AN2690 (Figure 1) are excellent examples of inhibitors that bind to the synthetic and editing active sites, respectively.

Figure 1. Structures of mupirocin and AN2690.

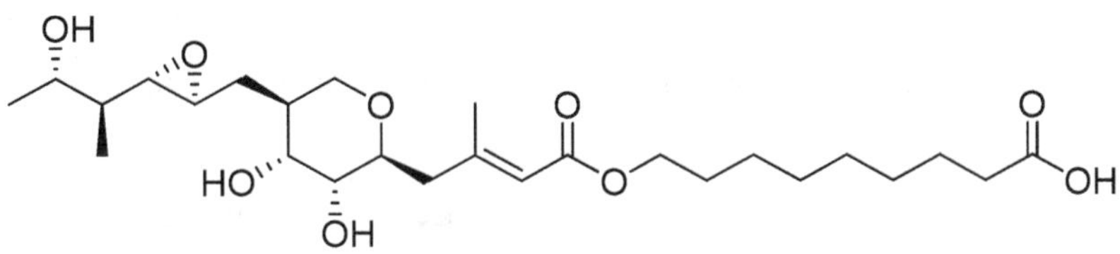

mupirocin (PA-A, **1**)

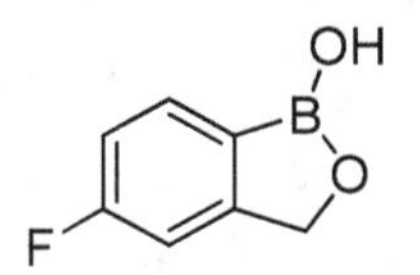

AN2690 (**2**)

Mupirocin (Bactroban, GSK, London, England), a natural product of *Pseudomonas fluorescens*, is the only aaRS inhibitor approved by the US Food and Drug Administration to this date [15]. It is a mixture of several pseudomonic acids, with pseudomonic acid A (PA-A) constituting greater than 90%. Mupirocin is primarily active against gram-positive pathogens, such as *Staphylococcus aureus* and *Streptococcus pyogenes*, and is used as a topical treatment for bacterial skin infections [16]. Mupirocin is targeted against IleRS. Crystal structures of IleRS bound with mupirocin and Ile-AMP show that mupirocin binding in the IleRS synthetic site is highly similar to Ile-AMP binding [16–18]. Therefore, mupirocin is a competitive inhibitor functioning by displacing endogenous Ile and ATP.

AN2690 (Tavaborole, Anacor, Palo Alto, CA, USA) is currently in Phase 3 clinical trials for treating onychomycosis. It is a fluorinated benzoxaborole that targets LeuRS [19]. The boron atom in the oxaborole ring of AN2690 binds to both the 2'- and the 3'-hydroxyl groups on the 3'-terminal adenosine. AN2690 occupies the non-cognate amino acid binding pocket in the editing domain of LeuRS. Therefore, by trapping $tRNA^{Leu}$ in the editing active site, such inhibitors prevent LeuRS catalytic turnover, inhibiting synthesis of leucyl-$tRNA^{Leu}$ and consequently blocking protein synthesis.

Traditional approaches for inhibitor discovery that have proved successful include serendipity, screening natural products and known active substances to identify the active components, drug metabolites, and observing side effects of existing medicines to identify potential involvement in other pathways. In the early 1990s, combinatorial chemistry was used to synthesize huge libraries of compounds and high-throughput screening of these libraries proved particularly successful [20]. However since 2000, computational methods such as virtual screening and structure-based drug design have become more popular in pharmaceutical research. *In silico* methods save time and money in the drug discovery process [20]. Virtual screening has been widely applied in the discovery of lead compounds [21–23]. It can be divided into docking-based and pharmacophore-based procedures. A classical docking-based virtual screening approach begins with the three dimensional (3D) structure of the target protein from the Protein Data Bank (PDB) [24] or from homology modeling. Small molecule structures from commercial databases are then docked into the binding pocket of the target protein. Scoring functions are then used to evaluate and rank the binding mode of each small molecule in the target protein binding site. Finally, high scoring molecules are tested for activity in inhibition or binding assays. Currently available docking software packages for virtual screening studies are represented by Glide [25,26], Gold [27], Dock [28], and AutoDock Vina [29].

Pharmacophore features are generally represented by points in 3D space. A pharmacophore feature could be comprised of functional groups such as hydrogen bond donors, hydrogen bond acceptors, cations, anions, aromatics and hydrophobic sites [30]. Pharmacophore features can be generated by identifying common chemical features from a set of bioactive compounds, or by observing important shared interactions in protein-ligand complex structures. There are several available programs for automatic generation of pharmacophore models including Catalyst [31], Phase [32] and LigandScout [33]. The generated pharmacophore can be used to screen small molecule databases to identify appropriate compounds.

Structure-based or rational drug design is now widely applied in most stages of the drug development process, from initial hit identification to lead optimization [34,35]. Several important drugs have been developed using this method, including human immunodeficiency virus-1 protease [36] and neuraminidase [37–39] inhibitors. Central to all structure-based discovery approaches is experimental determination of the 3D structure of the target protein or protein-ligand complex, or construction of a suitably accurate homology model.

2. Inhibitor Identification Using Virtual Screening

2.1. Leucyl-tRNA Synthetase Inhibitors

To discover inhibitors of *Trypanosoma brucei* LeuRS in order to develop drugs against human African trypanosomiasis, Zhao *et al.* (2012) constructed a homology model of the synthetic active site based on the crystal structure of *Pyrococcus horikoshii* LeuRS (1WKB [40]) using the *in silico* mutation method [41]. By analyzing the interactions of the substrate analog Leu-AMS and *T. brucei* LeuRS, pharmacophores I and II were generated and used to screen the SPECS database [42] using Catalyst (Figure 2). Hits that matched the pharmacophores well were docked using Glide, and the 2-pyrrolinone compound **3** was identified, and found to be active *in vitro*, with an IC_{50} of 170.3 μM

(Figure 2). Guided by the docking of compound **4** and the *T. brucei* LeuRS structure, various substituents at R^1, R^2, and R^3 were designed and synthesized. Structure-activity relationship studies generally corroborated the docking model, which showed that the R^2 phenyl group explored a new hydrophobic pocket, and the R^3 indolyl group was essential for the favorable interaction with the leucine recognition pocket. Finally, compound **5** was identified as the most potent inhibitor (IC_{50} = 31.9 μM).

Figure 2. Scheme used in the identification of LeuRS inhibitors. Pharmacophores I and II: the hydrophobic site is colored as orange sphere, hydrogen bond donor is colored as green arrow, hydrogen bond acceptor is colored as red arrow.

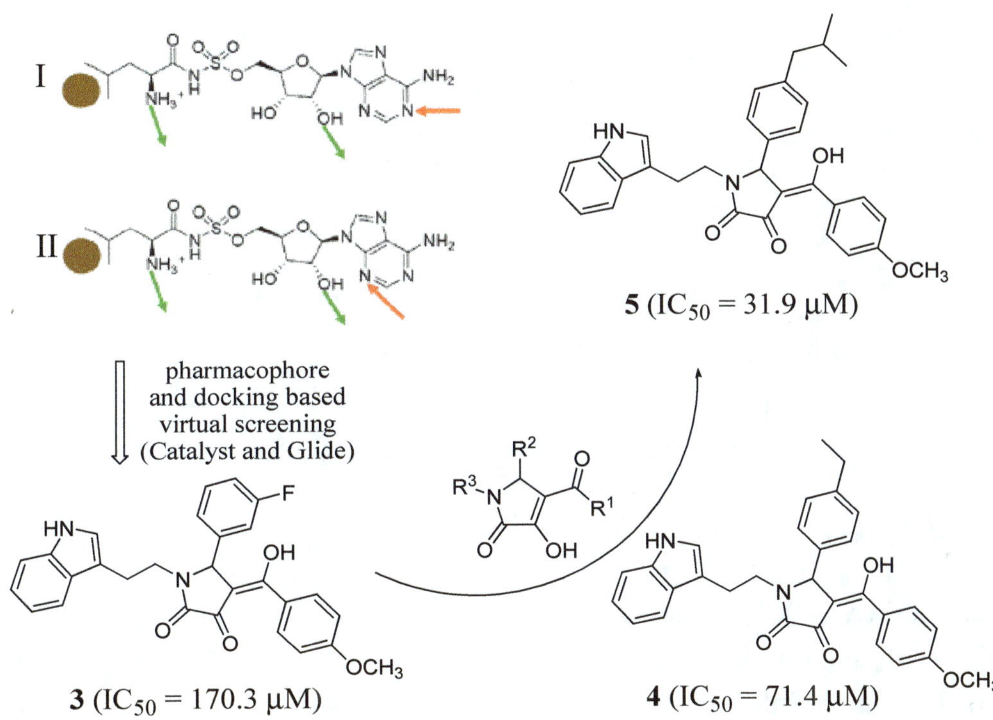

2.2. Tryptophanyl-tRNA Synthetase Inhibitors

Wu *et al.* (2007) applied a virtual screening approach to find new lead compounds to target *Staphylococcus epidermidis* TrpRS [43]. Figure 3 shows their inhibitor identification scheme. They first constructed a homology model of *S. epidermidis* TrpRS based on the crystal structures of *Bacillus stearothermophilus* TrpRS (1MAW, 1M83, 1MAU, 1MB2 [44]) using MODELLER (Accelrys, Inc. San Diego, CA, USA). Three compounds were identified as TrpRS inhibitors that arrested *S. epidermidis* growth from the SPECS database combining virtual screening, *in vitro* and *in vivo* experiments. The IC_{50} values of these compounds were 22.8, 42.2 and 18.3 μM as shown using the Kinase-Glo Luminescent Kinase assay, and these results were consistent with the results of the Pyrophosphate Reagent assay. All three compounds inhibited the growth of both *S. epidermidis* ATCC 12228 and ATCC 35984 strains with micromolar minimal inhibitory concentrations (MICs) (Table 2), and also exhibited low cytotoxicity with CC_{50} > 200 μM.

Table 2. Micromolar minimal inhibitory concentrations (MICs) of compounds **6**, **7** and **8**.

ID	MIC (µM)	
	***S. epidermidis* ATCC 12228**	***S. epidermidis* ATCC 35984**
6	6.25	6.25
7	25	25
8	100	100

Figure 3. Inhibitor identification scheme of Wu *et al.* (2007) for TrpRS [43].

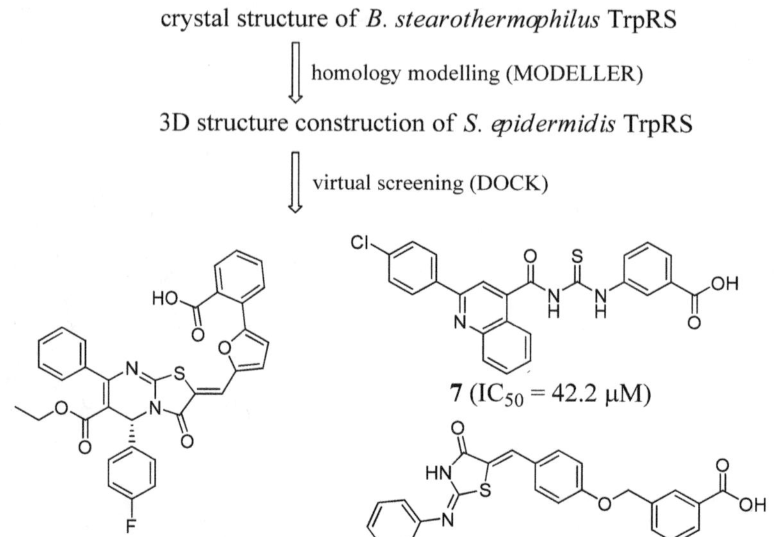

6 (IC_{50} = 22.8 µM) **8** (IC_{50} = 18.3 µM)

2.3. Asparaginyl-tRNA Synthetase Inhibitors

Sukuru and co-workers (2006) successfully identified seven diverse compounds that can inhibit the activity of AsnRS with micromolar affinity [45]. A template was generated using SLIDE to represent the active site of *B. malayi* AsnRS and its interactions with Asn-AMS based on X-ray crystal structure (2XGT [46]) (Figure 4). After screening the Cambridge Structural Database [47] and National Cancer Institute Plated Compounds Database [48] using SLIDE, they selected forty-five compounds for activity assays. NSC363624 is the most potent inhibitor (IC_{50} = 25 µM, Figure 4) which has a symmetric structure with two substituted triazine rings connected by a phenyl group.

Figure 4. Inhibitor identification scheme of Sukuru *et al.* (2006) for AsnRS [45].

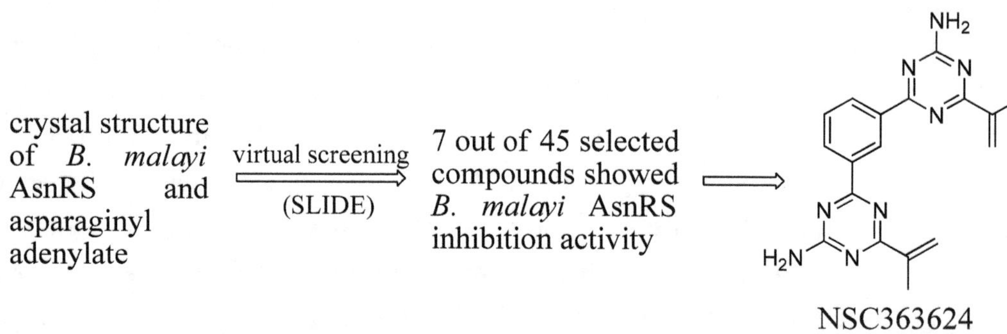

NSC363624
(**9**, IC_{50} = 25 µM)

2.4. Methionyl-tRNA Synthetase Inhibitors

Kim *et al.* (2006) performed a 2D-database search to discover new MetRS inhibitors [49]. Initially, they constructed a pharmacophore query based on methionyl adenylate (Figure 5) in which R^1 has hydrogen bonding characteristics similar to those of a carboxylic acid, amide, and sulfonamide, that may act as the amine or ring nitrogen of the adenine group. R^2 is an optionally substituted aryl or heteroaryl ring designed to mimic the lipophilic methionine side chain, and X^1 and X^2 are linkers. When the query structure was used to search a chemical database consisting of 508,143 commercially available compounds, four novel micromolar inhibitors of *Escherichia coli* MetRS were identified (Figure 5).

Figure 5. Inhibitor identification scheme of Kim *et al.* (2006) for MetRS [49].

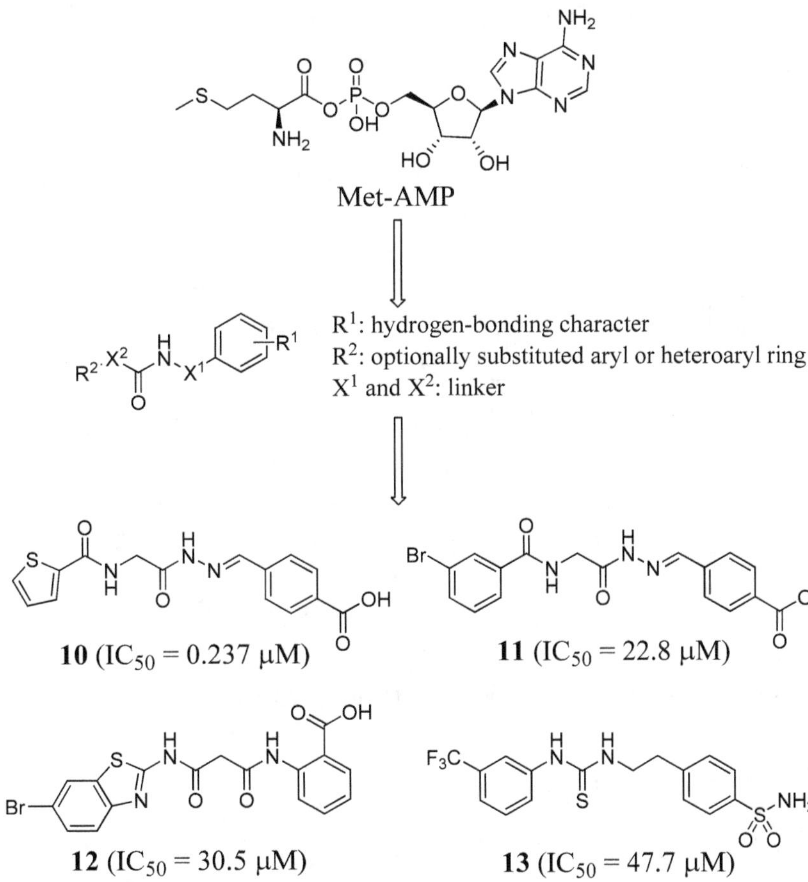

Bharatham *et al.* (2007) identified 246 potential MetRS inhibitors, and selected 29 based on structural diversity and wide coverage of the activity range to generate pharmacophore models using CATALYST [50]. The best ranking pharmacophore model contained four chemical features including a hydrogen bond donor, a hydrophobic aliphatic substituent, and two aromatic rings. When used to search the Maybridge database [51], two inhibitors (AW01179 and BTB00521, Figure 6) were identified as the top hits, although inhibition activity was estimated using HypoGen and not tested *in vitro* or *in vivo*.

Figure 6. Structures of AW01179 and BTB00521.

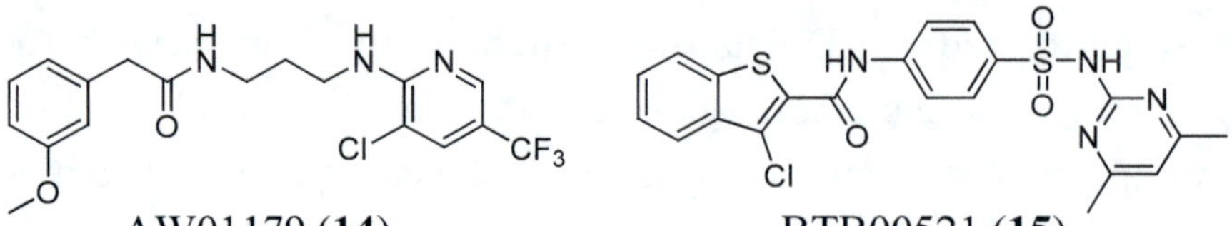

AW01179 (**14**) BTB00521 (**15**)

Finn *et al.* (2008) also attempted to identify novel MetRS inhibitors using pharmacophore-based virtual screening [52]. By analyzing crystal structures of *S. aureus* MetRS in complex with known inhibitors [52], a four point pharmacophore was generated using Catalyst which contained two hydrophobic regions, two hydrogen bond donors directed towards the carboxylate oxygens of Asp51, and an excluded volume. When this pharmacophore was used to search the ChemDiv diverse collection database [53], thirty-one compounds were identified. Enzyme assays established that twenty-two out of the thirty-one compounds demonstrated 50% or greater inhibition of *S. aureus* MetRS at 100 μM. The structures of the four most potent compounds and their IC_{50} values are shown in Figure 7.

Figure 7. Structures of MetRS inhibitors with IC_{50} < 10 μM.

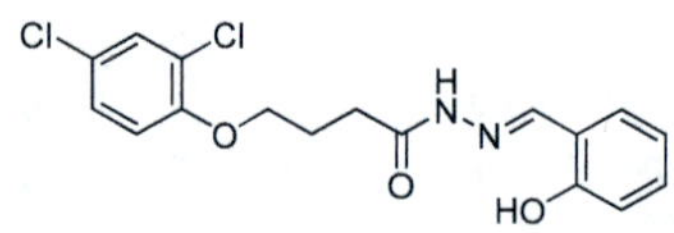

16 (IC_{50} = 4.8 μM)

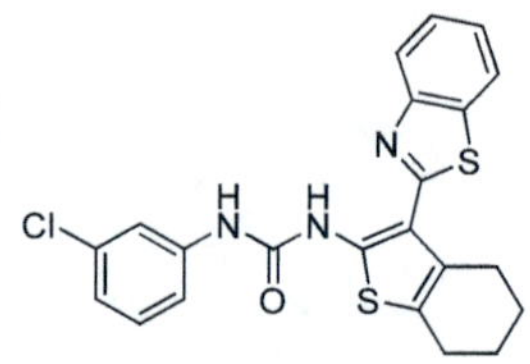

17 (IC_{50} = 6.3 μM)

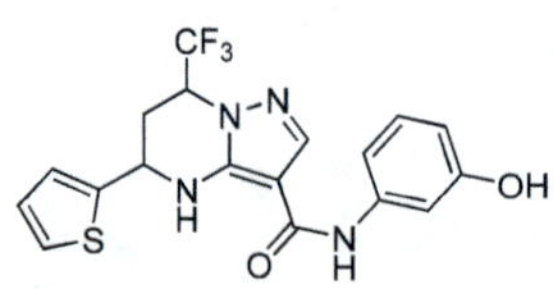

18 (IC_{50} = 5.7 μM)

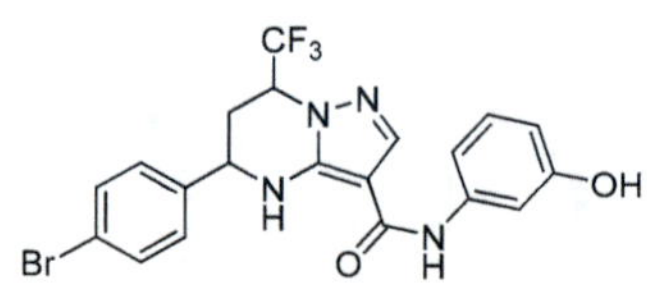

19 (IC_{50} = 8.3 μM)

3. Inhibitor Identification Using Structure-Based Drug Design

3.1. Leucyl-tRNA Synthetase Inhibitors

Ding *et al.* (2011) and Zhang *et al.* (2013) investigated *T. brucei* LeuRS using a structure-based drug design approach [12,54].

Figure 8. Inhibitor design scheme of Ding *et al.* (2011) for LeuRS. LeuRS is in cartoon representation and colored green, while compound **20** is in stick conformation and is colored salmon [12].

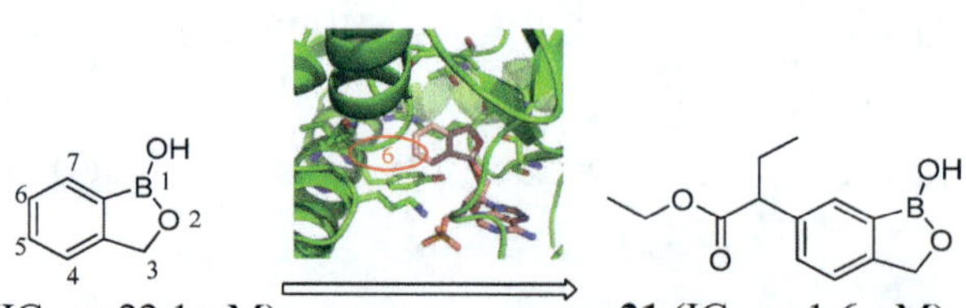

20 (IC_{50} = 22.1 μM) **21** (IC_{50} = 1.6 μM)

A 3D structure of the *T. brucei* LeuRS editing active site was constructed using homology modeling based on the crystal structure of *Candida albicans* LeuRS (2WFG [10]). The editing domain active site of the model was rather small and hydrophobic, and lined by nonpolar amino acid residues including Pro398, Ala443, Ile468, and Ala464. Compounds with smaller hydrophobic groups at the 6-position (compound **20**, IC_{50} = 22.1 μM; Figure 8) were more potent than those with hydrophilic and larger hydrophobic groups. The highest potency was shown by a compound with an ester at the 6-position (compound **21**, IC_{50} of 1.6 μM; Figure 8). 6-amide and 6-ketone derivatives were also developed to improve stability *in vivo*, and the 6-ketone analogs were comparable in potency to the 6-ester compounds.

By screening an in-house database of 500 compounds, Zhang *et al.* [54] identified a *T. brucei* LeuRS inhibitor with an *N*-(4-sulfamoylphenyl)thiourea core structure (compound **22**, IC_{50} = 174.5 μM; Figure 9). Computational studies suggested thiourea compounds bind to the synthetic active site rather than the editing active site. In order to capitalize on the intrinsic binding affinity for the leucyl-anchoring pocket, the leucyl group was introduced (compound **23**, IC_{50} = 31.4 μM; Figure 9). Superimposition of the docked pose of compound **23** and Leu-AMS showed a 1,4-substitution geometry at the central phenyl ring that could lead to significant deviation. Accordingly, a 1,3-substituted analog of **23** was designed and tested, and shown to result in a more potent *T. brucei* LeuRS inhibitor (compound **24**, IC_{50} = 1.1 μM; in Figure 9).

Figure 9. Structures of LeuRS inhibitors.

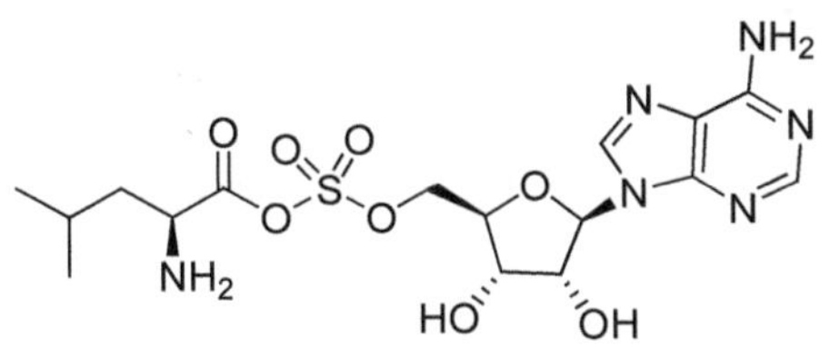

Leu-AMS

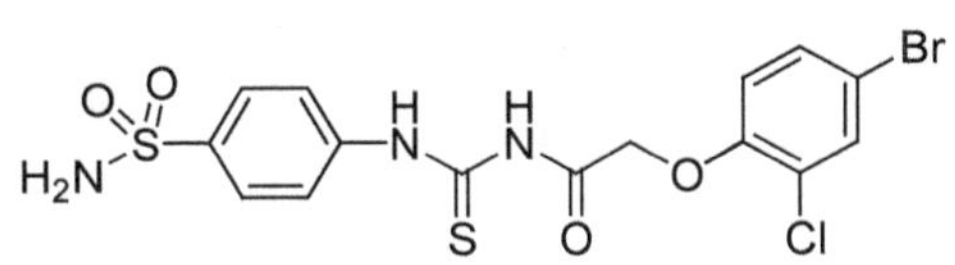

22 (IC_{50} = 174.5 μM)

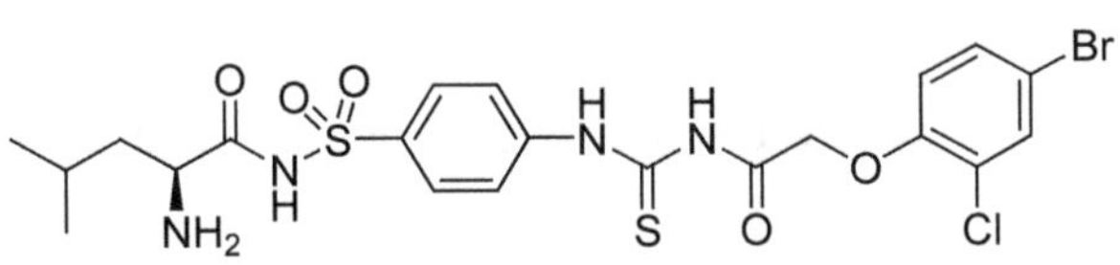

23 (IC_{50} = 31.4 μM)

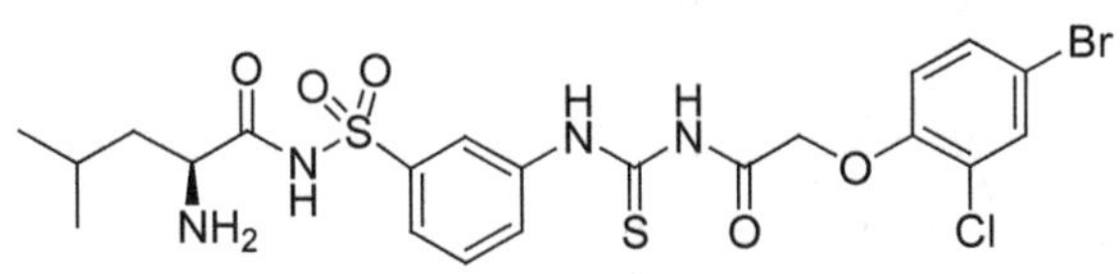

24 (IC_{50} = 1.1 μM)

3.2. Isoleucyl-tRNA Synthetase Inhibitors

As mentioned above, mupirocin is a drug that targets IleRS. Based on the IleRS-PA-A complex crystal structure, together with a detailed understanding of the reaction cycle of IleRS and characterization of the binding mode of the Ile-AMP reaction intermediate, Brown *et al.* (2000) designed and synthesized a series of novel IleRS inhibitors (Figure 10) [55]. They noticed that the binding site of PA-A overlaps with that of Ile-AMP, such that the dihydroxytetrahydropyran and ribose rings overlap but the epoxide containing group does not occupy the Ile-binding pocket which is lined with two tryptophans. They hypothesized that extra binding energy could be achieved by appropriate introduction of an Ile moiety to improve the potency of inhibitors. Based on this, four pharmacophores (amino acid side chain, linker, monate core and monate side chain) were generated (Figure 10). Through systematic optimization of these pharmacophores, compounds SB-234764 (K_d = 0.012 pM) and SB-236996 (K_d = 0.010 pM) were developed, which were much more potent than the starting compound (K_d = 140 pM).

Figure 10. Inhibitor design scheme of Brown *et al.* (2000) for IleRS [55].

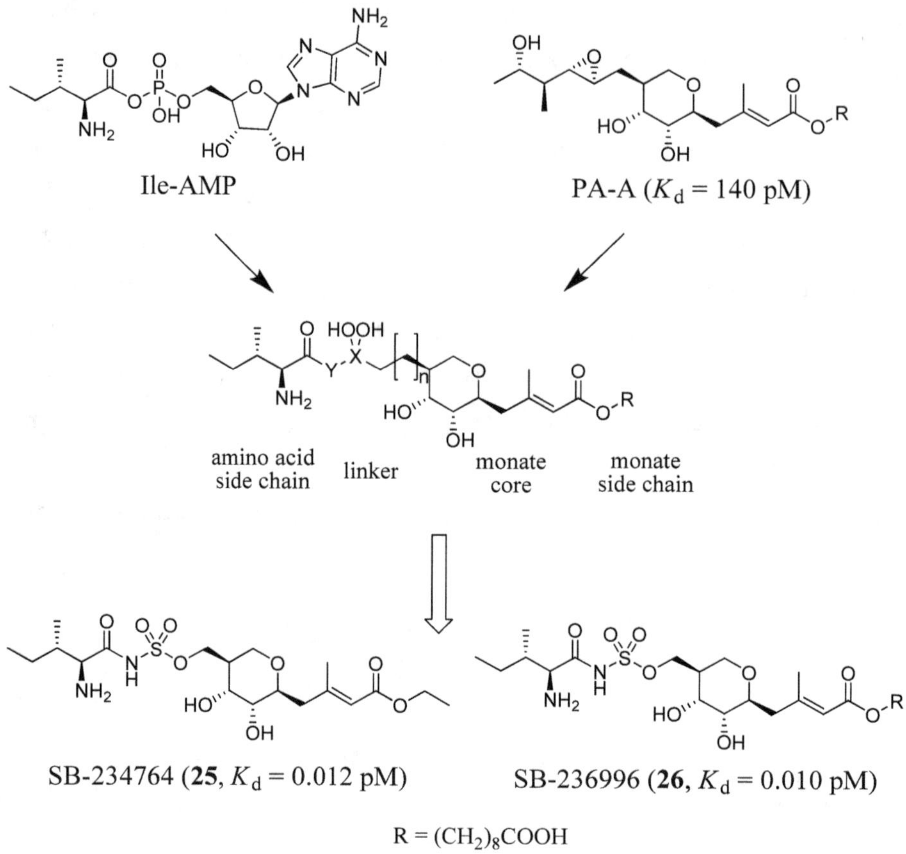

3.3. *Threonyl-tRNA Synthetase Inhibitors*

Teng *et al.* (2013) developed a series of potent and bacterial-selective ThrRS inhibitors through using crystal structures and structure-based drug design [56]. The crystal structure of the complex containing Thr-AMS (compound **27** in Table 2) bound to *E. coli* ThrRS was the starting point (1EVL [57]). Computational modeling using Discovery Studio suggested that compound **28** (Table 3) which has an aminoquinazoline fragment would bind to the ThrRS synthetic active site in a similar manner to Thr-AMS. Compound **28** was itself a potent inhibitor of *E. coli* ThrRS with a K_i of 2.9 nM, and the binding mode was confirmed by crystal structure. However, compound **28** also displayed potent human ThrRS inhibition activity. To explore the opportunity for selective bacterial ThrRS inhibition, compound **29** which lacks the H-bond critical for binding to Ser517 was expected to have weaker potency against human ThrRS. Indeed, **29** turned out to have a selectivity ratio of 270. Crystal structures of **29** in complex with both *E. coli* and human ThrRS demonstrated that the ATP sites of the two enzymes did bind **29** in two distinct modes.

Table 3. Inhibitors of threonyl-tRNA synthetase.

ID	Structure	ThrRS K_i (nM)		Selectivity ratio
		E. coli	Human	Human/*E.coli*
27	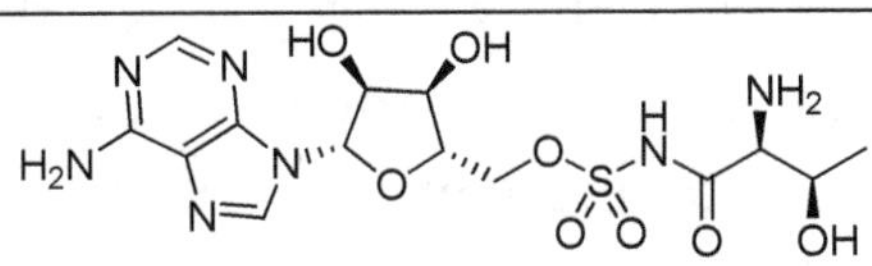	13.1	13.4	1
28	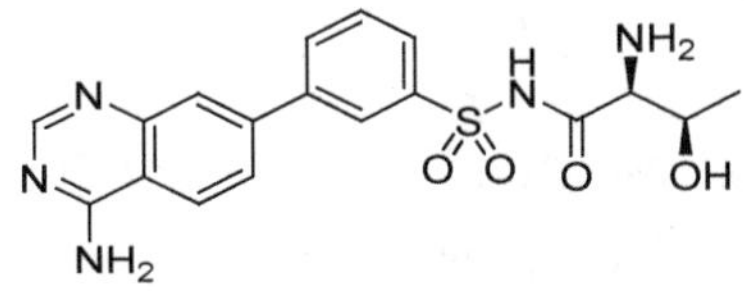	2.9	3.3	1.1
29	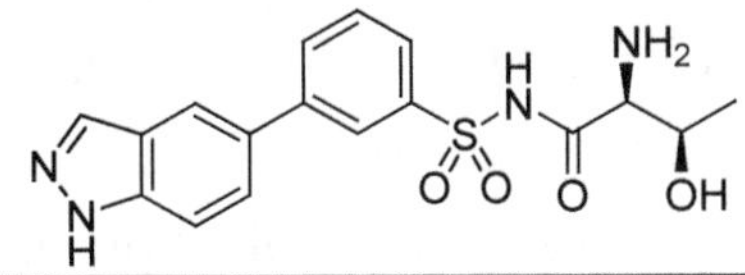	182	>50,000	>270

3.4. Methionyl-tRNA Synthetase Inhibitors

The Buckner group have carried out extensive work on the discovery of *T. brucei* MetRS inhibitors [58–61]. Based on the crystal structure of *Clostridium difficile* MetRS [62], they built a *T. bruce*i MetRS homology model, and docking of compound **30** (Figure 11) was the starting point for inhibitor design [59]. The binding mode of **30** showed that its aminoquinolinone ring forms hydrogen bonds through its NH groups with the carboxylate of Asp287. To capitalize on this interaction, urea- and guanidine-containing analogs (compounds **31** and **32** respectively; Figure 11) were investigated. Compound **31** was chosen as a template for further exploration to develop inhibitors with better permeability properties, due to its tighter binding. By modifying R^1, Ar and R^2, compound **34** was developed which was significantly more potent (IC_{50} = 220 nM) and compound **33** had higher cell permeability (Figure 11). Additionally, compound **31** was also found to have high cell permeability and was capable of penetrating the mouse blood-brain barrier.

Figure 11. Inhibitor design scheme of Shibata *et al.* (2012) for IleRS [59].

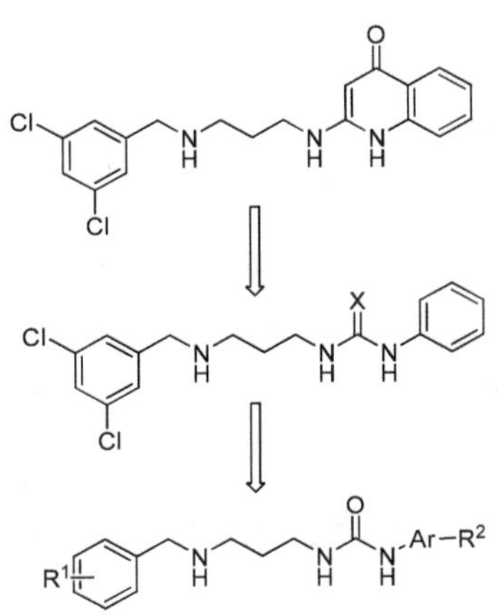

30

31: X = O (IC_{50} = 415 nM)
32: X = NH

33: R^1 = 3,5-diCl; Ar = 3-thiophene; R^2 = H (IC_{50} = 220 nM)
34: R^1 =3-Cl,5-OMe; Ar = 3-thiophene; R^2 = H (IC_{50} = 150 nM)

4. Conclusions

Aminoacyl-tRNA synthetases play a central role in protein synthesis by catalyzing the transfer of amino acids to their cognate tRNAs. These enzymes are clinically validated drug targets and they have been successfully targeted by anti-bacterial, anti-fungal and anti-parasitic agents. However, only one approved aaRS inhibitor is used clinically to date, illustrating the need for further aaRS inhibitor discovery and development.

All aaRSs from different species use the same reaction intermediates in the aminoacylation reaction, representing their conservatism in structures. However, some divergence has occurred throughout their evolution. For example, LeuRSs are evolved into bacterial and eukaryal/archaeal subtypes that possess a number of different residues in the active sites. Consequently, it is possible to take advantage of these active site variations to develop species-selective aaRS inhibitors, although it presents a challenge as in the development of any other selective inhibitors. The drug mupirocin which shows an 8000-fold selectivity for pathogenic IleRS over human IleRS is an excellent example of success.

Computational methods are commonly used in all areas of health science research. Among them, virtual screening and structure-based drug design have become established as powerful methods for identification and optimization of potential small molecule drugs. In this article, we reviewed the application of these approaches in the discovery and development of LeuRS, TrpRS, AsnRS, MetRS, IleRS and ThrRS inhibitors. This review demonstrates the wide use of computational methods for aaRS inhibitor discovery and development, which will surely assist the production of much needed novel antibiotics and other pharmaceutical agents.

Acknowledgments

We thank National Science Foundation of China (81222042 and 81102313); the Ministry of Science and Technology of China (2009CB918404 and 2012CB518001); China Postdoctoral Science Foundation Grant (2013M541525) and the E-Institutes of Shanghai Universities (EISU) Chemical Biology Division for financial support of this work.

References

1. Ibba, M.; Soll, D. Aminoacyl-tRNA synthesis. *Annu. Rev. Biochem.* **2000**, *69*, 617–650.
2. Eriani, G.; Delarue, M.; Poch, O.; Gangloff, J.; Moras, D. Partition of tRNA synthetases into two classes based on mutually exclusive sets of sequence motifs. *Nature* **1990**, *347*, 203–206.
3. Arnez, J.G.; Moras, D. Structural and functional considerations of the aminoacylation reaction. *Trends. Biochem. Sci.* **1997**, *22*, 211–216.
4. Schimmel, P. Development of tRNA synthetases and connection to genetic code and disease. *Protein Sci.* **2008**, *17*, 1643–1652.
5. Lincecum, T.L., Jr.; Tukalo, M.; Yaremchuk, A.; Mursinna, R.S.; Williams, A.M.; Sproat, B.S.; van den Eynde, W.; Link, A.; van Calenbergh, S.; Grotli, M.; *et al.* Structural and mechanistic basis of pre- and posttransfer editing by leucyl-tRNA synthetase. *Mol. Cell* **2003**, *11*, 951–963.

6. Jakubowski, H.; Goldman, E. Editing of errors in selection of amino acids for protein synthesis. *Microbiol. Rev.* **1992**, *56*, 412–429.
7. Beuning, P.J.; Musier-Forsyth, K. Hydrolytic editing by a class II aminoacyl-tRNA synthetase. *Proc. Natl. Acad. Sci. USA* **2000**, *97*, 8916–8920.
8. Dock-Bregeon, A.; Sankaranarayanan, R.; Romby, P.; Caillet, J.; Springer, M.; Rees, B.; Francklyn, C.S.; Ehresmann, C.; Moras, D. Transfer RNA-mediated editing in threonyl-tRNA synthetase. The class II solution to the double discrimination problem. *Cell* **2000**, *103*, 877–884.
9. Beaulieu, D.; Ohemeng, K.A. Patents on bacterial tRNA synthetase inhibitors: January 1996 to March 1999. *Exp. Opin. Ther. Patents* **1999**, *9*, 1021–1028.
10. Seiradake, E.; Mao, W.; Hernandez, V.; Baker, S.J.; Plattner, J.J.; Alley, M.R.; Cusack, S. Crystal structures of the human and fungal cytosolic Leucyl-tRNA synthetase editing domains: A structural basis for the rational design of antifungal benzoxaboroles. *J. Mol. Biol.* **2009**, *390*, 196–207.
11. Zhang, Y.K.; Plattner, J.J.; Freund, Y.R.; Easom, E.E.; Zhou, Y.; Gut, J.; Rosenthal, P.J.; Waterson, D.; Gamo, F.J.; Angulo-Barturen, I.; *et al.* Synthesis and structure-activity relationships of novel benzoxaboroles as a new class of antimalarial agents. *Bioorg. Med. Chem. Lett.* **2011**, *21*, 644–651.
12. Ding, D.; Meng, Q.; Gao, G.; Zhao, Y.; Wang, Q.; Nare, B.; Jacobs, R.; Rock, F.; Alley, M.R.; Plattner, J.J.; *et al.* Design, synthesis, and structure-activity relationship of *Trypanosoma brucei* leucyl-tRNA synthetase inhibitors as antitrypanosomal agents. *J. Med. Chem.* **2011**, *54*, 1276–1287.
13. Chopra, S.; Palencia, A.; Virus, C.; Tripathy, A.; Temple, B.R.; Velazquez-Campoy, A.; Cusack, S.; Reader, J.S. Plant tumour biocontrol agent employs a tRNA-dependent mechanism to inhibit leucyl-tRNA synthetase. *Nat. Commun.* **2013**, *4*, 1417.
14. Van de Vijver, P.; Ostrowski, T.; Sproat, B.; Goebels, J.; Rutgeerts, O.; van Aerschot, A.; Waer, M.; Herdewijn, P. Aminoacyl-tRNA synthetase inhibitors as potent and synergistic immunosuppressants. *J. Med. Chem.* **2008**, *51*, 3020–3029.
15. Ataide, S.F.; Ibba, M. Small molecules: Big players in the evolution of protein synthesis. *ACS Chem. Biol.* **2006**, *1*, 285–297.
16. Hurdle, J.G.; O'Neill, A.J.; Chopra, I. Prospects for aminoacyl-tRNA synthetase inhibitors as new antimicrobial agents. *Antimicrob. Agents Chemother.* **2005**, *49*, 4821–4833.
17. Hurdle, J.G.; O'Neill, A.J.; Ingham, E.; Fishwick, C.; Chopra, I. Analysis of mupirocin resistance and fitness in *Staphylococcus aureus* by molecular genetic and structural modeling techniques. *Antimicrob. Agents Chemother.* **2004**, *48*, 4366–4376.
18. Nakama, T.; Nureki, O.; Yokoyama, S. Structural basis for the recognition of isoleucyl-adenylate and an antibiotic, mupirocin, by isoleucyl-tRNA synthetase. *J. Biol. Chem.* **2001**, *276*, 47387–47393.
19. Rock, F.L.; Mao, W.; Yaremchuk, A.; Tukalo, M.; Crepin, T.; Zhou, H.; Zhang, Y.K.; Hernandez, V.; Akama, T.; Baker, S.J.; *et al.* An antifungal agent inhibits an aminoacyl-tRNA synthetase by trapping tRNA in the editing site. *Science* **2007**, *316*, 1759–1761.
20. Lavecchia, A.; di Giovanni, C. Virtual screening strategies in drug discovery: A critical review. *Curr. Med. Chem.* **2013**, *20*, 2839–2860.
21. Shoichet, B.K.; McGovern, S.L.; Wei, B.; Irwin, J.J. Lead discovery using molecular docking. *Curr. Opin. Chem. Biol.* **2002**, *6*, 439–446.

22. Reddy, A.S.; Pati, S.P.; Kumar, P.P.; Pradeep, H.N.; Sastry, G.N. Virtual screening in drug discovery—A computational perspective. *Curr. Protein. Pept. Sci.* **2007**, *8*, 329–351.
23. Zoete, V.; Grosdidier, A.; Michielin, O. Docking, virtual high throughput screening and *in silico* fragment-based drug design. *J. Cell. Mol. Med.* **2009**, *13*, 238–248.
24. PDB. Available online: http://www.pdb.org (accessed on 6 December 2013).
25. Friesner, R.A.; Banks, J.L.; Murphy, R.B.; Halgren, T.A.; Klicic, J.J.; Mainz, D.T.; Repasky, M.P.; Knoll, E.H.; Shelley, M.; Perry, J.K.; *et al.* Glide: A new approach for rapid, accurate docking and scoring. 1. Method and assessment of docking accuracy. *J. Med. Chem.* **2004**, *47*, 1739–1749.
26. Halgren, T.A.; Murphy, R.B.; Friesner, R.A.; Beard, H.S.; Frye, L.L.; Pollard, W.T.; Banks, J.L. Glide: A new approach for rapid, accurate docking and scoring. 2. Enrichment factors in database screening. *J. Med. Chem.* **2004**, *47*, 1750–1759.
27. Jones, G.; Willett, P.; Glen, R.C.; Leach, A.R.; Taylor, R. Development and validation of a genetic algorithm for flexible docking. *J. Mol. Biol.* **1997**, *267*, 727–748.
28. Moustakas, D.T.; Lang, P.T.; Pegg, S.; Pettersen, E.; Kuntz, I.D.; Brooijmans, N.; Rizzo, R.C. Development and validation of a modular, extensible docking program: DOCK 5. *J. Comput. Aided Mol. Des.* **2006**, *20*, 601–619.
29. Trott, O.; Olson, A.J. AutoDock Vina: Improving the speed and accuracy of docking with a new scoring function, efficient optimization, and multithreading. *J. Comput. Chem.* **2010**, *31*, 455–461.
30. Kim, K.H.; Kim, N.D.; Seong, B.L. Pharmacophore-based virtual screening: A review of recent applications. *Expert. Opin. Drug Discov.* **2010**, *5*, 205–222.
31. *Discovery Studio*; Accelrys Inc.: San Diego, CA, USA, 2013.
32. *Phase*; Schrödinger LLC: New York, NY, USA, 2013.
33. Wolber, G.; Langer, T. LigandScout: 3-D pharmacophores derived from protein-bound ligands and their use as virtual screening filters. *J. Chem. Inf. Model.* **2005**, *45*, 160–169.
34. Kalyaanamoorthy, S.; Chen, Y.P. Structure-based drug design to augment hit discovery. *Drug Discov. Today* **2011**, *16*, 831–839.
35. Zurcher, M.; Diederich, F. Structure-based drug design: Exploring the proper filling of apolar pockets at enzyme active sites. *J. Org. Chem.* **2008**, *73*, 4345–4361.
36. Reich, S.H.; Melnick, M.; Davies, J.F.; Appelt, K.; Lewis, K.K.; Fuhry, M.A.; Pino, M.; Trippe, A.J.; Nguyen, D.; Dawson, H.; *et al.* Protein structure-based design of potent orally bioavailable, nonpeptide inhibitors of human immunodeficiency virus protease. *Proc. Natl. Acad. Sci. USA* **1995**, *92*, 3298–3302.
37. Von Itzstein, M.; Wu, W.Y.; Kok, G.B.; Pegg, M.S.; Dyason, J.C.; Jin, B.; van Phan, T.; Smythe, M.L.; White, H.F.; Oliver, S.W.; *et al.* Rational design of potent sialidase-based inhibitors of influenza virus replication. *Nature* **1993**, *363*, 418–423.
38. Kim, C.U.; Lew, W.; Williams, M.A.; Liu, H.; Zhang, L.; Swaminathan, S.; Bischofberger, N.; Chen, M.S.; Mendel, D.B.; Tai, C.Y.; *et al.* Influenza neuraminidase inhibitors possessing a novel hydrophobic interaction in the enzyme active site: Design, synthesis, and structural analysis of carbocyclic sialic acid analogues with potent anti-influenza activity. *J. Am. Chem. Soc.* **1997**, *119*, 681–690.

39. Kim, C.U.; Lew, W.; Williams, M.A.; Wu, H.; Zhang, L.; Chen, X.; Escarpe, P.A.; Mendel, D.B.; Laver, W.G.; Stevens, R.C. Structure-activity relationship studies of novel carbocyclic influenza neuraminidase inhibitors. *J. Med. Chem.* **1998**, *41*, 2451–2460.
40. Fukunaqa, R.; Yokoyama, S. Crystal structure of leucyl-tRNA synthetase from the archaeon *Pyrococcus horikoshii* reveals a novel editing domain orientation. *J. Mol. Biol.* **2005**, *346*, 57–71.
41. Zhao, Y.; Wang, Q.; Meng, Q.; Ding, D.; Yang, H.; Gao, G.; Li, D.; Zhu, W.; Zhou, H. Identification of *Trypanosoma brucei* leucyl-tRNA synthetase inhibitors by pharmacophore- and docking-based virtual screening and synthesis. *Bioorg. Med. Chem.* **2012**, *20*, 1240–1250.
42. SPECS. Available online: http://www.specs.net (accessed on 6 December 2013).
43. Wu, Y.; Yu, K.; Xu, B.; Chen, L.; Chen, X.; Mao, J.; Danchin, A.; Shen, X.; Qu, D.; Jiang, H. Potent and selective inhibitors of *Staphylococcus epidermidis tr*yptophanyl-tRNA synthetase. *J. Antimicrob. Chemother.* **2007**, *60*, 502–509.
44. Retailleau, P.; Huang, X.; Yin, Y.; Hu, M.; Weinreb, V.; Vachette, P.; Vonrhein, C.; Bricogne, G.; Roversi, P.; Ilyin, V.; *et al.* Interconversion of ATP binding and conformational free energies by tryptophanyl-tRNA synthetase: Structures of ATP bound to open and closed, pre-transition-state conformations. *J. Mol. Biol.* **2003**, *325*, 39–63.
45. Sukuru, S.C.; Crepin, T.; Milev, Y.; Marsh, L.C.; Hill, J.B.; Anderson, R.J.; Morris, J.C.; Rohatgi, A.; O'Mahony, G.; Grotli, M.; *et al.* Discovering new classes of *Brugia malayi* asparaginyl-tRNA synthetase inhibitors and relating specificity to conformational change. *J. Comput. Aided Mol. Des.* **2006**, *20*, 159–178.
46. Crepin, T.; Peterson, F.; Haertlein, M.; Jensen, D.; Wang, C.; Cusack, S.; Kron, M. A hybrid structural model of the complete *Brugia malayi* cytoplasmic asparaginyl-tRNA synthetase. *J. Mol. Biol.* **2011**, *405*, 1056–1069.
47. Cambridge Structural Database. Available online: http://cds.dl.ac.uk/cds/datasets/crys/csd/csd.html (accessed on 6 December 2013).
48. National Cancer Institute Plated Compounds Database. Available online: http://cactus.nci.nih.gov/download/nci/ (accessed on 6 December 2013).
49. Kim, S.Y.; Lee, Y.S.; Kang, T.; Kim, S.; Lee, J. Pharmacophore-based virtual screening: The discovery of novel methionyl-tRNA synthetase inhibitors. *Bioorg. Med. Chem. Lett.* **2006**, *16*, 4898–4907.
50. Bharatham, N.; Bharatham, K.; Lee, K.W. Pharmacophore identification and virtual screening for methionyl-tRNA synthetase inhibitors. *J. Mol. Graph. Model.* **2007**, *25*, 813–823.
51. Maybridge. Available online: http://www.maybridge.com (accessed on 6 December 2013).
52. Finn, J.; Stidham, M.; Hilgers, M.; G, C.K. Identification of novel inhibitors of methionyl-tRNA synthetase (MetRS) by virtual screening. *Bioorg. Med. Chem. Lett.* **2008**, *18*, 3932–3937.
53. ChemDiv. Available online: http://eu.chemdiv.com (accessed on 6 December 2013).
54. Zhang, F.; Du, J.; Wang, Q.; Hu, Q.; Zhang, J.; Ding, D.; Zhao, Y.; Yang, F.; Wang, E.; Zhou, H. Discovery of *N*-(4-sulfamoylphenyl)thioureas as *Trypanosoma brucei* leucyl-tRNA synthetase inhibitors. *Org. Biomol. Chem.* **2013**, *11*, 5310–5324.
55. Brown, M.J.; Mensah, L.M.; Doyle, M.L.; Broom, N.J.; Osbourne, N.; Forrest, A.K.; Richardson, C.M.; O'Hanlon, P.J.; Pope, A.J. Rational design of femtomolar inhibitors of

isoleucyl tRNA synthetase from a binding model for pseudomonic acid-A. *Biochemistry* **2000**, *39*, 6003–6011.

56. Teng, M.; Hilgers, M.T.; Cunningham, M.L.; Borchardt, A.; Locke, J.B.; Abraham, S.; Haley, G.; Kwan, B.P.; Hall, C.; Hough, G.W.; *et al.* Identification of bacteria-selective threonyl-tRNA synthetase substrate inhibitors by structure-based design. *J. Med. Chem.* **2013**, *56*, 1748–1760.
57. Sankaranarayanan, R.; Dock-Bregeon, A.C.; Rees, B.; Bovee, M.; Caillet, J.; Romby, P.; Francklyn, C.S.; Moras, D. Zinc ion mediated amino acid discrimination by threonyl-tRNA synthetase. *Nat. Struct. Biol.* **2000**, *7*, 461–465.
58. Shibata, S.; Gillespie, J.R.; Kelley, A.M.; Napuli, A.J.; Zhang, Z.; Kovzun, K.V.; Pefley, R.M.; Lam, J.; Zucker, F.H.; Van Voorhis, W.C.; *et al.* Selective inhibitors of methionyl-tRNA synthetase have potent activity against *Trypanosoma brucei* infection in mice. *Antimicrob. Agents Chemother.* **2011**, *55*, 1982–1989.
59. Shibata, S.; Gillespie, J.R.; Ranade, R.M.; Koh, C.Y.; Kim, J.E.; Laydbak, J.U.; Zucker, F.H.; Hol, W.G.; Verlinde, C.L.; Buckner, F.S.; *et al.* Urea-based inhibitors of *Trypanosoma brucei* methionyl-tRNA synthetase: Selectivity and *in vivo* characterization. *J. Med. Chem.* **2012**, *55*, 6342–6351.
60. Koh, C.Y.; Kim, J.E.; Shibata, S.; Ranade, R.M.; Yu, M.; Liu, J.; Gillespie, J.R.; Buckner, F.S.; Verlinde, C.L.; Fan, E.; *et al.* Distinct states of methionyl-tRNA synthetase indicate inhibitor binding by conformational selection. *Structure* **2012**, *20*, 1681–1691.
61. Ranade, R.M.; Gillespie, J.R.; Shibata, S.; Verlinde, C.L.; Fan, E.; Hol, W.G.; Buckner, F.S. Induced resistance to methionyl-tRNA synthetase inhibitors in *Trypanosoma brucei* is due to overexpression of the target. *Antimicrob. Agents Chemother.* **2013**, *57*, 3021–3028.
62. Evans, R.; Green, L.; Sun, X.; Guiles, J.; Lorimer, D.; Burgin, A.; Janjic, N.; Jarvis, T.; Davies, D. Co-crystal structure of REP3123 bound to *Clostridium difficile* methionyl tRNA synthetase, poster F1−2114. In *Abstracts of the 47th Interscience Conference on Antimicrobial Agents and Chemotherapy*, Proceedings of the 47th Interscience Conference on Antimicrobial Agents and Chemotherapy, Chicago, IL, USA, 17–20 September 2007; American Society for Microbiology: Washington, DC, USA, 2007.

3

Biophysical Insights into the Inhibitory Mechanism of Non-Nucleoside HIV-1 Reverse Transcriptase Inhibitors

Grant Schauer [1,2], Sanford Leuba [1,2] and Nicolas Sluis-Cremer [3,*]

[1] Program in Molecular Biophysics and Structural Biology, Hillman Cancer Center, University of Pittsburgh, 5117 Centre Ave., Pittsburgh, PA 15213, USA; E-Mails: gds11@pitt.edu (G.S.); leuba@pitt.edu (S.L.)

[2] Department of Cell Biology, Hillman Cancer Center, University of Pittsburgh, 5117 Centre Ave., Pittsburgh, PA 15213, USA

[3] Department of Medicine, Division of Infectious Diseases, 3550 Terrace St., Pittsburgh, PA 15261, USA

* Author to whom correspondence should be addressed; E-Mail: nps2@pitt.edu

---

Abstract: HIV-1 reverse transcriptase (RT) plays a central role in HIV infection. Current United States Federal Drug Administration (USFDA)-approved antiretroviral therapies can include one of five approved non-nucleoside RT inhibitors (NNRTIs), which are potent inhibitors of RT activity. Despite their crucial clinical role in treating and preventing HIV-1 infection, their mechanism of action remains elusive. In this review, we introduce RT and highlight major advances from experimental and computational biophysical experiments toward an understanding of RT function and the inhibitory mechanism(s) of NNRTIs.

Keywords: HIV-1 reverse transcriptase; NNRTI; mechanism; biophysics; single-molecule

---

1. Introduction

HIV-1 reverse transcriptase (RT) is an RNA- or DNA-dependent DNA polymerase and also contains ribonuclease H (RNase H) activity, thereby containing all the necessary enzymatic activity for the multistep conversion of HIV-1 single stranded RNA (ssRNA) into double stranded DNA (dsDNA) for subsequent integration into the human genome. RT thus remains the prime target for new antiretroviral therapies. Non-nucleoside RT inhibitors (NNRTIs) are potent inhibitors of reverse transcription, but despite having been successfully used in the clinic for over 15 years, a comprehensive explanation for their inhibitory mechanism(s) has remained elusive.

2. Structure and Function of HIV-1 RT

2.1. Reverse Transcription

Reverse transcription [1,2] is initiated by RT at the 3'-end of cellular lysyl-$tRNA^{Lys,3}$, which hybridizes to the primer binding site (PBS) on the HIV-1 genome. During initiation, RNA-primed RT elongates until the 5'-end of the HIV-1 RNA is reached, forming minus-strand strong-stop DNA (Figure 1). Employing the RNase H activity of RT, the remaining HIV-1 genomic RNA is cleaved to allow the nascently synthesized DNA to circularize and hybridize with the repeat sequence (R) at the 3'-end of the HIV-1 ssRNA. After this strand transfer, the nascent DNA strand is further elongated by RT. RT hydrolyzes the remaining RNA, but leaves behind a purine-rich sequence, named the polypurine tract (PPT), which subsequently serves as a primer for the initiation of second strand DNA synthesis. RT then elongates the PPT primer. The RNase H activity of RT removes all remaining RNA, including the transfer-RNA (tRNA) primer and the PPT. Strand transfer takes place by PBS sequence homology. DNA polymerization and strand-displacement followed by further DNA elongation results in an integrase-competent dsDNA, which is flanked by Long Terminal Repeat (LTR) sequences at both ends.

Figure 1. Reverse transcription. Schematic of the multistep process of the conversion of viral RNA (red) into integrase-competent dsDNA (bottom) for insertion into the human genome. PBS, primer binding site.

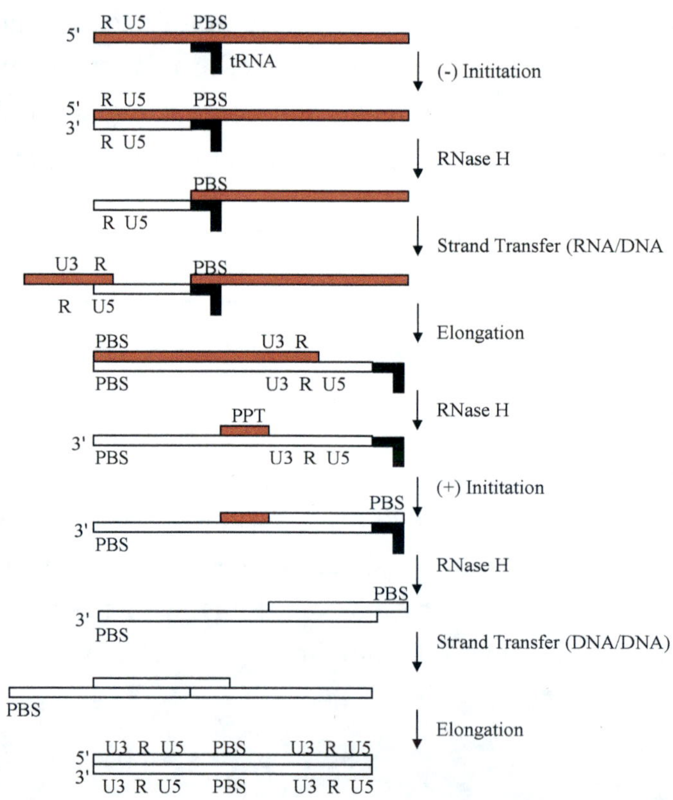

The resulting dsDNA is then a substrate for integrase, which catalyzes the insertion of dsDNA into the human genome [2].

2.2. Structural and Biophysical Studies of RT

With the ability to efficiently catalyze DNA polymerization on both RNA/DNA and DNA/DNA duplexes and also possessing RNase H activity, RT is an astonishingly versatile enzyme, due in large part to its modular structure. RT is a 110 kD heterodimer composed of two subunits: p66 (560 amino acids-long) and p51 (440 amino acids-long) [3]. Both subunits are a product of proteolytic processing of a gag-pol polyprotein by HIV-1 protease. In p51, the RNase-H domain (residues 440–560) has been cleaved, resulting in a structure that shares secondary structural elements with p66. However, since the overall tertiary structure is spatially configured differently than p66, p51 largely plays a scaffolding role for p66, the only subunit in the RT heterodimer to possess polymerase and RNase H activity [4,5].

X-ray crystallography, comprising the majority of biophysical studies on RT, has provided the community with an invaluable mechanistic understanding of RT. Beginning with the first structural elucidation of RT [5], there have since been hundreds of structures of RT solved (at the time of this writing, there were 245 structures of HIV-1 RT deposited in the Protein Database (PDB)). Additionally, a large body of mechanistic information complementary to the structural data has been gleaned from fluorescence-based kinetics studies. Like other DNA polymerases, RT is shaped like a right hand [5] (Figure 2).

Figure 2. Structure of reverse transcriptase (RT). RT is pictured as a solvent-accessible surface area model (taken from PDB ID: 1RTD). The DNA/DNA template/primer is shown as a cartoon. The p51 subdomain is colored in grey, and the p66 subdomain is subdivided into thumb (green), fingers (red), palm (purple), connection (wheat) and RNase H (blue) domains. The polymerase and RNase H active site residues are colored in yellow.

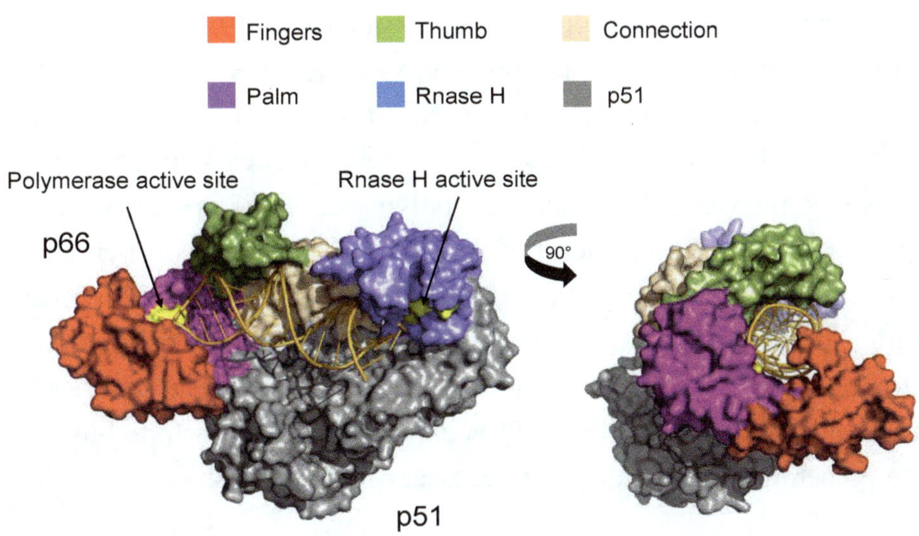

Its subdomains are accordingly named: fingers (residues 1–85 and 118–155), thumb (residues 237–318), palm (residues 86–117 and 156–236) and connection (319–426) subdomains (see Figure 2). In the *apo* structure of RT [6], the thumb and finger domains are nearly in contact, and the thumb resides in the nucleic acid binding cleft; however, to accommodate template/primer (T/P) substrates in the nucleic acid binding cleft, the thumb must extend outward [6,7]. Pre-steady-state kinetic analysis showed that binding to the T/P and deoxynucleotide triphosphate (dNTP) substrates is a two-step process involving initial entry of the T/P into the nucleic acid binding cleft and a slow step dependent on the conformational change of the enzyme to accommodate cognate dNTP, which can increase the overall binding affinity to the T/P substrate [8]. The resultant binary complex contains a T/P substrate simultaneously spanned by polymerase and RNase H active sites, approximately 18 base pairs apart.

2.3. Conformational Dynamics of Reverse Transcription

RT clamps down on the incoming nucleotide with its fingers [9], specifically, via the Lys65, Arg72, Asp113 and Ala114 residues of the β3-β4 loop [10]. Structures of RT complexed with a DNA/DNA template/primer that has been dideoxy terminated at the 3'-end (to create a dead-end complex in the presence of cognate dNTP) identified the binding site of dNTP: Asp113, Tyr115, Phe116 and Gln151 [9]. In agreement from pre-steady-state kinetics data and based on data from other polymerases [11], it was proposed that finger clamping of dNTP is a rate-limiting prerequisite to catalysis [12]. Pre-steady-state kinetic analysis showed that dNTP incorporation rates depend upon the composition of the T/P substrate and suggested that RT can bind T/P substrates in productive or nonproductive complexes in terms of dNTP incorporation and that RT must undergo an isomerization step in order for the RT-T/P complex to be converted into a productive one [13]. The presence of this dead-end complex was later directly visualized with single-pair Förster resonance energy transfer (FRET) via multiparameter fluorescence detection (MFD) [14,15]. It has further been shown, through a combination of stopped-flow fluorescence experiments and a molecular dynamics technique used to extend the timescales of simulations, called directional milestoning [16], that the rate-limiting step of finger bending is on the order of milliseconds and that the conformational change itself governs dNTP specificity [17]. The bending of the finger clamp precisely positions dNTP for incorporation, which is coordinated with the positioning of the growing primer by the "primer grip", composed of the β12-β13 hairpin [4]. Finally, formation of a nascent phosphodiester bond on a growing primer (*i.e.*, DNA polymerization) is coordinated by the catalytic triad in the palm subdomain (D110, D185, D186; part of the conserved YXDD motif, where Y and D stand for Tyrosine and Aspartate, respectively, and X stands for any amino acid), which is a conserved process amongst polymerases and is dependent upon two divalent metals [5]. The process of nucleotide addition ends in the release of pyrophosphate, which is accompanied by finger domain opening. In order for processive replication to continue, the T/P substrate needs to translocate one base pair with respect to RT, a process which is possibly linked to potential energy stored in the YMDD (Tyr-Met-Asp-Asp) loop, akin to a loaded springboard [18,19]. Consistent with this, atomic force microscopy (AFM) experiments demonstrated that finger clamping is likely responsible for generating the motor force necessary for RT translocation [20].

3. NNRTIs and Their Mechanism of Action

3.1. NNRTIs

NNRTIs are amphiphilic compounds that bind to a hydrophobic pocket in HIV-1 RT, called the NNRTI binding pocket (NNRTIBP), that is proximal to, but distinct from, the polymerase active site [21,22]. The NNRTIBP resides close to the p51-p66 interface and does not exist unless an NNRTI is bound [5,21]. NNRTIs typically bind with low nanomolar affinity and are potent inhibitors of reverse transcription [22]. In order of their approval from first to most recent, the five United States Federal Drug Adminstration (USFDA)-approved NNRTIs include nevirapine (NVP), delavirdine (DEL), efavirenz (EFV), etravirine (ETV) and rilpivirine (RIL) (Figure 3).

Figure 3. Chemical structures of USFDA-approved non-nucleoside RT inhibitors (NNRTIs).

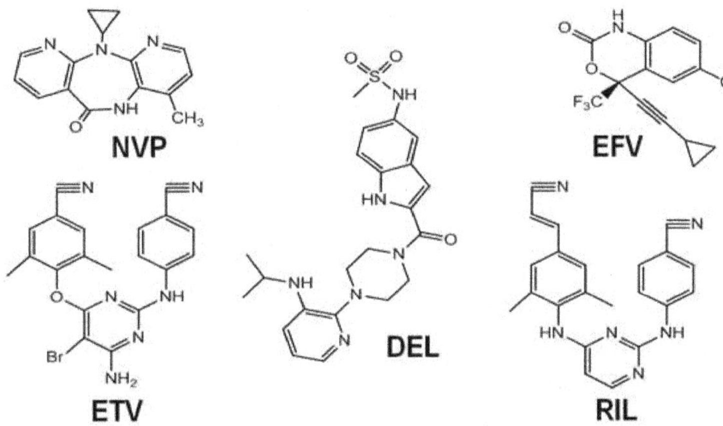

3.2. NNRTI Binding

Binding of NNRTI to RT is accompanied by rearrangements in the residues, Y181 and Y188, as well as in the primer grip motif, thereby forming the cavity of the NNRTIBP [5,6,21]. The fact that this pocket is not observed without the presence of NNRTIs has led some groups to conclude that an induced-fit mechanism [23] is responsible for NNRTI binding [24,25]; however, several other groups have demonstrated that conformational selection likely governs NNRTI binding [26–29]. Fourier transform ion cyclotron resonance mass spectrometry (FT-ICR MS) was originally used to show that p51 or p66 monomers undergo a population shift upon binding to EFV [28], and experiments monitoring EFV binding kinetics via tryptophan fluorescence were subsequently used to show that association of EFV to p51/p66 dimers was extremely slow, consistent with a conformational selection model [29]. Also consistent with this is data from nuclear magnetic resonance (NMR) experiments with RT that was site-specifically spin-labeled, which suggested that the primer grip was much more flexible than previously observed, presumably aiding in NNRTI binding [30]. The dynamic flexibility of the primer grip, comprising part of the NNRTIB, may provide evidence for conformational selection of NNRTIs. The likely important role of conformational selection has been further demonstrated using the Anisotropic Network Model [31] (ANM) combined with principle components analysis (PCA) methodology that both the unbound and NNRTI-bound conformation of RT intrinsically exist, but that NNRTI preferentially binds to conformations of RT within the structural ensemble, which are best predisposed toward NNRTI binding, shifting the population toward the NNRTI-bound form [27].

One obstacle to understanding the inhibitory mechanism of NNRTIs is that NNRTIs are practically insoluble in water at high concentrations (<10 mg/mL), such that traditional techniques for determining the equilibrium dissociation constant, K_d, are made difficult. For instance, traditional K_d determination with isothermal titration calorimetry (ITC), the "gold standard" for measuring reaction thermodynamics, is made difficult, since it typically requires higher concentrations of protein (and, therefore, drug) than attainable for NNRTIs. Most studies therefore use polymerase activity assays to indirectly infer K_d for NNRTIs; however, the disadvantage to this technique is that it only measures dNTP incorporation, such that, e.g., if NNRTIs primarily had an effect on nonproductive binding, it would be difficult to determine an accurate K_d. One alternative is surface plasmon resonance (SPR), which has been used to calculate K_d values for some NNRTIs [26,32,33]. Both SPR and ITC can be limited by extremely slow association and/or dissociation reactions, depending on the experiment [25,29]. One group circumvented the slow association of NNRTI to p51 monomers by extracting the K_d of EFV through the thermodynamic linkage between dimerization and NNRTI binding by combining kinetics from equilibrium dialysis, tryptophan fluorescence and native gel electrophoresis [29]. Another group recently surmounted the solubility barrier to ITC experiments by employing a technique using small, incremental injections of RT at regular intervals into the calorimeter (the Multiple Injection Method, or MIM [34]) containing fixed concentrations of NNRTI, demonstrating that, contrary to most reports, rather than acting on RT-DNA complexes, NNRTIs preferentially bind to free RT. This study showed that NNRTI-bound RT can then associate with DNA and form either a DNA-RT-NNRTI complex that exists in either a polymerase-competent incompetent configuration, both of which dNTP is unable to bind to [35].

3.3. Putative Mechanisms of Inhibition by NNRTIs

Interestingly, binding of NNRTIs does not appear to affect the ability of RT to bind incoming dNTPs [36,37]. Since the NNRTIBP is ~1 nm away from the polymerase active site and since binding of NNRTIs is noncompetitive with dNTP or template/primer binding [38], NNRTIs are considered classic allosteric inhibitors of HIV-1 RT DNA polymerization [21,39]. Despite their widespread use and remarkable efficacy, the precise mechanism(s) behind inhibition of reverse transcription by NNRTIs remains unclear [2,40–42], but proposed mechanisms have included restriction of thumb flexibility [5], distortion of the catalytic triad [43], primer grip repositioning [44], repositioning of the primer [45], enhancement of RNase H activity [46,47] and loosening of the thumb/finger clamp on template/primer substrates [48].

3.4. Molecular Arthritis

One persistent observation from the solved crystal structures of RT in the presence of NNRTIs has been the hyperextension of the thumb domain and a further opening of the finger domain accompanied by a marked decrease in flexibility, also known as "molecular arthritis [5]" (see Figure 4). Using hydrogen exchange mass spectrometry (HXMS), it was demonstrated that binding of EFV is allosterically coupled to regions of RT that are up to 6 nm away from the NRTIBP, primarily acting to stabilize the dynamics of these regions [39]. Elastic network models [31] and multicopy molecular dynamics simulations have shown that molecular arthritis appears to be closely related to the existence of a hinge region between the p51 and p66 subdomains [49,50]. Evaluating the slowest ANM modes, it was also shown that the presence of NNRTI changes the direction of finger and thumb motions, while the actual restriction of mobility was primarily seen in the presence of EFV, but not NVP [51]. Recently, however, one group using molecular dynamics simulations suggested that thumb mobility is not constrained as much as previously reported [52]. Notwithstanding, restriction of thumb flexibility has long been considered a key element in the inhibition mechanism [5], but whether molecular arthritis is directly and functionally tied to inhibition of polymerization is unclear.

Figure 4. Molecular arthritis. The structures of *apo* wild-type (WT) RT (left; PDB ID: 1DLO) and WT RT bound to efavirenz (right; PDB ID: 1FK9). Structures are represented by solvent-accessible surface area. The p51 subunit is colored in grey, and the p66 subunit is colored according to crystallographic B-factors, with blue being the least mobile and red the most mobile residues. EFV, efavirenz; NNRTIBP, NNRTI binding pocket.

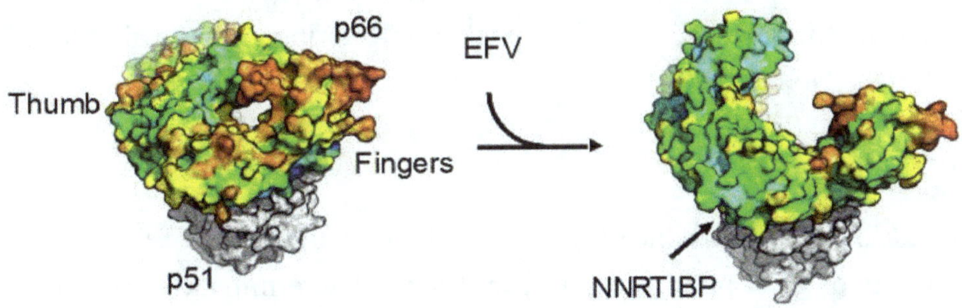

One possible effect of NNRTI-induced molecular arthritis is distortion of the catalytic triad [43]. In addition to the observation from crystal structures, repositioning of the catalytic triad into a polymerase incompetent configuration was directly observed in large, multicopy simulations of *apo* and NNRTI-bound RT [50]. Possibly linked to thumb hyperextension or simply to local rearrangements resulting from NNRTI binding, distortion of the precise constellation residues in the YXDD motif is presumed to be deleterious to polymerization activity. In a similar, but distinct, mechanism, it has also been postulated that NNRTI-induced molecular arthritis is responsible for repositioning the primer grip, eliminating the coordination of primer to the catalytic triad [44]. Furthermore, observation from crystal structures suggested that molecular arthritis is linked to the opening up of the RNase H domain, thereby providing access to degradation of DNA/RNA duplexes [53]. Consistent with this, EFV was shown to enhance RNase H activity [54].

3.5. Effect of NNRTIs on RT Dimerization

NNRTIs have been shown to enhance or stabilize the rate of RT heterodimerization, depending on the NNRTI studied, suggesting diversity in binding modes depending on chemical structure [55]. Perhaps counterintuitively, it has been demonstrated that some NNRTIs, such as EFV, act as potent chemical enhancers of HIV-1 RT heterodimerization [56,57]. To date, efavirenz was found to be the most potent enhancer of RT heterodimerization, whereas nevirapine has a weak effect, and delavirdine has no effect at all [56]. While some studies have demonstrated the effects of some potent NNRTIs, (e.g., efavirenz, dapivirine and etravirine) on the late stages of HIV replication [58,59], results from pre-steady-state kinetics experiments indicated that there does not appear to be a correlation between the impact of NNRTI-mediated enhancement of RT heterodimerization and the defects in RT polymerase function [60].

3.6. NNRTI Resistance Mutations

Due to the error-prone nature of RT, resistance mutations in and around the NNRTIBP arise during treatment regimens involving NNRTI therapy and are associated with virologic failure [61]. Because NNRTI resistance mutations are necessarily linked to RT function and to the mechanism of susceptibility of RT to NNRTIs, it is instructive to study them in order to better understand NNRTI mechanism [61]. It is generally understood that NNRTI resistance mutations can affect inhibitor binding by either altering one or more favorable interactions between the inhibitor and NNRTIBP (e.g., the Y181C mutation eliminates π-stacking interactions between this residue and the aromatic ring of the NNRTI [62]), by introducing steric barriers to NNRTI binding (e.g., G190E introduces a bulky side-chain, which may sterically interfere with NNRTI binding [63,64]) or by eliminating or altering the inter-residue network of the NNRTI-binding pocket, interfering with the ability of other residues in the pocket to envelop the NNRTI [21].

3.6.1. K103N

Although a discussion of all resistance mutations is beyond the scope of this review, particularly in light of the accessibility of growing RT crystals for structural determination, resulting in an abundance of solved structures of RT with various resistance mutations, we will discuss K103N as a prime example of an NNRTI-resistance mutation, since it is the most commonly reported clinical mutation [65]. K103N confers high level resistance to a wide range of NNRTIs, notably efavirenz (EFV) and nevirapine (NVP) (approximately 20-fold reduction in susceptibility) [66]. Despite its devastating effects on susceptibility to first-generation NNRTIs, the mechanism of resistance by K103N remains unclear. Observations from the first structures of K103N RT [62,67] indicated that neither K103 nor the mutant N103 residue make any contacts with EFV or NVP (although K103 can H-bond with delavirdine (DEL)) [65]. It was initially proposed that the K103N mutation was responsible for adding an extra H-bond with Y188 in the NNRTIBP, stabilizing the *apo* form of RT [67]; however, the energy of an H-bond is relatively weak compared to that of NNRTI binding, and surface plasmon resonance (SPR) studies showed that K103N actually facilitates entry and exit into the NNRTIBP [38]. It was later inferred from a K103N/Y181C double mutant that the extra H-bond was possibly mediated by a Na^+ ion and water molecules in the NNRTIBP [68]. Unlike first generation NNRTIs, the next generation diarylpyrimidine NNRTIs, etravirine (ETR) and rilpivirine (RIL), are effective against K103N RT [69–73]. At the moment, however, it is unclear what advantage these inhibitors have over first generation NNRTIs, although the crystal structures of K103N/Y181C RT and K103N/L101I RT suggest that the flexibility of ETR and RIL may allow the NNRTIBP to conform to these drugs, even in the presence of resistance mutations in the NNRTIBP [71]. It is also possible that the strengthened H-bond network conferred by K103N is broken by this new class of NNRTIs through a binding mode with extra favorable interactions [73]. Alternately, the diarylpyrimidine NNRTIs may be able to overcome K103N in a yet-to-be determined mechanism.

3.7. Structures of NNRTI-Bound RT-T/P Complexes

Until recently, there only existed structures of RT complexed with dsDNA template/primer (T/P) mimic substrates (mostly DNA/DNA, DNA/PPT) or structures of RT complexed with NNRTIs; however, there was no structure of a ternary RT-T/P-NNRTI complex. Recently, however, a structure of RT crosslinked to a DNA/DNA T/P substrate and bound to nevirapine (NVP) was elucidated [45], wherein the authors observed a distortion of the primer grip accompanied by thumb hyperextension and a 5.5 Å shift of the 3'-end primer away from the active site, dNTP binding site distortion and a reduction in RT-T/P contacts. By restricting the mobility of RT, it is possible that crosslinking of RT to the T/P substrate could lead to experimental artifacts. Soon after, structures of RT bound to several DNA/RNA T/P substrates in the presence of NVP and EFV were solved without the use of crosslinking agents [47]. The structure showed that NNRTIs pushed RT toward a degradative/RNase H-competent mode (as opposed to a polymerase competent mode) in the context of a hybrid substrate, which is consistent with the findings that EFV accelerates RNA degradation [54].

3.8. Insights into RT Dynamics from Time-Resolved Single-Molecule Experiments

Given the heterogeneity and complexity of the processes involved in reverse transcription and the suspected dynamic nature of RT-T/P interactions, there is great interest in observing single kinetic processes of reverse transcription. Due to the brief amount of time spent in the observation volume (~1 ms), single-molecule methods based on confocal fluorescence microscopy provide instantaneous, albeit detailed, structural information from FRET signals [14,15,74]. In contrast, time-resolved single-molecule techniques, relying on surface-tethering schemes, offer a glimpse into a broad range of dynamic processes for up to several minutes. One method to visualize RT dynamics on nucleic acid substrates is through direct visualization of the lengthening of flow-stretched, bead-tethered ssDNA [75] by RT as it completes primer extension [76]. Using this technique, different rates of primer extension and strand displacement were observed, and it was shown that RT dynamically switches between these modes, depending on the composition of the substrate [77]. Crucially, single-pair FRET experiments using total internal reflection fluorescence microscopy (TIRFM) [78] revealed that RT slides on nucleic acid substrates in a highly dynamic manner [48]. While cognate dNTP stabilized the polymerase competent configuration, NNRTI destabilized this configuration, enhancing the sliding dynamics of RT on the T/P. Remarkably, RT was shown to flip on template/primer substrates and can reside in different relative orientations based on the composition of the template/primer alone. It was further shown that RT bound T/P substrates in opposite orientations depending on whether it contains a DNA or RNA primer [77]. The relative orientation was a direct predictor of enzymatic activity; *i.e.*, when assayed for activity, the orientations were shown to be either polymerase-competent or RNase H-competent. When bound to a T/P with a PPT primer, RT could flip between these orientations, and the rate of flipping could be slowed by dNTP (directing RT toward a polymerase-competent configuration) or enhanced by NNRTIs (directing it toward an RNA-degradative orientation) [77]. Taken together, these results indicated that modulation of the grip on the T/P substrate itself by NNRTIs may alter RT-T/P dynamics, disfavoring the polymerase mode, suggesting that NNRTI-induced molecular arthritis affects the grip on the template/primer substrate. Intriguingly, it was recently demonstrated that the binding of the various dimeric isoforms of RT could be distinguished using a single-molecule fluorescence method, called Protein-Induced Fluorescence Enhancement (PIFE) [79], where the authors were able to monitor the binding of RT via the fluorescence intensity enhancement of a Cy3-labeled T/P resulting from the proximity of RT to Cy3 [80].

4. Conclusions and Future Directions

Combining results from many interdisciplinary biophysical experiments, a comprehensive picture of the mechanism(s) of action of NNRTIs is emerging, but is far from complete. Clearly, molecular arthritis plays a role in the inhibitory mechanism of NNRTI, but the exact link between modulation of structural dynamics as it relates to the attenuation of RT function remains unclear. Exciting new means of obtaining the crystal structures of T/P-bound RT in the presence of NNRTIs have emerged, and novel structures of T/P-bound, NNRTI-resistant RT mutants complexed to their relative NNRTIs are likely soon to follow. In X-ray crystallographic studies, crystal packing can lead to underestimation of flexibility, as shown for RT with Hydrogen-Deuterium Exchange Mass Spectrometry (HXMS) [81] and NMR [30], highlighting the necessity of using molecular dynamics simulations to probe the conformational dynamics of solved structures. Advances in sampling techniques for molecular simulations combined with the continually increasing computational power afforded by Moore's Law [82] will help bring these structures to life, providing unprecedented knowledge of the conformational dynamics of reverse transcription and its inhibition by NNRTIs. Advances in the techniques used to quantify the thermodynamics of NNRTI binding may help in understanding the evolutionary rationale for certain NNRTI resistance mutations. Novel spin-label probes for NMR experiments will permit measurement of the dynamics of new regions of RT in the presence or absence of NNRTIs. Although traditional biophysical experiments have afforded crucial insight into the mechanisms of reverse transcription and its inhibition by NNRTIs, single-molecule biophysical experiments offer many strategic advantages over bulk experiments when it comes to understanding the kinetic mechanisms of a dynamic enzyme like RT. For instance, since traditional bulk fluorescence spectroscopy experiments represent an ensemble average of many unsynchronized kinetic processes occurring simultaneously (not all of which are related to the observable of interest), data from these techniques may be difficult to interpret [83]. Furthermore, crystallographic structures represent an ensemble average of structural states at equilibrium, but cannot provide information on the transitional intermediates between these states. Eliminating bulk averaging and enabling the visualization of rare events not associated with low-energy structures, single-molecule techniques will likely be increasingly used to resolve signals from individual RT complexes without the need for synchronization [83]. It will be interesting to see whether the techniques involving the multiplexed observation of flow-stretched DNA molecules will be put to use to probe NNRTI function. Completely new types of information have been made available from single-pair FRET, and these experiments are likely to continue. Although experiments cannot provide the detailed structural and/or orientation information, it is nonetheless an extensible technique that will likely see more use in the near future. Since PIFE does not require exogenous labeling of RT, it does not require structural perturbation and also may be more accessible to some laboratories than FRET.

Furthermore, through observation of signals from T/P in the presence of unlabeled RT, single-molecule protein induced fluorescence enhancement (PIFE) experiments [80] allow the observation of physiologically relevant RT concentrations by providing a workaround to the "concentration problem" [83] seen in single-molecule FRET experiments, which are limited to the use of low nanomolar concentrations of labeled protein. Biophysical studies of RT structure, function and inhibition by NNRTI are all interrelated; e.g., the results from time-resolved single-molecule experiments must be informed by the structural data, the interpretation of which is, in turn, aided by new advances in computational simulation and biophysical techniques to measure protein dynamics. Combining these powerful methodologies, a great wave of enlightening structural information about HIV-1 RT is imminent, hopefully providing new strategies for rational drug design against this highly elusive target.

Acknowledgements

Our work on NNRTIs has been funded by grants AI081571, GM068406 and T32GM088119 from the National Institutes of Health (USA).

References

1. Hu, W.-S.; Hughes, S.H. HIV-1 reverse transcription. *Cold Spring Harb. Perspect. Med.* **2012**, *2*, doi:10.1101/cshperspect.a006882.
2. Sluis-Cremer, N.; Tachedjian, G. Mechanisms of inhibition of HIV replication by non-nucleoside reverse transcriptase inhibitors. *Virus Res.* **2008**, *134*, 147–156.
3. Divita, G.; Restle, T.; Goody, R.S. Characterization of the dimerization process of HIV-1 reverse transcriptase heterodimer using intrinsic protein fluorescence. *FEBS Lett.* **1993**, *324*, 153–158.
4. Jacobo-Molina, A.; Ding, J.; Nanni, R.G.; Clark, A.D.; Lu, X.; Tantillo, C.; Williams, R.L.; Kamer, G.; Ferris, A.L.; Clark, P. Crystal structure of human immunodeficiency virus type 1 reverse transcriptase complexed with double-stranded DNA at 3.0 A resolution shows bent DNA. *Proc. Natl. Acad. Sci. USA* **1993**, *90*, 6320–6324.
5. Kohlstaedt, L.A.; Wang, J.; Friedman, J.M.; Rice, P.A.; Steitz, T.A. Crystal structure at 3.5 Å resolution of HIV-1 reverse transcriptase complexed with an inhibitor. *Science* **1992**, *256*, 1783–1790.
6. Rodgers, D.W.; Gamblin, S.J.; Harris, B.A.; Ray, S.; Culp, J.S.; Hellmig, B.; Woolf, D.J.; Debouck, C.; Harrison, S.C. The structure of unliganded reverse transcriptase from the human immunodeficiency virus type 1. *Proc. Natl. Acad. Sci. USA* **1995**, *92*, 1222–1226.
7. Esnouf, R.; Ren, J.; Ross, C.; Jones, Y.; Stammers, D.; Stuart, D. Mechanism of inhibition of HIV-1 reverse transcriptase by non-nucleoside inhibitors. *Nat. Struct. Mol. Biol.* **1995**, *2*, 303–308.
8. KruhØfter, M.; Urbanke, C.; Grosse, F. Two step binding of HIV-1 reverse transcriptase to nucleic acid substrates. *Nucleic Acids Res.* **1993**, *21*, 3943–3949.
9. Huang, H.; Chopra, R.; Verdine, G.L.; Harrison, S.C. Structure of a covalently trapped catalytic complex of HIV-1 reverse transcriptase: Implications for drug resistance. *Science* **1998**, *282*, 1669–1675.
10. Grice, S.F.J.L. Human immunodeficiency virus reverse transcriptase: 25 Years of research, drug discovery, and promise. *J. Biol. Chem.* **2012**, *287*, 40850–40857.

11. Kati, W.M.; Johnson, K.A.; Jerva, L.F.; Anderson, K.S. Mechanism and fidelity of HIV reverse transcriptase. *J. Biol. Chem.* **1992**, *267*, 25988–25997.
12. Sarafianos, S.G.; Das, K.; Ding, J.; Boyer, P.L.; Hughes, S.H.; Arnold, E. Touching the heart of HIV-1 drug resistance: The fingers close down on the dNTP at the polymerase active site. *Chem. Biol.* **1999**, *6*, R137–R146.
13. Wöhrl, B.M.; Krebs, R.; Goody, R.S.; Restle, T. Refined model for primer/template binding by HIV-1 reverse transcriptase: Pre-steady-state kinetic analyses of primer/template binding and nucleotide incorporation events distinguish between different binding modes depending on the nature of the nucleic acid substrate. *J. Mol. Biol.* **1999**, *292*, 333–344.
14. Rothwell, P.J.; Berger, S.; Kensch, O.; Felekyan, S.; Antonik, M.; Wöhrl, B.M.; Restle, T.; Goody, R.S.; Seidel, C.A.M. Multiparameter single-molecule fluorescence spectroscopy reveals heterogeneity of HIV-1 reverse transcriptase:primer/template complexes. *Proc. Natl. Acad. Sci. USA* **2003**, *100*, 1655–1660.
15. Sisamakis, E.; Valeri, A.; Kalinin, S.; Rothwell, P.J.; Seidel, C.A.M. Accurate single-molecule FRET studies using multiparameter fluorescence detection. *Methods Enzymol.* **2010**, *475*, 455–514.
16. Faradjian, A.K.; Elber, R. Computing time scales from reaction coordinates by milestoning. *J. Chem. Phys.* **2004**, *120*, 10880–10889.
17. Kirmizialtin, S.; Nguyen, V.; Johnson, K.A.; Elber, R. How conformational dynamics of DNA polymerase select correct substrates: Experiments and simulations. *Structure* **2012**, *20*, 618–627.
18. Götte, M.; Rausch, J.W.; Marchand, B.; Sarafianos, S.; Le Grice, S.F.J. Reverse transcriptase in motion: Conformational dynamics of enzyme-substrate interactions. *Biochim. Biophys. Acta* **2010**, *1804*, 1202–1212.
19. Sarafianos, S.G.; Clark, A.D.; Das, K.; Tuske, S.; Birktoft, J.J.; Ilankumaran, P.; Ramesha, A.R.; Sayer, J.M.; Jerina, D.M.; Boyer, P.L.; *et al.* Structures of HIV-1 reverse transcriptase with pre- and post-translocation AZTMP-terminated DNA. *EMBO J.* **2002**, *21*, 6614–6624.
20. Lu, H.; Macosko, J.; Habel-Rodriguez, D.; Keller, R.W.; Brozik, J.A.; Keller, D.J. Closing of the fingers domain generates motor forces in the HIV reverse transcriptase. *J. Biol. Chem.* **2004**, *279*, 54529–54532.
21. Sluis-Cremer, N.; Temiz, N.A.; Bahar, I. Conformational changes in HIV-1 reverse transcriptase induced by nonnucleoside reverse transcriptase inhibitor binding. *Curr. HIV Res.* **2004**, *2*, 323–332.
22. De Béthune, M.-P. Non-nucleoside reverse transcriptase inhibitors (NNRTIs), their discovery, development, and use in the treatment of HIV-1 infection: A review of the last 20 years (1989–2009). *Antiviral Res.* **2010**, *85*, 75–90.
23. Zhou, H.-X. From induced fit to conformational selection: A continuum of binding mechanism controlled by the timescale of conformational transitions. *Biophys. J.* **2010**, *98*, L15–L17.
24. Basavapathruni, A.; Anderson, K.S. Reverse transcription of the HIV-1 pandemic. *FASEB J.* **2007**, *21*, 3795–3808.
25. Elinder, M.; Selhorst, P.; Vanham, G.; Oberg, B.; Vrang, L.; Danielson, U.H. Inhibition of HIV-1 by non-nucleoside reverse transcriptase inhibitors via an induced fit mechanism-importance of slow dissociation and relaxation rates for antiviral efficacy. *Biochem. Pharmacol.* **2010**, *80*, 1133–1140.

26. Geitmann, M.; Unge, T.; Danielson, U.H. Biosensor-based kinetic characterization of the interaction between HIV-1 reverse transcriptase and non-nucleoside inhibitors. *J. Med. Chem.* **2006**, *49*, 2367–2374.
27. Bakan, A.; Bahar, I. The intrinsic dynamics of enzymes plays a dominant role in determining the structural changes induced upon inhibitor binding. *Proc. Natl. Acad. Sci. USA* **2009**, *106*, 14349–14354.
28. Braz, V.A.; Barkley, M.D.; Jockusch, R.A.; Wintrode, P.L. Efavirenz binding site in HIV-1 reverse transcriptase monomers. *Biochemistry* **2010**, *49*, 10565–10573.
29. Braz, V.A.; Holladay, L.A.; Barkley, M.D. Efavirenz binding to HIV-1 reverse transcriptase monomers and dimers. *Biochemistry* **2010**, *49*, 601–610.
30. Zheng, X.; Mueller, G.A.; DeRose, E.F.; London, R.E. Solution characterization of [methyl-13C]methionine HIV-1 reverse transcriptase by NMR spectroscopy. *Antiviral Res.* **2009**, *84*, 205–214.
31. Chennubhotla, C.; Rader, A.J.; Yang, L.-W.; Bahar, I. Elastic network models for understanding biomolecular machinery: From enzymes to supramolecular assemblies. *Phys. Biol.* **2005**, *2*, S173–S180.
32. Radi, M.; Maga, G.; Alongi, M.; Angeli, L.; Samuele, A.; Zanoli, S.; Bellucci, L.; Tafi, A.; Casaluce, G.; Giorgi, G.; *et al.* Discovery of chiral cyclopropyl dihydro-alkylthio-benzyl-oxopyrimidine (S-DABO) derivatives as potent HIV-1 reverse transcriptase inhibitors with high activity against clinically relevant mutants. *J. Med. Chem.* **2009**, *52*, 840–851.
33. Geitmann, M.; Unge, T.; Danielson, U.H. Interaction kinetic characterization of HIV-1 reverse transcriptase non-nucleoside inhibitor resistance. *J. Med. Chem.* **2006**, *49*, 2375–2387.
34. Burnouf, D.; Ennifar, E.; Guedich, S.; Puffer, B.; Hoffmann, G.; Bec, G.; Disdier, F.; Baltzinger, M.; Dumas, P. kinITC: A new method for obtaining joint thermodynamic and kinetic data by isothermal titration calorimetry. *J. Am. Chem. Soc.* **2012**, *134*, 559–565.
35. Bec, G.; Meyer, B.; Gerard, M.-A.; Steger, J.; Fauster, K.; Wolff, P.; Burnouf, D.Y.; Micura, R.; Dumas, P.; Ennifar, E. Thermodynamics of HIV-1 reverse transcriptase in action elucidates the mechanism of action of non-nucleoside inhibitors. *J. Am. Chem. Soc.* **2013**, *135*, 9743–9752.
36. Rittinger, K.; Divita, G.; Goody, R.S. Human immunodeficiency virus reverse transcriptase substrate-induced conformational changes and the mechanism of inhibition by nonnucleoside inhibitors. *Proc. Natl. Acad. Sci. USA* **1995**, *92*, 8046–8049.
37. Spence, R.; Kati, W.; Anderson, K.; Johnson, K. Mechanism of inhibition of HIV-1 reverse transcriptase by nonnucleoside inhibitors. *Science* **1995**, *267*, 988–993.
38. Divita, G.; Mueller, B.; Immendoerfer, U.; Gautel, M.; Rittinger, K.; Restle, T.; Goody, R.S. Kinetics of interaction of HIV reverse transcriptase with primer/template. *Biochemistry* **1993**, *32*, 7966–7971.
39. Seckler, J.M.; Barkley, M.D.; Wintrode, P.L. Allosteric suppression of HIV-1 reverse transcriptase structural dynamics upon inhibitor binding. *Biophys. J.* **2011**, *100*, 144–153.
40. Esposito, F.; Corona, A.; Tramontano, E. HIV-1 reverse transcriptase still remains a new drug target: Structure, function, classical inhibitors, and new inhibitors with innovative mechanisms of actions. *Mol. Biol. Int.* **2012**, *2012*, 1–23.

41. Sarafianos, S.G.; Marchand, B.; Das, K.; Himmel, D.M.; Parniak, M.A.; Hughes, S.H.; Arnold, E. Structure and function of HIV-1 reverse transcriptase: Molecular mechanisms of polymerization and inhibition. *J. Mol. Biol.* **2009**, *385*, 693–713.
42. Ren, J.; Nichols, C.E.; Stamp, A.; Chamberlain, P.P.; Ferris, R.; Weaver, K.L.; Short, S.A.; Stammers, D.K. Structural insights into mechanisms of non-nucleoside drug resistance for HIV-1 reverse transcriptases mutated at codons 101 or 138. *FEBS J.* **2006**, *273*, 3850–3860.
43. Ren, J.; Esnouf, R.; Garman, E.; Somers, D.; Ross, C.; Kirby, I.; Keeling, J.; Darby, G.; Jones, Y.; Stuart, D. High resolution structures of HIV-1 RT from four RT-inhibitor complexes. *Nat. Struct. Biol.* **1995**, *2*, 293–302.
44. Das, K.; Ding, J.; Hsiou, Y.; Clark, A.D., Jr.; Moereels, H.; Koymans, L.; Andries, K.; Pauwels, R.; Janssen, P.A.; Boyer, P.L.; *et al.* Crystal structures of 8-Cl and 9-Cl TIBO complexed with wild-type HIV-1 RT and 8-Cl TIBO complexed with the Tyr181Cys HIV-1 RT drug-resistant mutant. *J. Mol. Biol.* **1996**, *264*, 1085–1100.
45. Das, K.; Martinez, S.E.; Bauman, J.D.; Arnold, E. HIV-1 reverse transcriptase complex with DNA and nevirapine reveals non-nucleoside inhibition mechanism. *Nat. Struct. Mol. Biol.* **2012**, *19*, 253–259.
46. Delviks-Frankenberry, K.A.; Nikolenko, G.N.; Pathak, V.K. The "connection" between HIV drug resistance and RNase H. *Viruses* **2010**, *2*, 1476–1503.
47. Lapkouski, M.; Tian, L.; Miller, J.T.; Le Grice, S.F.J.; Yang, W. Complexes of HIV-1 RT, NNRTI and RNA/DNA hybrid reveal a structure compatible with RNA degradation. *Nat. Struct. Mol. Biol.* **2013**, *20*, 230–236.
48. Liu, S.; Abbondanzieri, E.A.; Rausch, J.W.; Grice, S.F.J.L.; Zhuang, X. Slide into action: Dynamic shuttling of HIV reverse transcriptase on nucleic acid substrates. *Science* **2008**, *322*, 1092–1097.
49. Bahar, I.; Erman, B.; Jernigan, R.L.; Atilgan, A.R.; Covell, D.G. Collective motions in HIV-1 reverse transcriptase: Examination of flexibility and enzyme function. *J. Mol. Biol.* **1999**, *285*, 1023–1037.
50. Ivetac, A.; McCammon, J.A. Elucidating the inhibition mechanism of HIV-1 non-nucleoside reverse transcriptase inhibitors through multicopy molecular dynamics simulations. *J. Mol. Biol.* **2009**, *388*, 644–658.
51. Temiz, N.A.; Bahar, I. Inhibitor binding alters the directions of domain motions in HIV-1 reverse transcriptase. *Proteins Struct. Funct. Genet.* **2002**, *49*, 61–70.
52. Wright, D.W.; Sadiq, S.K.; de Fabritiis, G.; Coveney, P.V. Thumbs down for HIV: Domain level rearrangements do occur in the NNRTI-bound HIV-1 reverse transcriptase. *J. Am. Chem. Soc.* **2012**, *134*, 12885–12888.
53. Shaw-Reid, C.A.; Feuston, B.; Munshi, V.; Getty, K.; Krueger, J.; Hazuda, D.J.; Parniak, M.A.; Miller, M.D.; Lewis, D. Dissecting the effects of DNA polymerase and ribonuclease H inhibitor combinations on HIV-1 reverse-transcriptase activities. *Biochemistry* **2005**, *44*, 1595–1606.
54. Radzio, J.; Sluis-Cremer, N. Efavirenz accelerates HIV-1 reverse transcriptase ribonuclease H cleavage, leading to diminished zidovudine excision. *Mol. Pharmacol.* **2008**, *73*, 601–606.
55. Tachedjian, G.; Goff, S.P. The effect of NNRTIs on HIV reverse transcriptase dimerization. *Curr. Opin. Investig. Drugs* **2003**, *4*, 966–973.

56. Tachedjian, G.; Orlova, M.; Sarafianos, S.G.; Arnold, E.; Goff, S.P. Nonnucleoside reverse transcriptase inhibitors are chemical enhancers of dimerization of the HIV type 1 reverse transcriptase. *Proc. Natl. Acad. Sci. USA* **2001**, *98*, 7188–7193.
57. Venezia, C.F.; Howard, K.J.; Ignatov, M.E.; Holladay, L.A.; Barkley, M.D. Effects of efavirenz binding on the subunit equilibria of HIV-1 reverse transcriptase. *Biochemistry* **2006**, *45*, 2779–2789.
58. Tachedjian, G.; Moore, K.L.; Goff, S.P.; Sluis-Cremer, N. Efavirenz enhances the proteolytic processing of an HIV-1 pol polyprotein precursor and reverse transcriptase homodimer formation. *FEBS Lett.* **2005**, *579*, 379–384.
59. Figueiredo, A.; Moore, K.L.; Mak, J.; Sluis-Cremer, N.; de Bethune, M.-P.; Tachedjian, G. Potent nonnucleoside reverse transcriptase inhibitors target HIV-1 Gag-Pol. *PLoS Pathog.* **2006**, *2*, e119.
60. Xia, Q.; Radzio, J.; Anderson, K.S.; Sluis-Cremer, N. Probing nonnucleoside inhibitor-induced active-site distortion in HIV-1 reverse transcriptase by transient kinetic analyses. *Protein Sci.* **2007**, *16*, 1728–1737.
61. Menéndez-Arias, L. Molecular basis of human immunodeficiency virus type 1 drug resistance: Overview and recent developments. *Antiviral Res.* **2013**, *98*, 93–120.
62. Ren, J.; Nichols, C.; Bird, L.; Chamberlain, P.; Weaver, K.; Short, S.; Stuart, D.I.; Stammers, D.K. Structural mechanisms of drug resistance for mutations at codons 181 and 188 in HIV-1 reverse transcriptase and the improved resilience of second generation non-nucleoside inhibitors. *J. Mol. Biol.* **2001**, *312*, 795–805.
63. Huang, W.; Gamarnik, A.; Limoli, K.; Petropoulos, C.J.; Whitcomb, J.M. Amino acid substitutions at position 190 of human immunodeficiency virus type 1 reverse transcriptase increase susceptibility to delavirdine and impair virus replication. *J. Virol.* **2003**, *77*, 1512–1523.
64. Yap, S.-H.; Sheen, C.-W.; Fahey, J.; Zanin, M.; Tyssen, D.; Lima, V.D.; Wynhoven, B.; Kuiper, M.; Sluis-Cremer, N.; Harrigan, P.R.; *et al.* N348I in the connection domain of HIV-1 reverse transcriptase confers zidovudine and nevirapine resistance. *PLoS Med.* **2007**, *4*, e335.
65. Ren, J.; Stammers, D.K. Structural basis for drug resistance mechanisms for non-nucleoside inhibitors of HIV reverse transcriptase. *Virus Res.* **2008**, *134*, 157–170.
66. Clotet, B. Efavirenz: Resistance and cross-resistance. *Int. J. Clin. Pract. Suppl.* **1999**, *103*, 21–25.
67. Hsiou, Y.; Ding, J.; Das, K.; Clark, A.D., Jr.; Boyer, P.L.; Lewi, P.; Janssen, P.A.; Kleim, J.-P.; Rösner, M.; Hughes, S.H.; *et al.* The Lys103Asn mutation of HIV-1 RT: A novel mechanism of drug resistance. *J. Mol. Biol.* **2001**, *309*, 437–445.
68. Das, K.; Sarafianos, S.G.; Clark, A.D., Jr.; Boyer, P.L.; Hughes, S.H.; Arnold, E. Crystal structures of clinically relevant Lys103Asn/Tyr181Cys double mutant HIV-1 reverse transcriptase in complexes with ATP and non-nucleoside inhibitor HBY 097. *J. Mol. Biol.* **2007**, *365*, 77–89.
69. Andries, K.; Azijn, H.; Thielemans, T.; Ludovici, D.; Kukla, M.; Heeres, J.; Janssen, P.; de Corte, B.; Vingerhoets, J.; Pauwels, R.; *et al.* TMC125, a novel next-generation nonnucleoside reverse transcriptase inhibitor active against nonnucleoside reverse transcriptase inhibitor-resistant human immunodeficiency virus type 1. *Antimicrob. Agents Chemother.* **2004**, *48*, 4680–4686.
70. Das, K.; Bauman, J.D.; Clark, A.D.; Frenkel, Y.V.; Lewi, P.J.; Shatkin, A.J.; Hughes, S.H.; Arnold, E. High-resolution structures of HIV-1 reverse transcriptase/TMC278 complexes: Strategic flexibility explains potency against resistance mutations. *Proc. Natl. Acad. Sci. USA* **2008**, *105*, 1466–1471.

71. Azijn, H.; Tirry, I.; Vingerhoets, J.; de Béthune, M.-P.; Kraus, G.; Boven, K.; Jochmans, D.; van Craenenbroeck, E.; Picchio, G.; Rimsky, L.T. TMC278, a next-generation nonnucleoside reverse transcriptase inhibitor (NNRTI), active against wild-type and NNRTI-resistant HIV-1. *Antimicrob. Agents Chemother.* **2010**, *54*, 718–727.
72. Lansdon, E.B.; Brendza, K.M.; Hung, M.; Wang, R.; Mukund, S.; Jin, D.; Birkus, G.; Kutty, N.; Liu, X. Crystal structures of HIV-1 reverse transcriptase with etravirine (TMC125) and rilpivirine (TMC278): Implications for drug design. *J. Med. Chem.* **2010**, *53*, 4295–4299.
73. Rawal, R.K.; Murugesan, V.; Katti, S.B. Structure-activity relationship studies on clinically relevant HIV-1 NNRTIs. *Curr. Med. Chem.* **2012**, *19*, 5364–5380.
74. Kalinin, S.; Peulen, T.; Sindbert, S.; Rothwell, P.J.; Berger, S.; Restle, T.; Goody, R.S.; Gohlke, H.; Seidel, C.A.M. A toolkit and benchmark study for FRET-restrained high-precision structural modeling. *Nat. Methods* **2012**, *9*, 1218–1225.
75. Kim, S.; Blainey, P.C.; Schroeder, C.M.; Xie, X.S. Multiplexed single-molecule assay for enzymatic activity on flow-stretched DNA. *Nat. Methods* **2007**, *4*, 397–399.
76. Kim, S.; Schroeder, C.M.; Xie, X.S. Single-molecule study of DNA polymerization activity of HIV-1 reverse transcriptase on DNA templates. *J. Mol. Biol.* **2010**, *395*, 995–1006.
77. Abbondanzieri, E.A.; Bokinsky, G.; Rausch, J.W.; Zhang, J.X.; Le Grice, S.F.J.; Zhuang, X. Dynamic binding orientations direct activity of HIV reverse transcriptase. *Nature* **2008**, *453*, 184–189.
78. Fagerburg, M.V.; Leuba, S.H. Optimal practices for surface-tethered single molecule total internal reflection fluorescence resonance energy transfer analysis. *Methods Mol. Biol.* **2011**, *749*, 273–289.
79. Hwang, H.; Kim, H.; Myong, S. Protein induced fluorescence enhancement as a single molecule assay with short distance sensitivity. *Proc. Natl. Acad. Sci. USA* **2011**, *108*, 7414–7418.
80. Marko, R.A.; Liu, H.-W.; Ablenas, C.J.; Ehteshami, M.; Götte, M.; Cosa, G. Binding kinetics and affinities of heterodimeric *versus* homodimeric HIV-1 reverse transcriptase on DNA—DNA substrates at the single-molecule level. *J. Phys. Chem. B* **2013**, *117*, 4560–4567.
81. Seckler, J.M.; Howard, K.J.; Barkley, M.D.; Wintrode, P.L. Solution structural dynamics of HIV-1 reverse transcriptase heterodimer. *Biochemistry* **2009**, *48*, 7646–7655.
82. Moore, G.E. *Cramming More Components onto Integrated Circuits*; McGraw-Hill: New York, NY, USA, 1965.
83. Van Oijen, A.M. Single-molecule approaches to characterizing kinetics of biomolecular interactions. *Curr. Opin. Biotechnol.* **2011**, *22*, 75–80.

4

Fungal Beta-Glucosidases: A Bottleneck in Industrial use of Lignocellulosic Materials

Annette Sørensen [1,2], Mette Lübeck [1], Peter S. Lübeck [1] and Birgitte K. Ahring [1,2,*]

1 Section for Sustainable Biotechnology, Aalborg University Copenhagen, A C Meyers Vaenge 15, 2450 Copenhagen SV, Denmark; E-Mails: aso@bio.aau.dk (A.S.); mel@bio.aau.dk (M.L.); psl@bio.aau.dk (P.S.L.)

2 Bioproducts, Sciences & Engineering Laboratory, Washington State University, 2710 Crimson Way, Richland, WA 99354, USA

* Author to whom correspondence should be addressed; E-Mail: bka@bio.aau.dk

---

Abstract: Profitable biomass conversion processes are highly dependent on the use of efficient enzymes for lignocellulose degradation. Among the cellulose degrading enzymes, beta-glucosidases are essential for efficient hydrolysis of cellulosic biomass as they relieve the inhibition of the cellobiohydrolases and endoglucanases by reducing cellobiose accumulation. In this review, we discuss the important role beta-glucosidases play in complex biomass hydrolysis and how they create a bottleneck in industrial use of lignocellulosic materials. An efficient beta-glucosidase facilitates hydrolysis at specified process conditions, and key points to consider in this respect are hydrolysis rate, inhibitors, and stability. Product inhibition impairing yields, thermal inactivation of enzymes, and the high cost of enzyme production are the main obstacles to commercial cellulose hydrolysis. Therefore, this sets the stage in the search for better alternatives to the currently available enzyme preparations either by improving known or screening for new beta-glucosidases.

Keywords: beta-glucosidase; cellulase; biomass; hydrolysis; biofuels

---

1. Introduction

The ever-increasing energy consumption and the depletion of fossil resources have laid the foundation for a shift towards sustainable production of biofuels and bioproducts in biorefineries from renewable sources. Oil is currently the primary source of energy for the transportation sector and for production of chemicals and plastics. However, biorefineries are in the coming decades expected to supplement or replace oil refineries by maximizing biomass value, producing fuels and platform molecules for use as building blocks in the synthesis of chemicals and polymeric materials [1]. Biorefineries rely on the use of plant biomass in the form of dedicated energy crops or lignocellulosic agricultural residues as an abundant and inexpensive renewable energy resource [2]. Most biorefineries focus on production of a sugar platform of simple sugars that are released from biomass [1]. These sugars can then biologically or chemically be converted into fuels (e.g., ethanol, butanol and hydrocarbons), building block chemicals (e.g., different organic acids), as well as other high value bioproducts [3].

Plant biomasses are rich in lignocellulose which consists mainly of polysaccharides such as cellulose and hemicelluloses that together with the phenolic lignin polymer form a complex and rigid structure. The biomass composition depends on the plant/crop type, with cellulose being the most abundant component [4]. Cellulose is a long homogenous linear polymer of beta-D-glucosyl units linked by 1,4-beta-D-glucosidic bonds. The cellulose chains are assembled in larger rigid units held together by hydrogen bonds and weak van der Wall's forces. Through parallel orientation, the chains form a highly ordered crystalline structure, but are interspersed with amorphous regions of more disordered structure [5–7].

The complex structure of the cellulose fibrils embedded in an amorphous matrix of lignin and hemicellulose strengthen the plant cell wall and give plants a natural recalcitrance to biological degradation. Pretreatment is crucial as a first step for increasing the accessibility of the biomass polymers for the following enzyme hydrolysis. The different pretreatment methods available are plentiful—including alkali-, acid-, or organic—solvent pretreatment, steam-, ammonia fiber- or CO_2 explosion, and wet-oxidation [8–10]. The type of plant material as well as the severity of the pretreatment method applied will influence the characteristics of the lignocellulosic substrate for enzyme hydrolysis with regard to cellulose accessibility, degree of polymerization, hemicellulose content, lignin content, and other potential interfering compounds [11–13]. Such variation in biomass characteristics will influence the composition requirements for an optimal enzyme cocktail for the breakdown of different types of lignocellulosic biomasses [14]. Balanced enzyme cocktails and tailoring of enzymes for increased performance is of major importance for obtaining high yields of sugar monomers from hydrolysis, and especially the complete hydrolysis of cellulose is the main challenge that must be overcome. Among the cellulolytic enzyme complex, beta-glucosidases play a key role for the final conversion of cellobiose into glucose.

This review provides an overview of fungal beta-glucosidases in relation to industrial use of lignocellulosic materials. We discuss the significance of beta-glucosidases, how they represent the bottleneck in biomass conversion and the challenges in biomass hydrolysis in biorefineries.

2. Hydrolysis of Cellulose

The general biochemistry of cellulosic enzymatic hydrolysis has been reviewed extensively in previous literature [5,11,15,16] and will only be briefly discussed here.

The classical scheme for cellulose hydrolysis involves three main categories of enzymes: endo-1,4-beta-glucanases (EC 3.2.1.4), cellobiohydrolases (or exo-1,4-beta-glucanases) (EC 3.2.1.91), and beta-glucosidases (EC 3.2.1.21). Cellulose polymers are, through sequential and synergistic actions of these enzymes, degraded to glucose. The general consensus is that endo-glucanases randomly hydrolyze the internal 1,4-beta-linkages in primarily the amorphous regions of cellulose, rapidly decreasing the degree of polymerization. Cellobiohydrolases hydrolyze the cellulose polymer from the free ends, releasing cellobiose as product in a processive fashion, and finally, beta-glucosidases hydrolyze the cellobiose to glucose. Several fungal beta-glucosidases have furthermore been shown to produce glucose from larger cellodextrins, thus having the potential to increase the reaction rate and extent of cellulose hydrolysis [5,11,15,16].

The classical concept of cellulose hydrolysis described above has been agreed on for decades, but more recently, attention has been paid to accessory enzymes that are co-regulated or co-expressed by microbes during growth on cellulosic substrates. The crystalline chains in cellulose are tightly packed and additional factors are needed in order to make the substrate more accessible for the hydrolytic enzymes. Among such accessory enzymes are the GH61 proteins and the bacterial family 33 Carbohydrate Binding Modules (CBM33), which lack measurable hydrolytic activity, yet they are able to significantly enhance the activity of cellulases on pretreated biomass. Both proteins have flat substrate-binding surfaces and are capable of cleaving polysaccharide chains by oxidative reactions to disrupt the polymer packing, thereby increasing its accessibility [17–20].

The commercial viability of biorefineries has been burdened by the use of expensive enzymes needed to hydrolyze the biomass material after pretreatment [21,22]. It has been well established that producing higher concentration of sugars is an absolute necessity in an industrial setting as it lowers the heating requirements (lowering operating costs) and increases the volumetric efficiency (lowering capital costs) of the equipment [23]. Therefore, lowering the enzyme input and increasing the dry matter content during enzyme hydrolysis for higher cellulose conversion would be one of the most significant steps towards the direction of bioethanol production cost reduction and eventually leading to the commercialization of second generation biorefineries based on the lignocellulosic feedstock.

Several researchers have worked on using corn stover for the bioethanol production. Karr *et al.* used lime pretreatment followed by enzymatic hydrolysis at 5% solids concentration (SC) and 20 FPU (Spezyme CP and Novozym 188) and obtained 60% cellulose conversion [24]. Kim *et al.* introduced ammonia recycle percolation pretreatment followed by enzyme hydrolysis at 1% SC and 10 FPU (Spezyme CP and β-glucosidase (Sigma-Aldrich, St Louis, MO, USA)) and obtained 92% cellulose conversion [25,26]. This concentration of solids will, however, be far from an industrial process. Bura *et al.* used SO_2 catalyzed pretreatment followed by enzyme hydrolysis at 8% SC and 10 FPU (Spezyme CP, Novozym 188 and Multifect® Xylanase) and obtained 100% cellulose conversion [27]. However, again the solid concentration was far lower than needed for operating any industrial process. Using chemicals such as sulfur could further affect the down-stream processing of products, and for instance, sulfur will be attached to the solid fraction remaining after sugar extraction [28]. Recently,

Yang *et al.* used steam explosion pretreatment followed by enzyme hydrolysis at 25% SC and 20 FPU (Celluclast) and obtained 85% cellulose conversion [29]. Even though this study achieved high glucose concentrations, the amount of enzymes used was higher, affecting the applicability of the process.

Cellobiohydrolases and endoglucanases are often inhibited by cellobiose [30], making beta-glucosidases important for avoiding product inhibition through conversion of cellobiose to glucose, and thereby, avoiding decreased hydrolysis rates of cellulose over time. However, beta-glucosidases are often themselves inhibited by their product glucose [31,32] making beta-glucosidase the rate-limiting enzyme. Maintaining a high hydrolysis rate of cellulose ultimately requires highly efficient beta-glucosidases that tolerate glucose at high levels.

3. Enzymes: Past to Present

Trichoderma reesei is one of the most widely used species of filamentous fungi for the production of cellulolytic enzymes. The fungus was originally isolated during the Second World War where it was found to thrive on the US Army's tent canvas (cellulose). Since the 1950s, the original strain of *T. reesei* has been subjected to multiple rounds of strain improvement for enhanced cellulase production, including increasing enzyme titers and reducing the catabolite repression effect as well as protease activity [33]. The fungal enzyme product, however, lacks sufficient beta-glucosidase activity for complete and efficient industrial cellulose hydrolysis [5,34].

Enhancement of the beta-glucosidase activity of the *T. reesei* enzyme product has been achieved through displacement of the native promoter by homologous recombination with xylanase and cellulase promoters obtaining a 4–7.5-fold increase in beta-glucosidase activity [35]. Other ways of increasing the beta-glucosidase activity of *T. reesei* include heterologous expression of beta-glucosidase from other fungi [36–39] thus creating a single expression host for the production of all relevant enzymes for converting cellulosic biomass into monomeric sugars.

Beta-glucosidases are widely produced by different genera and species of the fungal kingdom including Ascomycetes and Basidiomycetes, where especially the ascomycete genus *Aspergillus* has been widely studied for beta-glucosidase production. *A. niger* has been setting the standard in commercial beta-glucosidase production [40], but within the last few years more research papers have been published on efficient beta-glucosidases e.g., from other *Aspergillus* species and from the *Penicillium* genus [38,41,42].

Commercial enzyme preparations for cellulosic biomass hydrolysis were initially prepared as separate fungal fermentation products that needed to be combined for efficient hydrolysis, e.g., Celluclast (a *T. reesei* cellobiohydrolase and endo-glucanase product) and Novozym188 (an *A. niger* beta-glucosidase product) by Novozymes. More recently, the enzyme companies, Novozymes and Genencor, have replaced these two preparations with single products that contain the full array of enzymes for cellulosic biomass hydrolysis. Whether the products originate from strain improvement of the production strain to express all enzymes or if the products are mixes based on two or more fermentations is not disclosed by the companies. The optimal hydrolysis conditions of most commercial cellulosic enzymes are temperatures around 50 °C and a pH around 5. The enzyme loading must be optimized based on the biomass.

The current trend for the major enzyme companies is to team up with cellulosic biorefinery companies to specifically meet their needs in hydrolysis, working on optimizing the enzymes for a particular biomass and pretreatment method. Recently, Genencor has partnered with DuPont (http://biosciences.dupont.com), DSM with Poet (www.poetdsm.com), and Novozymes with Mossi & Ghisolfi Group (www.novozymes.com/en/news/news-archive/Pages/novozymes-partner-to-open-largest-cellulosic-ethanol-plant-in-2012.aspx), building the world's first commercial-scale cellulosic ethanol plant in Crescentino, Italy.

With enzymes being an expensive part of biomass processing, it would be of great interest to make enzyme production part of the processes within the biorefinery. Other research therefore looks at producing enzymes on-site to cut away the profit enzyme companies include in their pricing. By efficiently implementing enzyme production within the biorefinery, completing the value chain can be achieved by using streams within the biorefinery as fungal growth medium for enzyme production, and directly using this product (enzymes, fungus, and medium) in hydrolysis of biomass. This has already been shown for different fungi, e.g., *T. reesei* cultured on pretreated wheat straw [43], *A. niger* and *A. saccharolyticus* [44] cultured on the fiber waste fraction left after hydrolysis and fermentation, and *A. japonicus* cultured on castor bean meal waste for the biodiesel production [45].

Evaluating the overall production cost, the price of enzymes typically contributes to a large part of the total cost [37]. Efficient enzymes for lignocellulose degradation are, therefore, of high demand. As most of the currently used pretreatment methods remove lignin from the sugar polymers and in many cases also hydrolyze most of the hemicellulose, the main target for enzyme treatment is cellulose decomposition into glucose with beta-glucosidases being key enzymes in terms of complete cellulose hydrolysis.

4. The Bottleneck Enzyme: Beta-Glucosidase

Beta-glucosidases are most commonly classified based on either substrate specificity or nucleotide sequence identity. Beta-glucosidases hydrolyze the *O*-glycosyl linkage of terminal, non-reducing beta-D-glucosyl residues with release of beta-D-glucose, e.g., the bond in cellobiose. A wide specificity for beta-D-glucosides is found and there are examples of beta-glucosidases hydrolyzing beta-D-galactosides, alpha-L-arabinosides, beta-D-xylosides, or beta-D-fucosides [46]. Based on substrate specificity, beta-glucosidases have traditionally been divided into cellobiases (high specificity towards cellobiose), aryl- beta-glucosidases (high specificity towards substrates such as *p*-nitrophenyl-beta-D-glucopyranoside (pNPG)), or broad specificity beta-glucosidases [31,47]. Most beta-glucosidases are placed in the last category.

A classification based on substrate specificity cannot sufficiently accommodate enzymes that act on several substrates; the best accommodation for this is the classification system proposed by Henrissat (1991) which is based on sequence and structural features [48]. The strength of this system especially lies in the investigation of the active site of the enzymes, with significant similarity of sequences being a strong indication of similarity in the fold of the structure, and analysis of the primary structure can thereby assign potential conserved active-site residues. Fungal beta-glucosidases are primarily placed in the family 3 glycosyl hydrolases with the active site signature pattern defined as written below, where the aspartate (D) is the active site residue involved in catalysis (underlined) [46,49].

GH3 active site signature:
[LIVM](2) – [KR] – X – [EQKRD] – X(4) – G – [LIVMFTC] – [LIVT] – [LIVMF] – [ST] – D – X(2) – [SGADNIT]

Structural information is valuable for protein engineering purposes to improve enzyme activity and stability. Only a few GH3 beta-glucosidase structures have been solved and published: *Hordeum vulgare* (barley) [50], *Kluyveromyces marxianus* (a yeast) [51], *Thermotoga neapolitana* (a hyperthermophilic bacterium) [52], *Pseudoalteromonas sp.* (a marine bacterium) [53], a compost microbial community [54], and only recently, one crystal structure from a filamentous fungus: the *Aspergillus aculeatus* beta-glucosidase BGL1 [55]. Furthermore, the BGL1 of *T. reesei* is in the protein database PDB, but the accompanying research article has not been published. Homology modeling has been the method of choice for obtaining structural information from fungal beta-glucosidases which have no available crystal structures. The beta-glucosidases from *Aspergillus saccharolyticus* and *Penicillium purpugenum* were modeled prior to the availability of other fungal beta-glucosidase crystal structures, and even though the sequence identity was relatively low to the template structures used, it was obvious that the residues important for substrate binding and catalysis were conserved and that the distance between the catalytic residues is similar to that of other solved beta-glucosidases [38,56].

The solved structure of the fungal *A. aculeatus* BGL1 consists of three domains: a catalytic TIM (triosephosphateisomerase) barrel-like domain, an α/β sandwich domain, and a FnIII (fibronectin type III) domain. These domains are connected with two linker regions. The active site and the catalytic residues of AaBGL1 are located at the domain interface between the barrel and the α/β sandwich domains [55]. Hydrolysis of beta-1,4-glycosidic bonds by beta-glucosidases is carried out by an overall retaining double-displacement mechanism [57]. Two catalytic carboxylic acid residues at the active site facilitate the reaction with one carboxylic acid acting as a nucleophile and the other as an acid/base catalyst [58]. The catalytic nucleophile of GH3 family enzymes is always present at a specific structural location just after the $\beta7$ strand of the TIM barrel domain, however, the position and identities for the acid/base catalyst are not completely conserved but are rather phylogenetically variable, and thus, less readily divined [55,59]. The topology of the active sites of all glycoside hydrolases falls into three general classes: (i) pocket or crater, (ii) cleft or groove, and (iii) tunnel. Beta-glucosidases and non-processive exo-acting enzymes have a pocket or crater topology that is well suited for recognition of a saccharide non-reducing extremity [60], with the depth and shape of the pocket or crater reflecting the number of sub-sites that contribute to substrate binding and the length of the leaving group [61]. Hydrolytic activity towards cellodextrins is commonly reported for fungal beta-glucosidases [11], and compared to other beta-glucosidases, the structure of the *A. aculeatus* BGL1 active site has a long cleft extending from sub-site +1 which appears to be a more suitable binding pocket for cellooligosaccharides [55]. Meanwhile, the catalytic pocket of *A. saccharolyticus* BGL1 is wider than other beta-glucosidases as it is missing a loop structure by the active site. Amino acids at this loop have been described to have weak H-bonds with glucose at the -1 sub-site, thus the deletion of this loop may plan an important role in altering substrate accessibility as well as rapid release of the product from the enzymes [38].

Based on genomic data, fungal beta-glucosidases are often reported to have several putative glycosylation sites based on their predicted amino acid sequence. The crystal structure of the *A.*

aculeatus BGL1 was found to be highly glycosylated by many large N-glycan chains, which is believed to facilitate increased resistance to proteolytic attack and contributes to protein stability [55].

5. Beta-Glucosidases in Biomass Hydrolysis: The Challenges

In relation to industrial biomass conversion, a good beta-glucosidase facilitates efficient hydrolysis at specified operating conditions. Key points to consider when evaluating a beta-glucosidase are hydrolysis rate, inhibitors, and stability, with product inhibition and thermal instability often being a restriction for maintaining high conversion rates throughout the hydrolysis. It is obvious that activity and stability varies among different beta-glucosidases. Previous papers have listed beta-glucosidases and their properties [47,62]; as an addition to this, in Table 1 we here present a list of some more recently characterized fungal beta-glucosidases.

Table 1. Beta-glucosidases and their properties.

Microorganism	Activity			Temp.	pH	Inhibition			Ref.
	Substrate	Km (mM)	Vmax (U/mg)	Opt. (°C)	Opt.	Substrate	Inhibitor	Ki (mM)	
Aspergillus fumigatus Z5	pNPG			60	6.0				[42]
Aspergillus niger NRRL 599	pNPG	3.11	20.83	60	4.8				[63]
Aspergillus saccharolyticus	pNPG	1.9	45	58	4.2				[38]
Aspergillus terreus NRRL 265	pNPG Cellobiose	2.5 3.7		60	5.0	pNPG	Glucose	13.6	[64]
Daldinia eschscholzii	pNPG	1.52	3.21	50	5.0	pNPG	Glucose	0.79	[65]
Fomitopsis palustris FFPRI 0507	pNPG Cellobiose	0.117 4.81		70	4.5	pNPG	Glucose	0.35	[66]
Fusarium solani	pNPG	1	55.6	65	4.5				[67]
Humicola insolens	pNPG Cellobiose	0.16 0.51	18.1 86.0		6.2				[68]
Neosartorya ficheri NRRL181	pNPG	68	886	40	6.0				[69]
Peniclillium funiculosum NCL1	pNPG	0.057	1920	60	4.0–5.0	pNPG	Glucose	1.5	[70]
Phoma sp. KCTC11825BP	pNPG Cellobiose	0.3 3.2		60	4.5	pNPG	Glucose	1.7	[71]
Tolypocladium cylinndrosporum	pNPG	0.85	85.23	60	2.4	pNPG	Glucose	39.5	[72]
Trichoderma koningii AS3.2774	pNPG	2.67		50	5.0				[73]
Trichoderma reesei	pNPG			70	5.0				[74]

Fungi naturally produce a broad array of lignocellulosic enzymes, and with more and more full genome sequences available, it becomes evident just how many different enzymes their genome encode for. However, the genetic code itself does not necessarily imply that the fungus is optimally expressing the needed enzymes for efficient biomass hydrolysis. For example, the amount and types of cellulases (GH5, 6, 7, 12, 45, 61) and associated hemicellulase activities (GH10, 11, 26, 29, 39, 62, 67, 74, 93) are relatively small in the genome of *T. reesei* compared with other ascomycetes, even though the fungus is one of the most efficient cellulose degraders known [75]. Function can be predicted from the genetic code, but profound expertise does not yet exist in linking gene sequence to the actual activity and efficiency of the encoded enzyme [76]. The pathway to this has been initiated through homology modeling based on known 3D enzyme structures. Structures of most enzyme families have been resolved, including beta-glucosidases as mentioned previously, and with more templates becoming available, homology modeling can predict the folds and activity of gene sequences. This knowledge is useful for enzyme optimization using protein engineering methods such as site-directed mutagenesis, e.g., for higher thermal stability [22]. However, most current research has focused on testing the performance of individual enzymes heterologously expressed and purified, free from contaminating activities, against each other, and studies on optimally balanced enzyme cocktails have been undertaken to identify the best combination and ratio of de novo enzyme mixtures for biomass hydrolysis [22,77]. However, among the difficulties in expressing the enzymes heterologously for studying their activities are that different hosts might alter the original glycosylation pattern in the enzyme, thereby seriously altering their activity and/or stability [78].

One area of focus that must be addressed is how to perform such enzyme screenings in high through-put systems on actual biomass samples [79]. In practical terms, when studying the activity and kinetics of beta-glucosidases, it is important to consider the substrate that is being used, as substrate specificity of beta-glucosidases varies [47,62,80–82] and the choice of substrate will influence the kinetic data obtained. Several different substrates with varying sensitivity and ease of use can be applied for the determination of beta-glucosidase activity. Some enzyme testing is currently done on artificial or purified substrates rather than complex biomasses. However, data obtained using synthetic biomass substrates or single purified components have little value and limited applicability in predicting and modeling real biomass hydrolysis [11]. Those substrates can be valuable in terms of studying specific activities, but conclusions should never be extended to actual biomass hydrolysis as it is often the case that activities are found to be lower due to reduced substrate accessibility as well as enzyme inhibitors.

High conversion rates are essential for efficient conversion of biomass. Accumulation of glucose during hydrolysis can significantly lower the rate of cellulose hydrolysis through inhibition by blocking the active site or preventing the hydrolyzed substrate from leaving [83]. In case of product inhibition (glucose), the effect is naturally increased during the course of the reaction as more and more glucose is formed, and for beta-glucosidases the end-product is generally not removed during hydrolysis so the actual reaction rate will differ more and more from maximum reaction rate. High tolerance of beta-glucosidases towards glucose accumulation is, therefore, of great importance. A broad range of data on inhibition by glucose is described in the literature, with several of the published K_i values collected in a table in the Handbook of Carbohydrate Engineering [47]. The K_i values reported range from below one to thousands. Even within the same fungal species, there is great variation in the extent of product inhibition reported for different beta-glucosidases [47]. Compounds other than glucose are potentially present in biomass that can be inhibitory and influence the activity of beta-glucosidases, including (but not exclusively) other simple sugars, sugar derivatives, amines, and phenols [84].

A decrease in the rate of glucose formation can also be caused by transglycosylation events as the enzyme reaction is a reversible process. Other than inhibiting the reaction by occupying the active site, glucose can also be considered to take part in transglycosylation, thus using the active site capacity in non-hydrolyzing action which will decrease the overall rate of hydrolysis. Transglycosylation is obviously an unwanted event in biomass hydrolysis, but it is frequently reported for beta-glucosidases [62]; especially at high substrate concentrations, the transglycosylation is observed [85]. Targeted mutagenesis aiming at displacing essential amino acids involved in transglycosylation could potentially reduce this mechanism [86].

Enzyme performance in actual biomass hydrolysis is affected by several factors including temperature, pH, and solids loading. First of all, the condition of the biomass is defined by the pretreatment method applied. Many pretreatment methods rely on high temperatures and acidic conditions to make the biomass accessible for enzyme hydrolysis. Enzymes will, depending on extremity and time of exposure be inactivated by pH and temperature variations. Ionic groups are involved in enzyme catalysis, such as the acid-base catalyst in the beta-glucosidase active site, and the protonation state of the carboxylic acid residue catalyst and the carboxylate nucleophile is essential for the enzymatic reaction, therefore, a pH change could impair the catalytic mechanism [87]. Beta-glucosidases perform well at pH 4–5 [47,62], but at pH much lower than that, a significant decrease in activity is found. Therefore, in most cases, the pretreated biomass must be pH adjusted to some degree as the acidity is usually beyond this. Regarding temperature, according to the van't Hoff rule, reaction rates double with every 10 degrees Celsius increase of temperature, which applies to all chemical reactions including enzyme catalyzed reactions. However, when reaching high temperatures, protein stability will be affected, leading to denaturation, and thus irreversible inactivation of the enzyme. Mesophilic fungi that typically grow at 24–27 degrees Celsius are often times reported to produce beta-glucosidases with temperature optima around 60–75 degrees Celsius, and only moderate increases in thermal stability are seen in enzymes derived from thermophilic fungi [88]. Temperature and pH optimum for microbial beta-glucosidases have been reported in different reviews [47,62,88], but for biomass hydrolysis processes that typically run for the duration of several hours or even days, the stability of the enzyme at specified temperatures is important. Several papers claim to have discovered thermostable beta-glucosidases, however, often the activity was only verified at the high temperature for a short duration of time [88].

In industrial biofuel production, the pretreatment of biomass needs to be performed at very high dry matter content, above 20% (w/w), in order to increase product concentrations and decrease reactor volumes and distillation costs [23]. Most studies have, however, shown that hydrolysis rates decrease with increasing dry matter content in biomass hydrolysis. Suggested explanations for this are inefficient means of mixing, product inhibition, lignin or hemicellulose derivatives, or inhibition by adsorption of the enzymes to the biomass surface. Based on different correlation studies, it has been found that the adsorption effect best describes the decrease in hydrolysis with increase in solids' loading [23]. It has further been recognized that enzyme performance is reduced by interaction with lignin or lignin–carbohydrate complex; however, of the cellulase and xylanase enzymes tested, beta-glucosidase was the least affected by lignin [89]. Attempts have been made to deal with this issue by adding non-enzyme proteins to the hydrolysis that will be absorbed by the biomass instead of the active enzymes [90]. Another more advanced solution would be to engineer the hydrophobicity of the surface amino acids on the enzymes to make them less prone to adsorption by the biomass.

Beta-glucosidases act on soluble substrates and are with regards to biomass hydrolysis highly dependent on the action of cellobiohydrolases and endoglucanases to provide substrate, as the beta-glucosidases cannot access the insoluble cellulose fibers. Meanwhile, cellobiohydrolases and endoglucanases are highly dependent on beta-glucosidases to maintain efficient hydrolysis by relieving product inhibition. Therefore, a balanced enzyme cocktail is essential for efficient hydrolysis of biomass. The optimal ratio of the enzymes will depend on the specific activity of the enzymes used, the condition of the biomass substrate (sugar accessibility) as well as physical reaction conditions [91]. The total amount of enzyme required directly reflects on cost. Economics of enzymatic hydrolysis has long been a topic of discussion and concern for the feasibility of lignocellulosic biomass conversion. Properties of new enzymes are continuously reported in literature as well as research on optimizing the enzyme cocktails for biomass hydrolysis with emphasis on using reduced enzyme loadings yet obtaining same hydrolysis efficiency. One strategy for resolving this is the minimal enzyme cocktail concept which concerns identification of the minimal number, the minimal levels, and the optimal combination of the best performing mono-active enzymatic activities to achieve degradation to monomeric sugar units [14]. Ideally, based on minimal enzyme cocktail concept studies, rather than using purified enzymes, a selected enzyme producing microorganism should be genetically modified with distinct promoters for each enzyme gene to facilitate optimal expression of each enzyme component. It should be ensured that the enzymes are correctly post-processed by such host microorganisms so that they are correctly folded and have optimal activity, stability, *etc*. Furthermore, such microorganisms should not have intra- or extracellular proteolytic activity that would affect enzyme expression.

6. New and Improved Beta-Glucosidases

In order to optimize the use of different biomasses, it is important to identify new beta-glucosidases with improved abilities on the specific biomasses as well as with improved abilities such as stability and high conversion rates. As already discussed, product inhibition impairing yields, thermal inactivation of enzymes, and high cost of enzyme production are main obstacles of commercial cellulose hydrolysis and therefore set the stage in the search for better alternatives to the currently available enzyme preparations. The choice stands between screening for new beta-glucosidases and improving known beta-glucosidases.

The number of fungal species on earth is estimated to 1.5 million of which as little as approximately 5% are known [92,93], a statement that calls for a more directed effort for unraveling the potential of unknown species found in nature. The identification and characterization of new fungal species are often encountered in literature. Within the black Aspergilli, to which several efficient beta-glucosidase products belong, several new species have been identified within recent years [94–102]. Screening for new enzymes can be performed at the genomic as well as the proteomic level—in either case, it can be a mixed gene or protein pool or a sample representing a specific species. The number of organisms being fully genome-sequenced is constantly increasing, and along with it, the sequences for new genes. Comparative searches in databases can reveal new beta-glucosidase sequences, but to know if they are better than current standard, they must be cloned, expressed, and assayed. Using a metagenomics approach, environmental DNA has been screened for beta-glucosidase activity with the findings of novel beta-glucosidases [103–105]. As another approach, screening secreted fungal proteins for new and improved beta-glucosidases has been reported with success, generally finding black Aspergilli to be superior [106,107].

Through genetic changes, enzymes can be tailored to obtain improved abilities. The changes can either be random by classical methods of mutagenesis or specifically targeted improvements aided by the solved crystal structures.

The increased activity obtained from classical mutagenesis is most often due to changes at the regulatory level of enzyme expression leading to increased production of the gene of interest or decreased expression of conflicting genes and is therefore minded on production strain improvements, rather than changes to the enzyme itself for improved activity. One good example was the use of a combination of UV irradiation and nitrosomethyl guanidine treatment to develop the *T. reesei* strain RutC30 with improved total protein production and activity; one of the best existing *T. reesei* cellulase mutants [108].

Mutation, recombination, and selection set the stage for functional evolution in nature. Directed evolution mimics natural evolution by combining reiterative random mutagenesis and recombination with screening or selection for enzyme variants with improved properties [109–111]. Compared to classical mutagenesis, directed evolution targets a specific gene of choice with random changes being performed delimited to the gene of choice, followed by evaluation of the mutants [112]. Several publications exist on such strategy for non-fungal beta-glucosidases. For example, several single amino acid substitutions generated through error prone PCR were found to contribute to increased thermal resistance of *Paenibacillus polymyxa* beta-glucosidase that were then further recombined by gene shuffling [113]. The improvements of the final best clone were attributed to three mutations leading to formation of salt bridges and amino acids less prone to oxidation [114]. A similar approach of combining error prone PCR and gene shuffling was performed on *Pyrococcus furiosus* beta-glucosidase, generating an improvement of low temperature cellobiose hydrolysis [115]. More recently, gene shuffling of beta-glucosidases from *Thermobifida fusca* and *Paebibacillus polymxyxa* resulted in a mutant with increased thermostability compared to both parental enzymes, reported as a 144-fold increase in half-life of inactivation, and a 94% increase in k_{cat} towards cellobiose [116].

To perform more advanced mutagenesis, such as rational design, bioinformatics is a prerequisite. Protein structure can guide the fine-tuning of e.g., the active site by rational design by only a few specific mutations. A great amount of knowledge is available on the protein engineering possibilities for improving activity as well as stability [117]. With only a few filamentous fungal beta-glucosidase structures recently having been solved, most rational design has been performed on non-fungal beta-glucosidases. However, in a recent study, specific amino acids were mutated in the outer channel of the active site of a *T. reesei* beta-glucosidase to significantly improve activity as well as increase the thermostability [118].

7. Conclusions

Fungal beta-glucosidases are important enzymes in efficient hydrolysis of cellulosic biomass, as they relieve the inhibition of the cellobiohydrolases and endoglucanases by reducing cellobiose accumulation. They are key enzymes in the final part of biomass hydrolysis for producing the monomer sugars for the production of biofuels and platform molecules that can serve as building blocks in the synthesis of chemicals and polymeric materials. They are often the bottleneck in the process, and the most important challenge to overcome is product inhibition. To have a profitable biomass conversion process, the hydrolysis must yield high glucose concentrations and the beta-glucosidases must, therefore, not be inhibited by their product but maintain high conversion rates at high glucose concentrations.

References

1. Cherubini, F. The biorefinery concept: Using biomass instead of oil for producing energy and chemicals. *Energy Conver. Manag.* **2010**, *51*, 1412–1421.
2. Knauf, M.; Moniruzzaman, M. Lignocellulosic biomass processing: A perspective. *Int. Sugar J.* **2004**, *106*, 147–150.
3. Werpy, T.; Petersen, G.; Aden, A.; Bozell, J.; Holladay, J.; White, J.; Manheim, A. *Top Value added Chemicals from Biomass, Volume 1: Results of screening for potential candidates from sugars and synthesis gas*; U.S. Department of Energy, Oak Ridge, TN, USA, **2004**; available at heep://www.eere.energy.gov/biomass/pdfs/35523.pdf (access on 2 September 2013).
4. U.S. Department of Energy. *Biomass Feedstock Composition and Property Database*; Available online at: http://www.afdc.energy.gov/biomass/progs/search1.cgi (access on 30 August 2013).
5. Lynd, L.R.; Weimer, P.J.; van Zyl, W.H.; Pretorius, I.S. Microbial cellulose utilization: Fundamentals and biotechnology. *Microbiol. Mol. Biol. Rev.* **2002**, *66*, 506–577.
6. Beguin, P.; Aubert, J.P. The biological degradation of cellulose. *FEMS Microbiol. Rev.* **1994**, *13*, 25–58.
7. Berg, J.M.; Tymoczko, J.L.; Stryer, L. *Biochemistry*, 5th ed.; W.H. Freeman and Company: New York, NY, USA, 2002.
8. Alvira, P.; Tomas-Pejo, E.; Ballesteros, M.; Negro, M.J. Pretreatment technologies for an efficient bioethanol production process based on enzymatic hydrolysis: A review. *Bioresour. Technol.* **2010**, *101*, 4851–4861.
9. Mosier, N.; Wyman, C.; Dale, B.; Elander, R.; Lee, Y.Y.; Holtzapple, M.; Ladisch, M. Features of promising technologies for pretreatment of lignocellulosic biomass. *Bioresour. Technol.* **2005**, *96*, 673–686.
10. Sun, Y.; Cheng, J.Y. Hydrolysis of lignocellulosic materials for ethanol production: A review. *Bioresour. Technol.* **2002**, *83*, 1–11.

11. Zhang, Y.-P.; Himmel, M.E.; Mielenz, J.R. Outlook for cellulase improvement: Screening and selection strategies. *Biotechnol. Adv.* **2006**, *24*, 452–481.

12. Kabel, M.A.; Bos, G.; Zeevalking, J.; Voragen, A.G.J.; Schols, H.A. Effect of pretreatment severity on xylan solubility and enzymatic breakdown of the remaining cellulose from wheat straw. *Bioresour. Technol.* **2007**, *98*, 2034–2042.

13. Chang, V.S.; Holtzapple, M.T. Fundamental factors affecting biomass enzymatic reactivity. *Appl. Biochem. Biotechnol.* **2000**, *84–86*, 5–37.

14. Meyer, A.S.; Rosgaard, L.; Sorensen, H.R. The minimal enzyme cocktail concept for biomass processing. *J. Cereal. Sci.* **2009**, *50*, 337–344.

15. Zhang, Y.-P.; Lynd, L.R. Toward an aggregated understanding of enzymatic hydrolysis of cellulose: Noncomplexed cellulase systems. *Biotechnol. Bioeng.* **2004**, *88*, 797–824.

16. Wang, M.; Liu, K.; Dai, L.; Zhang, J.; Fang, X. The structural and biochemical basis for cellulose biodegradation. *J. Chem. Technol. Biotechnol.* **2013**, *88*, 491–500.

17. Harris, P.V.; Welner, D.; McFarland, K.C.; Re, E.; Poulsen, J.N.; Brown, K.; Salbo, R.; Ding, H.; Vlasenko, E.; Merino, S.; Xu, F.; Cherry, J.; Larsen, S; Leggio, L.L. Stimulation of lignocellulosic biomass hydrolysis by proteins of glycoside hydrolase family 61: Structure and function of a large, enigmatic family. *Biochemistry (N Y)* **2010**, *49*, 3305–3316.

18. Langston, J.A.; Shaghasi, T.; Abbate, E.; Xu, F.; Vlasenko, E.; Sweeney, M.D. Oxidoreductive cellulose depolymerization by the enzymes cellobiose dehydrogenase and glycoside hydrolase 61. *Appl. Environ. Microbiol.* **2011**, *77*, 7007–7015.

19. Quinlan, R.J.; Sweeney, M.D.; lo Leggio, L.; Otten, H.; Poulsen, J.N.; Johansen, K.S.; Krogh, K.B.R.M.; Jørgensen, C.I.; Tovborg, M.; Anthonsen, A.; Tryfona, T.; Walter, C.P.; Dupree, P.; Xu, F.; Davies, G.J.; Walton, P.H. Insights into the oxidative degradation of cellulose by a copper metalloenzyme that exploits biomass components. *Proc. Natl. Acad. Sci. USA* **2011**, *108*, 15079–15084.

20. Vaaje-Kolstad, G.; Westereng, B.; Horn, S.J.; Liu, Z.; Zhai, H.; Sorlie, M.; Eijsink, V.G.H. An oxidative enzyme boosting the enzymatic conversion of recalcitrant polysaccharides. *Science* **2010**, *330*, 219–222.

21. Lynd, L.R.; Laser, M.S.; Bransby, D.; Dale, B.E.; Davison, B.; Hamilton, R.; Himmel, M.; Keller, M.; McMillan, J.D.; Sheehan, J.; Wyman, C.E. How biotech can transform biofuels. *Nat. Biotechnol.* **2008**, *26*, 169–172.

22. Banerjee, G.; Scott-Craig, J.S.; Walton, J.D. Improving enzymes for biomass conversion: A basic research perspective. *Bioenergy Res.* **2010**, *3*, 82–92.

23. Kristensen, J.B.; Felby, C.; Jorgensen, H. Yield-determining factors in high-solids enzymatic hydrolysis of lignocellulose. *Biotechnol. Biofuels* **2009**, *2*, 11.

24. Kaar, W.; Holtzapple, M. Using lime pretreatment to facilitate the enzymic hydrolysis of corn stover. *Biomass Bioenergy* **2000**, *18*, 189–199.

25. Kim, S.; Holtzapple, M. Lime pretreatment and enzymatic hydrolysis of corn stover. *Bioresour. Technol.* **2005**, *96*, 1994–2006.

26. Kim, T.; Lee, Y. Pretreatment of corn stover by soaking in aqueous ammonia. *Appl. Biochem. Biotechnol.* **2005**, *124*, 1119–1131.

27. Bura, R.; Chandra, R.; Saddler, J. Influence of xylan on the enzymatic hydrolysis of steam-pretreated corn stover and hybrid poplar. *Biotechnol. Prog.* **2009**, *25*, 315–322.
28. Zhu, J.Y.; Zhu, W.; OBryan, P.; Dien, B.S.; Tian, S.; Gleisner, R.; Pan, X.J. Ethanol production from SPORL-pretreated lodgepole pine: Preliminary evaluation of mass balance and process energy efficiency. *Appl. Microbiol. Biotechnol.* **2010**, *86*, 1355–1365.
29. Yang, J.; Zhang, X.; Yong, Q.; Yu, S. Three-stage enzymatic hydrolysis of steam-exploded corn stover at high substrate concentration. *Bioresour. Technol.* **2011**, *102*, 4905–4908.
30. Murphy, L.; Bohlin, C.; Baumann, M.J.; Olsen, S.N.; Sorensen, T.H.; Anderson, L.; Borch, K.; Westh, *P.* Product inhibition of five *Hypocrea jecorina* cellulases. *Enzyme Microb. Technol.* **2013**,*52*, 163–169.
31. Shewale, J.G. Beta-Glucosidase - its role in cellulase synthesis and hydrolysis of cellulose. *Int. J. Biochem.* **1982**, *14*, 435–443.
32. Xiao, Z.Z.; Zhang, X.; Gregg, D.J.; Saddler, J.N. Effects of sugar inhibition on cellulases and beta-glucosidase during enzymatic hydrolysis of softwood substrates. *Appl. Biochem. Biotechnol.* **2004**, *113–116*, 1115–1126.
33. Peterson, R.; Nevalainen, H. *Trichoderma reesei* RUT-C30 - thirty years of strain improvement. *Microbiology* **2012**, *158*, 58–68.
34. Reczey, K.; Brumbauer, A.; Bollok, M.; Szengyel, Z.; Zacchi, G. Use of hemicellulose hydrolysate for beta-glucosidase fermentation. *Appl. Biochem. Biotechnol.* **1998**, *70–72*, 225–235.
35. Rahman, Z.; Shida, Y.; Furukawa, T.; Suzuki, Y.; Okada, H.; Ogasawara, W.; Morikawa, Y. Application of *Trichoderma reesei* cellulase and xylanase promoters through homologous recombination for enhanced production of extracellular beta-glucosidase I. *Biosci. Biotechnol. Biochem.* **2009**, *73*, 1083–1089.
36. Murray, P.; Aro, N.; Collins, C.; Grassick, A.; Penttila, M.; Saloheimo, M.; Tuohy, M. Expression in *Trichoderma reesei* and characterisation of a thermostable family 3 beta-glucosidase from the moderately thermophilic fungus *Talaromyces emersonii*. *Protein Expr. Purif.* **2004**, *38*, 248–257.
37. Merino, S.T.; Cherry, J. Progress and challenges in enzyme development for Biomass utilization. *Biofuels* **2007**, *108*, 95–120.
38. Sørensen, A.; Ahring, B.K.; Lübeck, M.; Ubhayasekera, W.; Bruno, K.S.; Culley, D.E.; Lübeck, P.S. Identifying and characterizing the most significant beta-glucosidase of the novel species *Aspergillus saccharolyticus*. *Can. J. Microbiol.* **2012**, *58*, 1035–1046.
39. Nakazawa, H.; Kawai, T.; Ida, N.; Shida, Y.; Kobayashi, Y.; Okada, H.; Tani, S.; Sumitani, J. Kawaguchi, T.; Morikawa, Y.; Ogasawara, W. Construction of a recombinant *Trichoderma reesei* strain expressing *Aspergillus aculeatus* beta-glucosidase 1 for efficient biomass conversion. *Biotechnol. Bioeng.* **2012**, *109*, 92–99.
40. Dekker, R.F.H. Kinetic, inhibition, and stability properties of a commercial beta-D-glucosidase (cellobiase) preparation from *Aspergillus niger* and its suitability in the hydrolysis of lignocellulose. *Biotechnol. Bioeng.* **1986**, *28*, 1438–1442.

41. Krogh, K.B.R.; Morkeberg, A.; Jorgensen, H.; Frisvad, J.C.; Olsson, L. Screening genus *Penicillium* for producers of cellulolytic and xylanolytic enzymes. *Appl. Biochem. Biotechnol.* **2004**, *113–116*, 389–401.
42. Liu, D.; Zhang, R.; Yang, X.; Zhang, Z.; Song, S.; Miao, Y.; Shen, Q. Characterization of a thermostable beta-glucosidase from *Aspergillus fumigatus* Z5, and its functional expression in *Pichia pastoris* X33. *Microb. Cell Factories* **2012**, *11*, 25.
43. Gyalai-Korpos, M.; Mangel, R.; Alvira, P.; Dienes, D.; Ballesteros, M.; Reczey, K. Cellulase production using different streams of wheat grain- and wheat straw-based ethanol processes. *J. Ind. Microbiol. Biotechnol.* **2011**, *38*, 791–802.
44. Sørensen, A.; Teller, P.J.; Lübeck, P.S.; Ahring, B.K. Onsite enzyme production during bioethanol production from biomass: Screening for suitable fungal strains. *Appl. Biochem. Biotechnol.* **2011**, *164*, 1058–1070.
45. Herculano, P.N.; Porto, T.S.; Moreira, K.A.; Pinto, G.A.S.; Souza-Motta, C.M.; Porto, A.L.F. Cellulase production by *Aspergillus japonicus* URM5620 Using Waste from Castor Bean (Ricinus communis L.) Under Solid-State Fermentation. *Appl. Biochem. Biotechnol.* **2011**, *165*, 1057–1067.
46. Bairoch, A. The ENZYME database in 2000. *Nucleic Acids Res.* **2000**, *28*, 304–305.
47. Eyzaguirre, J.; Hidalgo, M.; Leschot, A. Beta-Glucosidases from Filamentous Fungi: Properties, Structure, and Applications. In *Handbook of Carbohydrate Engineering*; Taylor and Francis Group, LLC: Boca Raton, FL 33487, USA, 2005; pp. 645–685.
48. Henrissat, B. A classification of glycosyl hydrolases based on amino-acid-sequence similarities. *Biochem. J.* **1991**, *280*, 309–316.
49. Bairoch, A. Prosite - a dictionary of sites and patterns in proteins. *Nucleic Acids Res.* **1992**, *20*, 2013–2018.
50. Varghese, J.N.; Hrmova, M.; Fincher, G.B. Three-dimensional structure of a barley beta-D-glucan exohydrolase, a family 3 glycosyl hydrolase. *Structure* **1999**, *7*, 179–190.
51. Yoshida, E.; Hidaka, M.; Fushinobu, S.; Koyanagi, T.; Minami, H.; Tamaki, H.; Kitaoka, M.; Katayama, T.; Kumagai, H. Purification, crystallization and preliminary X-ray analysis of beta-glucosidase from *Kluyveromyces marxianus* NBRC1777. *Acta Crystallogr. Sect. F-Struct. Biol. Cryst. Commun.* **2009**, *65*, 1190–1192.
52. Pozzo, T.; Pasten, J.L.; Karlsson, E.N.; Logan, D.T. Structural and functional analyses of beta-glucosidase 3B from *Thermotoga neapolitana*: A thermostable three-domain representative of glycoside hydrolase 3. *J. Mol. Biol.* **2010**, *397*, 724–739.
53. Nakatani, Y.; Cutfield, S.M.; Cowieson, N.P.; Cutfield, J.F. Structure and activity of exo-1,3/1,4-beta-glucanase from marine bacterium *Pseudoalteromonas sp* BB1 showing a novel C-terminal domain. *Febs. J.* **2012**, *279*, 464–478.
54. McAndrew, R.P.; Park, J.I.; Heins, R.A.; Reindl, W.; Friedland, G.D.; D'haeseleer, P.; Northen, T.; Sale, K.L.; Simmons, B.A.; Adams, P.D. From soil to structure, a novel dimeric beta-glucosidase belonging to the glycoside hydrolase family 3 isolated from compost using metagenomic analysis. *J. Biol. Chem.* **2013**, *288*, 14985–14992.

55. Suzuki, K.; Sumitani, J.; Nam, Y.; Nishimaki, T.; Tani, S.; Wakagi, T.; Kawaguchi, T.; Fushinobu, S. Crystal structures of glycoside hydrolase family 3 β-glucosidase 1 from *Aspergillus aculeatus*. *Biochem. J.* **2013**, *452*, 211–221.
56. Jeya, M.; Joo, A.; Lee, K.; Tiwari, M.K.; Lee, K.; Kim, S.; Lee, J. Characterization of beta-glucosidase from a strain of *Penicillium purpurogenum* KJS506. *Appl. Microbiol. Biotechnol.* **2010**, *86,* 1473–1484.
57. Sinnott, M.L. Catalytic mechanisms of enzymatic glycosyl transfer. *Chem. Rev.* **1990**, 90, 1171–1202.
58. McCarter, J.D.; Withers, S.G. Mechanisms of enzymatic glycoside hydrolysis. *Curr. Opin. Struct. Biol.* **1994**, *4*, 885–892.
59. Thongpoo, P.; McKee, L.S.; Araujo, A.C.; Kongsaeree, P.T.; Brumer, H. Identification of the acid/base catalyst of a glycoside hydrolase family 3 (GH3) beta-glucosidase from *Aspergillus niger* ASKU28. *Biochim. Biophys. Acta-Gen.* **2013**, *1830*, 2739–2749.
60. Davies, G.; Henrissat, B. Structures and mechanisms of glycosyl hydrolases. *Structure* **1995**, 3, 853–859.
61. Davies, G.J.; Wilson, K.S.; Henrissat, B. Nomenclature for sugar-binding subsites in glycosyl hydrolases. *Biochem. J.* **1997**, *321*, 557–559.
62. Bhatia, Y.; Mishra, S.; Bisaria, V.S. Microbial beta-glucosidases: Cloning, properties, and applications. *Crit. Rev. Biotechnol.* **2002**, *22*, 375–407.
63. Zahoor, S.; Javed, M.M.; Aftab, S.; Latif, F.; Ikram-ul-Haq. Metabolic engineering and thermodynamic characterization of an extracellular beta-glucosidase produced by *Aspergillus niger*. *Afr. J. Biotechnol.* **2011**, *10*, 8107–8116.
64. Elshafei, A.M.; Hassan, M.M.; Morsi, N.M.; Elghonamy, D.H. Purification and some kinetic properties of beta-glucosidase from *Aspergillus terreus* NRRL 265. *Afr. J. Biotechnol.* **2011**, *10*, 19556–19569.
65. Karnchanatat, A.; Petsom, A.; Sangvanich, P.; Piaphukiew, J.; Whalley, A.J.S.; Reynolds, C.D.; Sihanonth, P. Purification and biochemical characterization of an extracellular beta glucosidase from the wood-decaying fungus *Daldinia eschscholzii* (Ehrenb. :Fr.) Rehm. FEMS Microbiol lett. 2007, *270*, 162–170.
66. Yoon, J.; Kim, K.; Cha, C. Purification and characterization of thermostable beta-glucosidase from the brown-rot basidiomycete *Fomitopsis palustris* grown on microcrystalline cellulose. *J. Microbiol.* **2008**, *46*, 51–55.
67. Bhatti, H.N.; Batool, S.; Afzal, N. Production and characterization of a novel beta-glucosidase from *Fusarium solani*. *Int. J. Agric. Biol.* **2013**, *15*, 140–144.
68. Kalyani, D.; Lee, K.; Tiwari, M.K.; Ramachandran, P.; Kim, H.; Kim, I.; Jeya, M.; Lee, J. Characterization of a recombinant aryl beta-glucosidase from *Neosartorya fischeri* NRRL181. *Appl. Microbiol. Biotechnol.* **2012**, *94*, 413–423.
69. Moreira Souza, F.H.; Nascimento, C.V.; Rosa, J.C.; Masui, D.C.; Leone, F.A.; Jorge, J.A.; Furriel, R.P.M. Purification and biochemical characterization of a mycelial glucose- and xylose-stimulated beta-glucosidase from the thermophilic fungus *Humicola insolens*. *Process Biochem.* **2010**, *45*, 272–278.

70. Ramani, G.; Meera, B.; Vanitha, C.; Rao, M.; Gunasekaran, P. Production, purification, and characterization of a beta-glucosidase of *Penicillium funiculosum* NCL1. *Appl. Biochem. Biotechnol.* **2012**, *167*, 959–972.

71. Choi, J.; Park, A.; Kim, Y.J.; Kim, J.; Cha, C.; Yoon, J. Purification and characterization of an extracellular beta-glucosidase produced by *Phoma sp* KCTC11825BP isolated from rotten mandarin peel. *J. Microbiol. Biotechnol.* **2011**, *21*, 503–508.

72. Zhang, Y.B.; Yuan, L.J.; Chen, Z.J.; Fu, L.; Lu, J.H.; Meng, Q.F.; He, H.; Yu, X.X.; Lin, F.; Teng, L.R. Purification and characterization of beta-glucosidase from a newly isolated strain *Tolypocladium cylindrosporum* Syzx4. *Chem. Res. Chin. Univ.* **2011**, *27*, 557–561.

73. Lin, J.; Pillay, B.; Singh, S. Purification and biochemical characteristics of beta-D-glucosidase from a thermophilic fungus, *Thermomyces lanuginosus* SSBP. *Biotechnol. Appl. Biochem.* **1999**, *30*, 81–87.

74. Chen, P.; Fu, X.Y.; Ng, T.B.; Ye, X.Y. Expression of a secretory beta-glucosidase from *Trichoderma reesei* in *Pichia pastoris* and its characterization. *Biotechnol. Lett.* **2011**, *33*, 2475–2479.

75. Kubicek, C.P.; Herrera-Estrella, A.; Seidl-Seiboth, V.; Martinez, D.A.; Druzhinina, I.S.; Thon, M.; Zeilinger, S.; Casas-Flores, S.; Horwitz, B.A.; Mukherjee, P.K.; Mukherjee, M.; Kredics, L.; Alcaraz, L.D.; Aerts, A.; Antal, Z.; Atanasova, L.; Cervantes-Badillo, M.G.; Challacombe, J.; Chertkov, O.; McCluskey, K.; Coulpier, F.; Deshpande, N.; von Döhren, H.; Ebbole, D.J.; Esquivel-Naranjo, E.U.; Fekete, E.; Flipphi, M.; Glaser, F.; Gómez-Rodríguez, E.Y.; Gruber, S.; Han, C.; Henrissat, B.; Hermosa, R.; Hernández-Oñate, M.; Karaffa, L.; Kosti, I.; Le Crom, S.; Lindquist, E.; Lucas, S.; Lübeck, M.; Lübeck, P.S.; Margeot, A.; Metz, B.; Misra, M.; Nevalainen, H.; Omann, M.; Packer, N.; Perrone, G.; Uresti-Rivera, E.E.; Salamov, A.; Schmoll, M.; Seiboth, B.; Shapiro, H.; Sukno, S.; Tamayo-Ramos, J.A.; Tisch, D.; Wiest, A.; Wilkinson, H.H.; Zhang, M.; Coutinho, P.M.; Kenerley, C.M.; Monte, E.; Baker, S.E; Grigoriev, I.V. Comparative genome sequence analysis underscores mycoparasitism as the ancestral life style of *Trichoderma*. *Genome Biol.* **2011**, *12*, R40.

76. Yan, S.; Wu, G. Prediction of optimal pH in hydrolytic reaction of beta-glucosidase. *Appl. Biochem. Biotechnol.* **2013**, *169*, 1884–1894.

77. Banerjee, S.; Mudliar, S.; Sen, R.; Giri, B.; Satpute, D.; Chakrabarti, T.; Pandey, R.A. Commercializing lignocellulosic bioethanol: Technology bottlenecks and possible remedies. *Biofuels Bioprod. Bioref.* **2010**, *4*, 77–93.

78. Jeoh, T.; Michener, W.; Himmel, M.E.; Decker, S.R.; Adney, W.S. Implications of cellobiohydrolase glycosylation for use in biomass conversion. *Biotechnol. Biofuels* **2008**, *1*, 10.

79. Banerjee, G.; Car, S.; Scott-Craig, J.S.; Borrusch, M.S.; Bongers, M.; Walton, J.D. Synthetic multi-component enzyme mixtures for deconstruction of lignocellulosic biomass. *Bioresour. Technol.* **2010**, *101*, 9097–9105.

80. Riou, C.; Salmon, J.M.; Vallier, M.J.; Gunata, Z.; Barre, P. Purification, characterization, and substrate specificity of a novel highly glucose-tolerant beta-glucosidase from *Aspergillus oryzae*. *Appl. Environ. Microbiol.* **1998**, *64*, 3607–3614.

81. Langston, J.; Sheehy, N.; Xu, F. Substrate specificity of *Aspergillus oryzae* family 3 beta-glucosidase. *Biochim. Biophys. Acta-Proteins Proteomics* **2006**, *1764*, 972–978.

82. Korotkova, O.G.; Semenova, M.V.; Morozova, V.V.; Zorov, I.N.; Sokolova, L.M.; Bubnova, T.M.; Okunev, O.N.; Sinitsyn, A.P. Isolation and properties of fungal beta-glucosidases. *Biochemistry (Mosc)* **2009**, 74, 569–577.
83. Andric, P.; Meyer, A.S.; Jensen, P.A.; Dam-Johansen, K. Reactor design for minimizing product inhibition during enzymatic lignocellulose hydrolysis: I. Significance and mechanism of cellobiose and glucose inhibition on cellulolytic enzymes. *Biotechnol. Adv.* **2010**, *28*, 308–324.
84. Dale, M.P.; Ensley, H.E.; Kern, K.; Sastry, K.A.R.; Byers, L.D. Reversible inhibitors of beta-glucosidase. *Biochemistry (N Y)* **1985**, *24*, 3530–3539.
85. Bohlin, C.; Praestgaard, E.; Baumann, M.J.; Borch, K.; Praestgaard, J.; Monrad, R.N.; Westh, P. A comparative study of hydrolysis and transglycosylation activities of fungal beta-glucosidases. *Appl. Microbiol. Biotechnol.* **2013**, *97*, 159–169.
86. Frutuoso, M.A.; Marana, S.R. A single amino acid residue determines the ratio of hydrolysis to transglycosylation catalyzed by beta-glucosidases. *Protein Peptide Lett.* **2013**, *20*, 102–106.
87. McIntosh, L.P.; Hand, G.; Johnson, P.E.; Joshi, M.D.; Korner, M.; Plesniak, L.A.; Ziser, L.; Wakarchuk, W.W.; Withers, S.G. The pK_a of the general acid/base carboxyl group of a glycosidase cycles during catalysis: A C-13-NMR study of *Bacillus circuluns* xylanase. *Biochemistry (N Y)* **1996**, *35*, 9958–9966.
88. Yeoman, C.J.; Han, Y.; Dodd, D.; Schroeder, C.M.; Mackie, R.I.; Cann, I.K.O. Thermostable enzymes as biocatalysts in the biofuel industry. *Adv. Appl. Microbiol.* **2010**, *70*, 1–55.
89. Berlin, A.; Balakshin, M.; Gilkes, N.; Kadla, J.; Maximenko, V.; Kubo, S.; Saddler, J. Inhibition of cellulase, xylanase and beta-glucosidase activities by softwood lignin preparations. *J. Biotechnol.* **2006**, *125*, 198–209.
90. Yang, B.; Wyman, C.E. BSA treatment to enhance enzymatic hydrolysis of cellulose in lignin containing substrates. *Biotechnol. Bioeng.* **2006**, *94*, 611–617.
91. Selig, M.J.; Hsieh, C.C.; Thygesen, L.G.; Himmel, M.E.; Felby, C.; Decker, S.R. Considering water availability and the effect of solute concentration on high solids saccharification of lignocellulosic biomass. *Biotechnol. Prog.* **2012**, *28*, 1478–1490.
92. Hawksworth, D.L. The fungal dimension of biodiversity - Magnitude, significance, and conservation. *Mycol. Res.* **1991**, *95*, 641–655.
93. Hawksworth, D.L. The magnitude of fungal diversity: The 1.5 million species estimate revisited. *Mycol. Res.* **2001**, *105*, 1422–1432.
94. Mares, D.; Andreotti, E.; Maldonado, M.E.; Pedrini, P.; Colalongo, C.; Romagnoli, C. Three new species of *Aspergillus* from Amazonian forest soil (Ecuador). *Curr. Microbiol.* **2008**, 57, 222–229.
95. De Vries, R.P.; Frisvad, J.C.; van de Vondervoort, P.J.I.; Burgers, K.; Kuijpers, A.F.A.; Samson, R.A.; Visser, J. *Aspergillus vadensis*, a new species of the group of black Aspergilli. *Antonie Van Leeuwenhoek* **2005**, *87*, 195–203.
96. Samson, R.A.; Houbraken, J.A.M.P.; Kuijpers, A.F.A.; Frank, J.M.; Frisvad, J.C. New ochratoxin A or sclerotium producing species in *Aspergillus* section *Nigri*. *Stud. Mycol.* **2004**, *50*, 45–61.

97. Perrone, G.; Varga, J.; Susca, A.; Frisvad, J.C.; Stea, G.; Kocsube, S.; Tóth, B.; Kozakiewicz, Z.; Samson, R.A. *Aspergillus uvarum* sp nov., an uniseriate black Aspergillus species isolated from grapes in Europe. *Int. J. Syst. Evol. Microbiol.* **2008**, *58*, 1032–1039.
98. Noonim, P.; Mahakarnchanakul, W.; Varga, J.; Frisvad, J.C.; Samson, R.A. Two novel species of *Aspergillus* section *Nigri* from Thai coffee beans. *Int. J. Syst. Evol. Microbiol.* **2008**, *58*, 1727–1734.
99. Noonim, P.; Mahakarnchanakul, W.; Nielsen, K.F.; Frisvad, J.C.; Samson, R.A. Isolation, identification and toxigenic potential of ochratoxin A-producing *Aspergillus* species from coffee beans grown in two regions of Thailand. *Int. J. Food Microbiol.* **2008**, *128*, 197–202.
100. Varga, J.; Kocsube, S.; Toth, B.; Frisvad, J.C.; Perrone, G.; Susca, A.; Meijer, M.; Samson, R.A. *Aspergillus brasiliensis* sp nov., a biseriate black *Aspergillus* species with world-wide distribution. *Int. J. Syst. Evol. Microbiol.* **2007**, *57*, 1925–1932.
101. Serra, R.; Cabanes, F.J.; Perrone, G.; Castella, G.; Venancio, A.; Mule, G.; Kozakiewicz, Z. *Aspergillus ibericus*: A new species of section *Nigri* isolated from grapes. *Mycologia* **2006**, *98*, 295–306.
102. Sørensen, A.; Lubeck, P.S.; Lubeck, M.; Nielsen, K.F.; Ahring, B.K.; Teller, P.J.; Frisvad, J.C. *Aspergillus saccharolyticus* sp nov., a black *Aspergillus* species isolated in Denmark. *Int. J. Syst. Evol. Microbiol.* **2011**, *61*, 3077–3083.
103. Kim, S.; Lee, C.; Kim, M.; Yeo, Y.; Yoon, S.; Kang, H.; Koo, B. Screening and characterization of an enzyme with beta-glucosidase activity from environmental DNA. *J. Microbiol. Biotechnol.* **2007**, *17*, 905–912.
104. Jiang, C.; Ma, G.; Li, S.; Hu, T.; Che, Z.; Shen, P.; Yan, B.; Wu, B. Characterization of a novel beta-glucosidase-like activity from a soil metagenome. *J. Microbiol.* **2009**, *47*, 542–548.
105. Jiang, C.; Hao, Z.; Jin, K.; Li, S.; Che, Z.; Ma, G.; Wu, B. Identification of a metagenome-derived beta-glucosidase from bioreactor contents. *J. Mol. Catal. B-Enzym.* **2010**, *63*, 11–16.
106. Sternberg, D.; Vijayakumar, P.; Reese, E.T. Beta-glucosidase - microbial-production and effect on enzymatic-hydrolysis of cellulose. *Can. J. Microbiol.* **1977**, *23*, 139–147.
107. Sørensen, A.; Lübeck, P.S.; Lübeck, M.; Teller, P.J.; Ahring, B.K. Beta-Glucosidases from a new *Aspergillus* species can substitute commercial beta-glucosidases for saccharification of lignocellulosic biomass. *Can. J. Microbiol.* **2011**, 57, 638–650.
108. Montenecourt, B.S.; Eveleigh, D.E. Selective screening methods for the isolation of high yielding cellulase mutants of *Trichoderma reesei*. In *Hydrolysis of Cellulose: Mechanisms of Enzymic and Acid Catalysis*; Brown, R.D., Jurasek, L., Eds.; American Chemical Society: Washington DC, USA, 1979; pp. 289–301.
109. Chirumamilla, R.R.; Muralidhar, R.; Marchant, R.; Nigam, P. Improving the quality of industrially important enzymes by directed evolution. *Mol. Cell. Biochem.* **2001**, *224*, 159–168.
110. Cherry, J.R.; Fidantsef, A.L. Directed evolution of industrial enzymes: An update. *Curr. Opin. Biotechnol.* **2003**, *14*, 438–443.
111. Tobin, M.B.; Gustafsson, C.; Huisman, G.W. Directed evolution: The 'rational' basis for 'irrational' design. *Curr. Opin. Struct. Biol.* **2000**, *10*, 421–427.
112. Antikainen, N.M.; Martin, S.F. Altering protein specificity: Techniques and applications. *Bioorg. Med. Chem.* **2005**, *13*, 2701–2716.

113. Gonzalez-Blasco, G.; Sanz-Aparicio, J.; Gonzalez, B.; Hermoso, J.A.; Polaina, J. Directed evolution of beta-glucosidase A from *Paenibacillus polymyxa* to thermal resistance. *J. Biol. Chem.* **2000**, *275*, 13708–13712.
114. Arrizubieta, M.J.; Polaina, J. Increased thermal resistance and modification of the catalytic properties of a beta-glucosidase by random mutagenesis and *in vitro* recombination. *J. Biol. Chem.* **2000**, *275*, 28843–28848.
115. Lebbink, J.H.G.; Kaper, T.; Bron, P.; van der Oost, J.; de Vos, W.M. Improving low-temperature catalysis in the hyperthermostable *Pyrococcus furiosus* beta-glucosidase CelB by directed evolution. *Biochemistry (N Y)* **2000**, *39*, 3656–3665.
116. Pei, X.; Yi, Z.; Tang, C.; Wu, Z. Three amino acid changes contribute markedly to the thermostability of beta-glucosidase BglC from *Thermobifida fusca*. *Bioresour. Technol.* **2011**, *102*, 3337–3342.
117. Steiner, K.; Schwab, H. Recent Advances in rational approaches for enzyme engineering. *Comput. Struct. Biotechnol. J.* **2012**, *2*, doi:10.5936/csbj.201209010.
118. Lee, H.; Chang, C.; Jeng, W.; Wang, A.H.-J; Liang, P. Mutations in the substrate entrance region of beta-glucosidase from *Trichoderma reesei* improve enzyme activity and thermostability. *Protein Eng. Des. Sel.* **2012**, *25*, 733–740.

5

Angling for Uniqueness in Enzymatic Preparation of Glycosides

Antonio Trincone

Institute of Biomolecular Chemistry, National Research Council, Via Campi Flegrei, 34, Pozzuoli 80078, Naples, Italy; E-Mail: antonio.trincone@icb.cnr.it

---

Abstract: In the early days of biocatalysis, limitations of an enzyme modeled the enzymatic applications; nowadays the enzyme can be engineered to be suitable for the process requirements. This is a general bird's-eye view and as such cannot be specific for articulated situations found in different classes of enzymes or for selected enzymatic processes. As far as the enzymatic preparation of glycosides is concerned, recent scientific literature is awash with examples of uniqueness related to the features of the biocatalyst (yield, substrate specificity, regioselectivity, and resistance to a particular reaction condition). The invention of glycosynthases is just one of the aspects that has thrust forward the research in this field. Protein engineering, metagenomics and reaction engineering have led to the discovery of an expanding number of novel enzymes and to the setting up of new bio-based processes for the preparation of glycosides. In this review, new examples from the last decade are compiled with attention both to cases in which naturally present, as well as genetically inserted, characteristics of the catalysts make them attractive for biocatalysis.

Keywords: biocatalysis; glycosides; enzymes; glycosidases; glycosynthases

---

1. Introduction

In a brilliant, recently published analysis of the research-guided development in the field of biocatalysis during the last century, different authors recognized three historical waves of innovations that totally changed the field of biocatalysis to the present industrially accomplished level [1]. In a nutshell, while in the past limitations of an enzyme modeled the enzymatic process, today the enzyme can be engineered to be suitable for the process requirements. However this general bird's-eye view cannot be specific for articulated contexts in which each single class of enzyme or selected enzymatic process is at the present state. In another similar general bird's-eye analysis, Riva identified a long wave of successes still far from reaching the end in biocatalysis [2], due to the difficulties encountered in the shift from 'classical' processes to biobased ones. It is clear that exploiting natural catalysts to obtain selective transformations of non-natural substrates is far from being fully explored; among many others, the cases represented by the new concept of 'third generation biorefineries' [2] (producing chemicals from biomasses), or by the new glycoside hydrolases and other enzymes found in marine environments [3] have re-fostered new research trends in the field. As a matter of example, although investigation into hemicellulases as biorefining enzymes has been slow, as reported in a recent analysis [4], xylan-related biocatalysis has continued to make steady progress in many areas, including the discovery and characterization of a wide range of hemicellulases. Talking more specifically about biobased glycosynthesis, these studies are opening new prospects for the use of pentose sugars as building blocks for engineered pentosides as non-ionic surfactants or prebiotic food/feed ingredients.

Carbohydrates are involved in a broad range of functions in cell living systems. Structural roles and energy storage as functions were recognized during the first half of the last century while attainments in glycobiology and glycochemistry, during the last twenty years, have further revealed that carbohydrate parts of biomolecules (glycoproteins, glycolipids, *etc.*) are involved in important biological functions mainly related to cell recognition events [5–7]. It should not be neglected that carbohydrates are important molecules also in the technological domain. Synthetic carbohydrate-containing polymers have a wide range of applications in medical biotechnology [8]. A number of novel dietary carbohydrates produced by enzymatic syntheses have been introduced into food technology during the last decade [9]. In innovative fine chemical manufacturing solutions, straightforward synthesis of products is of interest (e.g., chromophoric oligosaccharides of strictly defined structure as valuable tools for the kinetic analysis of hydrolytic activities and for characterization of new exo- or endo-glycosidases). Finally, in cosmetics, prodrug action of enzymatically glycosylated natural lipophilic antioxidants is currently under consideration.

In general glycosylation is considered to be an important and quite special method for the structural modification of a compound. It allows the conversion of a lipophilic compound into a hydrophilic one changing pharmacokinetic properties or creating drug delivery systems. It could also be generalized that in a glycoside, the type of aglycon determines the application: long alkyl chains allow glycosides to possess useful properties as surfactants and emulsifiers; aglycon based on unsaturated alkyl chains are said to be valuable, as glycosides, for fungal infections or as antimicrobial agents; glycosides of peptides and steroids are used in antitumor formulations and cardiac-related drugs, respectively; and glycosides of flavors and fragrances are used as "controlled release" compounds [10,11].

Sugar units have more than one site through which the chains are extended. Each of these sites frequently shows very similar reactivity, thus the masking of reactive centers by protecting groups is essential in order to direct coupling through the right position. For this reason protection and deprotection steps of functional groups are in use extensively in the arsenal of the synthetic carbohydrate chemist; moreover ensuring the correct stereochemistry of the glycosidic linkage formed entails additional difficulty. Carbohydrate related synthetic chemistry can still be considered one of the well-explored branches of organic chemistry and very rich in significant and spectacular successes, although important alternative biomethodologies for assembling glycosidic linkages are presently known and acknowledged. It is worth noting that in comparison with chemical methods, enzymatic glycosylation is particularly useful for the modification of complex biologically active substances, when generally harsh conditions or use of toxic (heavy metals) catalysts are undesirable. Enzymes may represent an imperative choice in fields such as agriculture and food or cosmetics where chemical strategies are not acceptable [12]. In a very recent report detailing different examples of enzymatic glycosylation of small molecules, the authors concluded that challenging substrates require tailored catalysts, and the progress in the field of enzyme engineering and screening of new catalytic activities are both expected to result in new applications of biocatalytic glycosylations in various industrial sectors [13].

The enzymes responsible for the synthesis of glycosidic linkage have been recognized as transglycosylases and named glycosyltransferases, specifying the glycosyl donor and the reaction product. These enzymes transfer sugar moieties from activated donors to specific acceptors, forming stereochemically specific glycosidic bonds, and are responsible *in vivo* for the synthesis of most cell-surface glycoconjugates, using eight common sugar nucleotides as activated donors (Leloir pathway). Sugar phosphates act as donors for other glycosyltransferases (non-Leloir pathway). Another widespread group of enzymes, named glycoside hydrolases (glycosidases), exists; they are involved in the carbohydrate metabolism being responsible for the hydrolysis of glycosidic linkages; they can act as exo- or endo-glycosidases and are involved in a series of important biological events such as energy uptake, in processes inherent cell wall metabolism, in glycan processing during *in vivo* glycoprotein synthesis, *etc.* Based on historical grounds, glycoside hydrolases were implicated in most experimental observations during the early studies into the biological synthesis of glycosidic linkages at the beginning of the last century. Hence, the concepts of enzymatically promoted synthesis by both hydrolysis-reversal and glycosyl transfer soon appeared [14]. By the end of the 1980s, several research projects [15] testified the importance of different and interesting glycoside hydrolases, especially from the marine environment; their main application was centered on the structural identification efforts that faced the complexity of oligosaccharide structures before the instrumental exploit of 2D NMR and MS spectroscopy.

Different wild-type glycosidases and their modified versions are enzymes deserving new expectations in research and development today. Significant progress has been made in recent years for the application of these enzymes: even while the major breakthrough was the invention of glycosynthases, protein engineering, metagenomics and reaction engineering led to the discovery of an expanding number of novel enzymes and to the setting up of new bio-based processes for the preparation of important glycosides. This review will compile different examples where glycoside hydrolases are the key enzymes in the process.

2. Natural Enzymes for the Synthesis of Glycosidic Linkages

In chemical terms, considering both hydrolytic or synthetic aspects of esterases, glycosidases, phosphatases, transglycosidases and peptidases, the enzymatic mechanisms are based on displacement reactions and could be grouped together. This line of thought proved to be highly productive in historical terms, allowing the collection and rationalization of the amount of mechanistic data especially for glycosidases and transglycosylases. The stereochemistry of the mechanisms of glycoside hydrolases was analyzed by Koshland [16] more than 60 years ago and allowed the classification of inverting and retaining enzymes according to the anomeric configuration found in the product with respect to that in the starting substrate. Very recently, it has become clear that other mechanisms have evolved, such as the one based on elimination [17]. In 2010, in an interesting review on diversity of catalytic base nucleophile of glycoside hydrolases, it was reported that a variety of systems are used to replace this function, including substrate-assisted catalysis, a network of several residues, and the use of non-carboxylate residues or exogenous nucleophiles [18].

Glycosyltransferases-mediated reactions are thought to proceed via an oxocarbenium-ion-like transition state as proposed for glycosidase reactions on the basis of solid structural, mechanistic and *ab initio* molecular orbital calculations data [19]. Glycosyltransferases are catalysts for natural glycosylation reactions, known as "Leloir" glycosyl transferases (GT). Glycoside phosphorylases (GP), requiring glycosyl phosphates and transglycosidases (TG), employing non-activated carbohydrates (e.g., sucrose), are additional examples of synthetic enzymes. However glycoside hydrolases (GH) can also be used for synthetic purposes under either kinetic (transglycosylation) or thermodynamic (reverse hydrolysis) control.

In this paragraph new examples related to the transfer of glycosyl residues between two oxygen nucleophiles are compiled with attention to those cases in which the natural characteristics of the catalyst make it attractive for biocatalysis (importance of molecular skeleton of substrates, yield, regioselectivity, resistance to particular reaction condition, *etc.*).

2.1. Interesting Transfer of Glycosyl Residues in Natural Enzymes for the Synthesis of Glycosidic Linkages

Hyaluronic acid, synthesized by hyaluronan synthases, is a biopolymer abundant in extracellular matrices that is degraded by hyaluronidases. Biocompatibility and biodegradability of this polymer and of related small derivatives are of great interest in pharmaceutics in a number of molecular devices for drug delivery. Bovine testicular hyaluronidase (BTH) is a commercially available hyaluronidase preparation that has long been considered a prototype of mammalian hyaluronidases. Presumably all mammalian hyaluronidases can catalyze hydrolysis as well as transglycosylation reactions of hyaluronic acid fragments (Figure 1). In the case of BTH, the hydrolysis is favored at acidic pH values, while transglycosylation occurs preferentially at neutral pH and at low NaCl concentrations. The availability of recombinant expression systems for the production of purified human hyaluronidases PH-20 and Hyal-1 has facilitated the first detailed analysis of the enzymatic reaction products for these two enzymes. HA hexasaccharide, which is generally accepted as being the minimum substrate of BTH, is not a substrate of recombinant human PH-20 and Hyal-1 as recently demonstrated [20] although BTH and PH-20 belong to the same type of hyaluronidase. Interestingly, HA octasaccharide can be used as the substrate of Hyal-1 at pH 3.5. The substrate was converted quickly at concentration between 25 μM to 1 mM; above 1 mM weak substrate inhibition was observed. The study of transfer reactions, selectivities and yields are all features of interest for a possible use of these enzymes in biocatalytic steps for the manipulation of important biomolecules [21].

Figure 1. Tetrasaccharidic moiety of a hyaluronic acid chain and point of attach of hyaluronase for the hydrolysis and transglycosylation reactions.

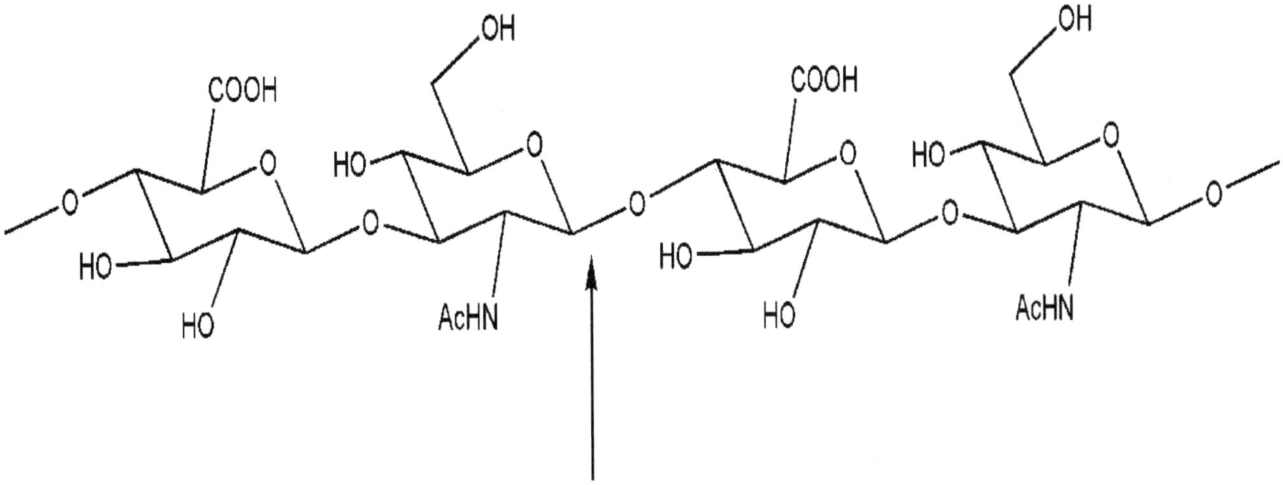

Additional new examples of this enzyme can be derived from other environments: the venoms of two classes of fish, freshwater stingray (members of the genus Potamotrygon) and stonefish (members of the genus Synanceia), contain, along with proteinaceous toxins, also hyaluronidases. These proteins are considered as spreading factors that facilitate the tissue diffusion of toxins by degrading hyaluronan; owing to this quick action it can possess very interesting features for biocatalysis [22].

Sensitive and reliable enzymatic tools for the analysis of glycan chains are needed. The O-linked glycans are attached to serine or threonine through the GalNAc residue at the reducing end. The hydrolysis of the O-glycosidic α-linkage between GalNAc of the disaccharide Gal-β-1,3-GalNAc- and threonine or serine, is catalyzed by endo-α-N-acetylgalactosaminidase (EC 3.2.1.97). Most of the enzymes of this type are strictly specific for the disaccharidic structure and have no action on substrates with longer or different glycosyl chains. Using the protein sequence of a known endo-α-GalNAcase from *B. longum*, four potential sequences were found from BLAST (Basic Local Alignment Search Tool) search. Cloned and expressed proteins were purified and characterized [23]. Substrate specificity was investigated on aryl substrates as indicated in Figure 2 or by using natural glycoproteins. All three new enzymes are active on Core 1 substrate (Gal-β-1,3-GalNAc) while only two of them (EngEF and EngPA) were active after 24 h on Core 3 disaccharide (Glc-β-1,3-GalNAc). Interestingly, these enzymes acted also as transglycosylating agents; when reacted with simple alkanols (from methanol to nonanol) as acceptors, Core 1 and Core 3 acted as donor disaccharides in test reactions. Although the yields as judged by TLC analysis were low (at 0.8-1.6 mM donor and 13% v/v alkanols), positive reactions were observed up to 4–5 alkanol carbon atoms. As stated by the authors in the concluding remarks, the action of EngEF and EngPA enzymes acting on Core 3 in addition to Core 1 O-glycans could make these enzymes powerful tools for the release of O-glycan sugars from glycoproteins. More importantly they can also be used as templates in future protein engineering experiments for a possible creation of endo-α-GalNAcases capable of acting on O-linked glycans, regardless of their sugar composition.

Figure 2. Substrates and percentages of products liberated at the link indicated by each single enzyme.

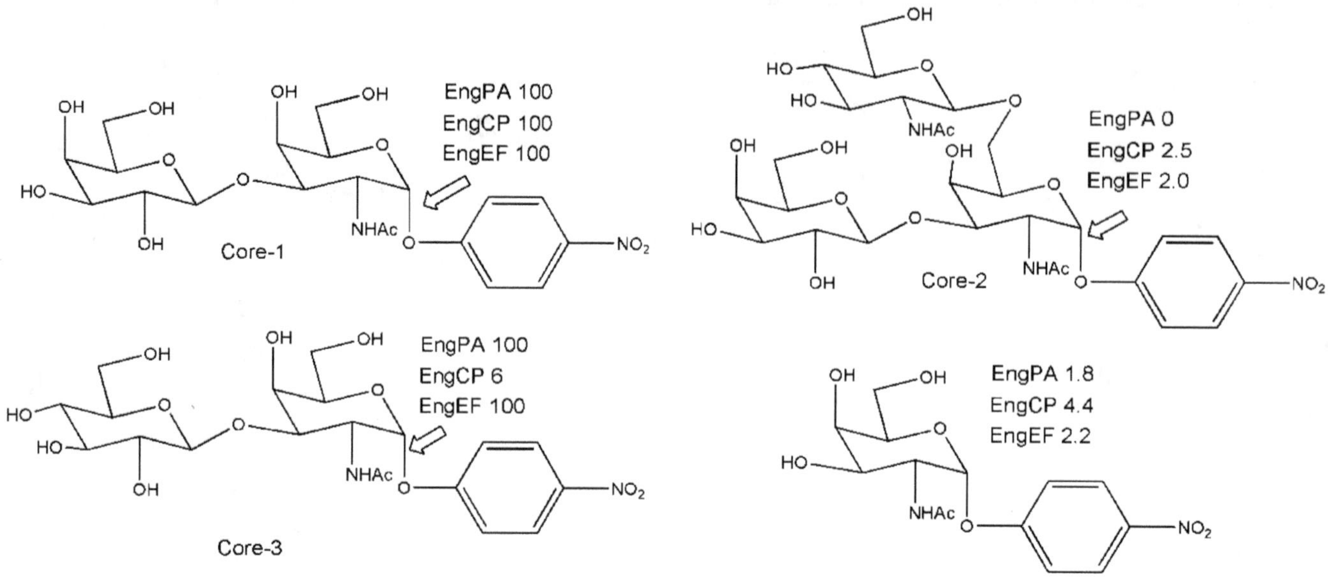

One of the last outstanding results in transglycosylation reactions in the last decade is, without any doubt, the recognition that endo-β-N-acetylglucosaminidase can transglycosylate a large oligosaccharide onto various glycosyl acceptors. In 2001, the Shoda group reported on the synthesis of a novel disaccharide possessing a 1,2-oxazoline moiety and tested it with a series of enzymes for transglycosylation activity. Typical oxazoline substrate 1, as depicted in Figure 3, reacted with Endo-M or Endo-A from *Mucor hiemalis* or *A. protophormiae*, respectively and the acceptor GlcNAc-β-1-O-pNP for the synthesis of a trisaccharidic derivative in high yield [24]. These interesting biocatalysts (Endo-M or Endo-A), which belong to GH family 85 were inactive when they have to use acceptors capped with an α-1,6-fucose (1,6-fucosyl-GlcNAc derivative as acceptors for transglycosylation). But more recently, the Huang group, screening various endo-β-N-acetylglucosaminidases using appropriate synthetic oxazoline donors and compound 2, Figure 3, as acceptor substrate, found that endo-β-N-acetylglucosaminidases from *Flavobacterium meningosepticum* (including Endo-F2 and Endo-F3), were able to glycosylate α-1,6-fucosylated GlcNAc derivative to provide natural, core-fucosylated complex type N-glycopeptides [25]. The product(s) were isolated in high yield by HPLC and this efficiency of transglycosylation is quite impressive, given the fact that only two-fold of the donor substrate was used.

Starting from the known consideration that extracellular glycosidases from fungi, (used to access low molecular weight sugars by their hydrolytic action on polymeric substrates), must possess high flexibility in substrate specificity, the group of Kren recently investigated new enzymes in which the 4-hydroxy moiety of the pyranose ring of the substrate is not essential for binding to the enzyme active site. In their report they emphasize also the transglycosylation activity of β-N-acetylhexosaminidase from *Talaromyces flavus*. The results reported on the production of three novel 4′-deoxy-disaccharides prepared in high yields (52% total disaccharide fraction) starting with phenyl 2-acetamido-2,4-dideoxy-substrate. The conditions of the transglycosylation reaction were first optimized on an analytical scale by varying the concentrations and ratios of the reaction components and identifiying the following reaction conditions as the most efficient: 75 mM donor, 300 mM acceptor, incubation for 5–6 h at 35 °C [26].

Figure 3. Typical oxazoline substrate 1 reacting with Endo-M or Endo-A and capped α-1,6-fucose (1,6-fucosyl-GlcNAc derivative) used as acceptors for transglycosylation by Endo-F2 and Endo-F3 [25].

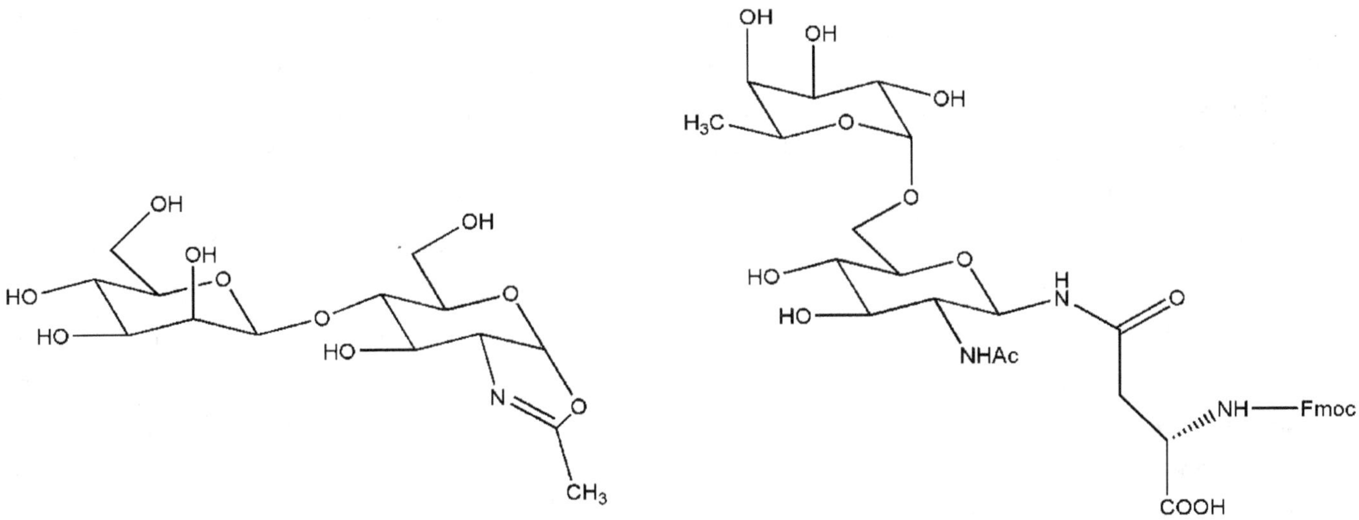

1 2

The enzymatic formation of β-D-fucosides is hardly described in the scientific literature. An extraordinarily broad substrate specificity for both hydrolysis and transglycosylation was exhibited by a glycosidase isolated from the China white jade snail. Acceptor specificity for monosaccharides and transfer efficiency have both been investigated for this promising enzyme [27] and from the results obtained, the authors indicated a very high transfucosylation efficiency using p-nitrophenyl derivative of β-D-fucose at 10 mM and acceptors (at 20–100 mM) such as glucose (88% yield) and xylose (93% yield); the interglycosidic linkage formed with glucose is β-1,6 thus they proposed this biocatalyst as a useful candidate for disaccharide synthesis. The Fuc-β-Xyl disaccharide formed in other reactions is a building block for the synthesis of asterosaponins of marine origin, although in natural compound the interglycosidic linkage is different (Fuc-β-(1-2)-Xyl) [28].

In screening a suitable biocatalyst for galactosylation of nucleosides, Ye *et al.* very recently found that β-galactosidases from *Kluyveromyces lactis* (Sigma, USA) and a crude glycosidase extract of apple seeds, had high hydrolytic activities toward oNPGal, but could not catalyze transglycosylation reactions between oNPGal and 2β-deoxynucleosides [29]. Indeed, only very few reports have appeared in the literature in the last 10 years leading with successful glycosidase-mediated nucleoside glycosylation: (i) after low-yielding results using β-galactosidase from *Aspergillus oryzae* [30], the first successful approach, which appeared in 2007, was (ii) the convenient synthesis of β-galactosyl derivatives of antiviral and anticancer nucleosides, utilizing the purified β-galactosidase activity from the hepatopancreas of *Aplysia fasciata* [31] and the second valid method is (iii) the regioselective galactosylation of floxuridine (FUdR) catalyzed by a commercial β-galactosidase from bovine liver with a high yield (75%) and an excellent 5-β-regioselectivity (>99%) [32] using o-nitrophenyl β-D-galactoside as glycosyl donor. Such desirable products were synthesized with satisfactory yields (41–68%) and moderate to high 5β-regioselectivities (87–100%), also by using the crude enzyme extract.

In the hepatopancreas and visceral mass of the mollusc *Aplysia fasciata* a wide range of glycoside hydrolases (α-glucosidase, β-galactosidase, β-glucosidase, β-mannosidase and others) were found and successfully used for the hydrolysis and synthesis of glycosidic bonds. The formation of α-D-oligoglucosides from maltose (up to tetra- and pentasaccharides) was studied in detail [33] and it was also demonstrated that the enzyme was able to form mono- and di- and poly-glucosides of different externally added acceptors [3 and references cited therein]. This activity has been applied in the very recent chemoenzymatic synthesis of α-6-sulfoquinovosyl-1,2-O-diacylglycerols, a class of natural lipids that have attracted biomedical attention as possible antitumors, antivirals, and immunomodulators. This synthesis started with the enzymatically controlled transfer of glucose from maltose (10:1 molar ratio) to (*rac*)-1,2-O-isopropylidene glycerol. The 1,2 glycerol acetonide conversion of 59% in 29 h was obtained and reaction products were exclusively 3-α-glycosyl derivatives of 1,2-O-isopropylidene glycerol. In particular 3-α-glucosyl-1,2-O-isopropylidene glycerol was produced as the major component (30% yield) together with a mixture of di- and tri-saccharide analogues (23 and 6% yields, respectively). Under similar conditions the transglycosidase of *Aspergillus niger* and the α-glucosidase of *Bacillus stearothermophilus* gave 3-α-glucosyl-1,2-O-isopropylidene glycerol with yields below 2%. This comparison suggests that marine enzymes may offer a viable option for the synthesis of different types of glycolipids [34].

There are other glycoside hydrolase activities that are hardly used in synthesis. A thermostable α-L-arabinofuranosidase was reported in 2002 for its ability to perform transarabinosylation reactions for the synthesis of different α-L-arabinofuranosides of various alcohols. The same enzyme was later reported as a useful biocatalyst for the synthesis of p-nitrophenyl α-L-arabinofuranosyl-(1-2)-α-L-arabinofuranoside, p-nitrophenyl β-D-xylopyranosyl-(1-2)-β-D-xylopyranoside, p-nitrophenyl β-D-xylopyranosyl-(1-3)-β-D-xylopyranoside and benzyl α-L-arabinofuranosyl-(1-2)- α-D-xylopyranoside. Such a disaccharidic motif could be used as the reference compound for the analysis of hemicellulase action and for raising antibodies to well-defined motifs for immunochemical-based analyses of plant cell walls [35]. The same group reported also the synthesis of galactofuranosides using the above catalyst [36]. β-D-galactofuranose is a cell wall constituent in several pathogenic species including *Mycobacterium tuberculosis* and *leprae*, agents of tuberculosis and leprosy, respectively. The synthesis of galactofuranose analogues acting as inhibitors of the biosynthetic enzymes (UDP-galactopyranose mutase and galactofuranosyl transferases) or the synthesis of galactofuranosides that could be used for the elaboration of vaccines, is important for this aspect. Authors reported also on the chemical synthesis of p-nitrophenyl galactofuranoside used as donor and the low yield obtained (16%) could act as an additional center of interest for the synthetic aspect of this kind of enzyme.

An interesting case of the effect of solvents in transglycosylation reactions has been studied recently for the synthesis of N-acetyl-D-lactosamine, Gal-β-1,4-GlcNAc (LacNAc), catalyzed by the β-galactosidase from *Thermus thermophilus* (TTP0042). The authors studied the percentages of all products formed (self-condensation products, hydrolytic originating galactose and transfer regioisomers). The low yield (about 20%) of LAcNAc when the reaction is performed in buffer is increased by up to 91% when the process takes place in the presence of glycerol based solvents. According to conformational studies of the enzyme structure by circular dichroism and fluorescence, the authors concluded that in the presence of these solvents the enzyme modifies secondary and tertiary structure and this may also be the cause of some additional regioselectivity changes observed [37]. Unfortunately no molecular details were furnished, indeed no clear relationship between the very different solvent structures used and the regioselectivity change was envisaged.

A new and interesting fungal diglycosidase was isolated from *Acremonium* sp. It has an α-rhamnosyl-β-glucosidase activity with transglycosylation potential of the entire disaccharide (rutinose) moiety from natural products to different acceptors. This enzyme allowed the synthesis of the diglycoconjugated fluorogenic substrate 4-methylumbelliferyl-rutinoside. The synthesis was performed in one step from the corresponding aglycone, 4-methylumbelliferone, and hesperidin as rutinose donor with a 16% yield regarding the sugar acceptor [38]. Despite the low yield, the fluorogenic substrate formed, 4-methylumbelliferyl rutinoside (Figure 4), which is not commercially available, is important for the study of diglycosidases since it allows the detection of enzymes specific for rutinose mobilization. Although the activity of this enzyme could seem quite specific for a broad interest, we must bear in mind its potential use for industrial processing of plant-based foods and the ability to transglycosylate rutinosyl units from abundant and inexpensive by-products of citrus industry.

Figure 4. Transrutinosylation performed by diglycosidase isolated from Acremonium sp.

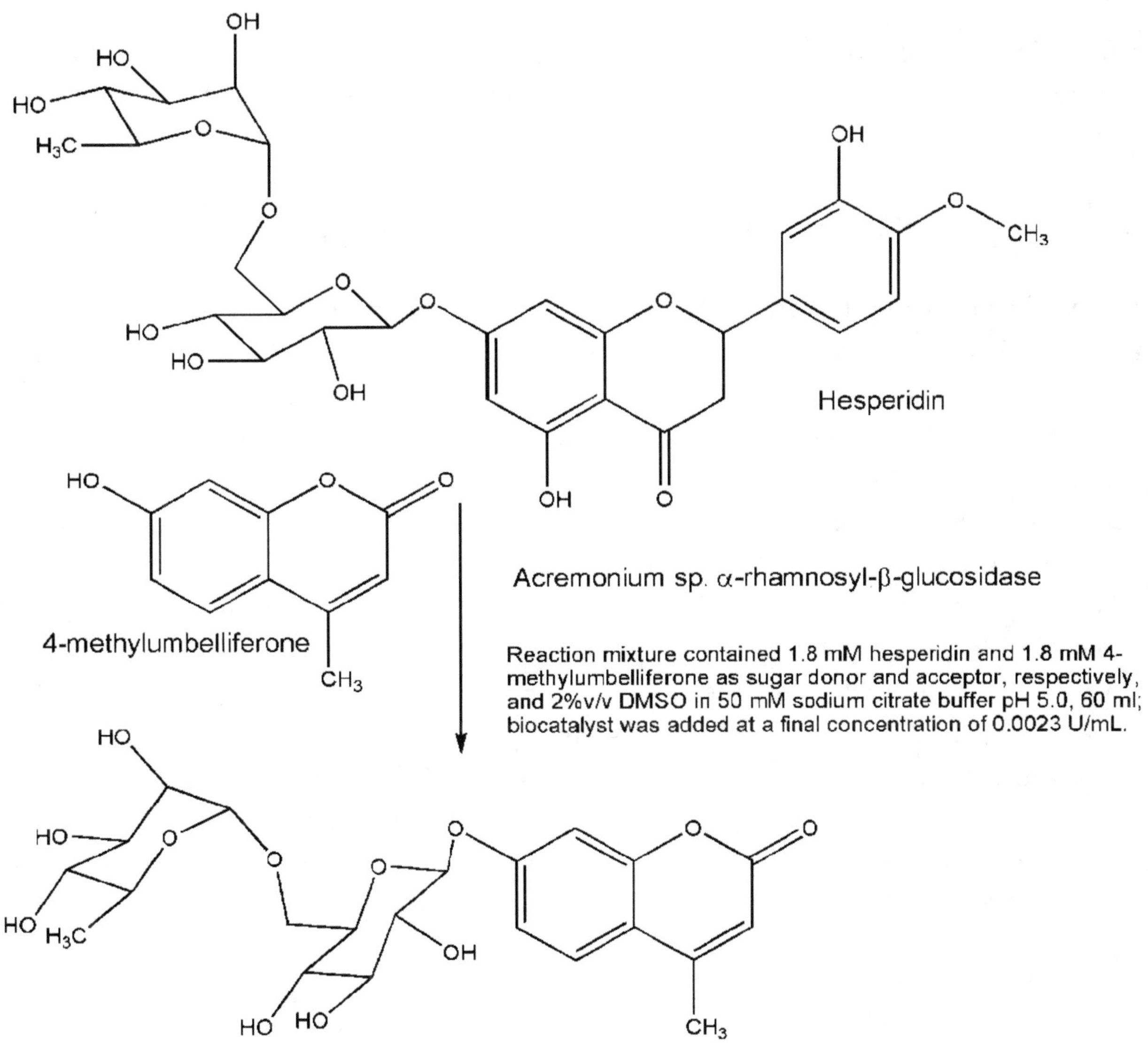

3. Engineered Enzymes

From 2004 to 2013 PubMed indexed 78 hits having the words glycosynthase(s) in the title, including review articles that accounted for ca. 20%. Similar figures can be found in Scopus and in Web of Knowledge. Using data published up to the mid-term of 2003, Perugino *et al.* compiled a review [39] accounting for 11 different glycoside hydrolases from bacteria, eukarya and archaea, that were modified as efficient glycosynthases. In 2012, in an update report from the same lab [40] a quadrupled value was found. Although these lists were declared not exhaustive, a similar number is also reported in the review of 2011 from Wong labs [41]. Indeed a number of excellent reviews periodically survey the most recent achievements in the field of glycosynthases from different perspectives, such as their production and application to the synthesis of glycoconjugates. The obvious success is due to quantitative yields that can be reached in reactions using these active-site modified glycoside hydrolases. The glycosynthase approach is certainly of great help while keeping molecular diversity offered by different natural glycoside hydrolases, including features such as resistance to temperature, organic solvent, *etc.,* making this aspect very interesting for the exploitation of these enzymes in biocatalysis. A survey of all these examples will not be repeated here, where instead just one case of a spectacular glycosphingolipids synthesis is reported below.

The focus in the remaining part of this paragraph, dedicated to engineered enzymes, is on those cases where genetic modifications could help in different manner than increasing yield by glycosynthase philosophy.

3.1. Preparative Glycosphingolipids Synthesis Operated by Glycosynthase

Pharmacological interest for glycosphingolipids, a class of glycolipids based on the aminodiol sphingosine, is high since these compounds have numerous biological functions in processes of human physiology [42]. Gangliosides are a subclass of glycosphingolipids containing also sialic acid in the oligosaccharide moiety, composed of different sugars. Certain gangliosides are involved in viral infections mechanisms. Commercial sources for glycosphingolipids rely on isolation of compounds from natural sources, such as bovine brain and canine blood. This is impractical for large-scale preparation, as well as posing risks for contamination and requiring great effort for extensive purification. A spectacular example of application of glycosynthase technology for high yield production of different gangliosides has been recently reported [43]. The work is based on the transformation of a natural hydrolytic enzyme, the endoglycoceramidase II (EGC II) from *Rhodococcus* strain M-777, in a glycosynthase. This glycosynthase version of the enzyme with Ala, Ser or Gly as substituents of active site nucleophile, is not capable of performing hydrolysis, but it can react in synthetic mode using α-anomer of different oligosaccharidic fluorides (di- to pentasaccharides) in the presence of acceptors such as D-erythro-sphingosine or sphingosine analogues (Figure 5). Yields from 70% to 100% were obtained for the lyso product on 300 mg scale.

Figure 5. Reaction yields for EGCII glycosynthase with various glycosyl fluoride.

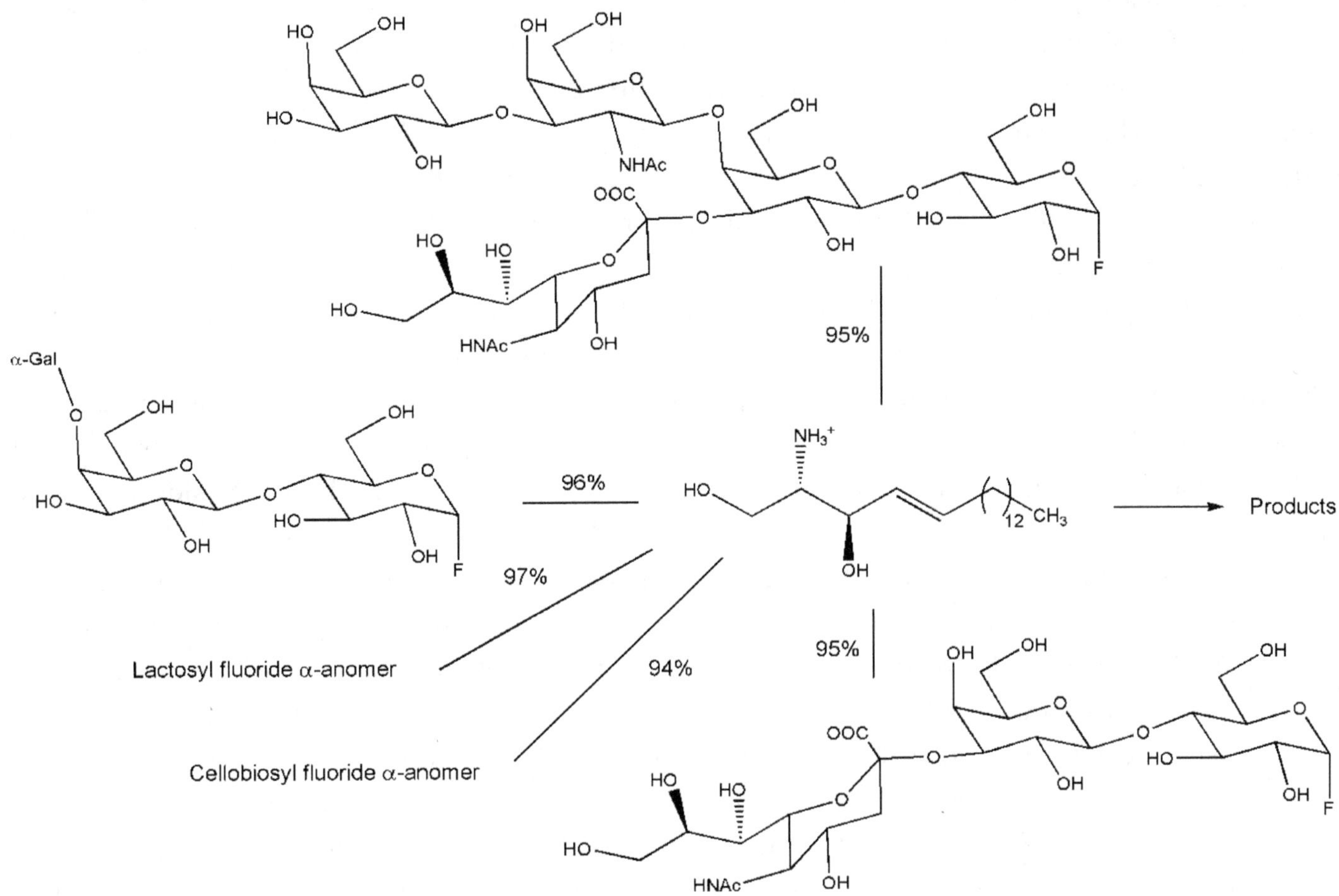

Reactants so structurally different could pose problems, but the solubility of the hydrochloride salt of the sphingosine in aqueous buffer (25 mM after sonication at 37 °C) is considered by the authors a benefit for possible molecular diversity that can be obtained with successive acylation steps with structurally diverse acyl chains. In fact, N-acyl substituents in ceramides affects biological activity of these compounds to some extent, and the acyl substituent has also been identified as a convenient position for conjugation of non-lipid substituents, including fluorescent tags [44].

3.2. Engineering Glycoside Hydrolases not at the Active Site

Engineering the active-site environs is a technique that has been used to enhance the transglycosylation activity of glycosidases and to modify other features of interest in biocatalysis. When structural architecture of the active site is not known, a site-specific "randomization" approach could be used in which an amino acid residue near the catalytic site can be replaced by each of all other amino acids to screen mutant enzyme(s) of interest. This strategy was used to obtain α-glucosidase variant(s) for increasing the production of theanderose, a trisaccharide obtained from sucrose by α-glucosylation to the C6 of glucose. An amino acid residue (Gly273 or Thr272) near the putative catalytic site (Glu271) of this *Bacillus* α-glucosidase was replaced by all other naturally-occurring amino acids thus increasing the transglucosylation activity. The highest specificity for theanderose formation (*i.e.*, the highest content of theanderose in the reaction product) was obtained with the isoleucine containing mutant (T272I), which showed 1.74 times higher productivity (per sucrose-hydrolyzing unit) of theanderose than that of the wild-type enzyme. The authors concluded, however, that elucidation of three-dimensional structures should help to understand the details of the mechanism eliciting the specificity obtained [45]. More recently, the transglycosylation/hydrolysis ratio was shown to be 3- and 8-fold increased with two mutants of a different enzyme, the *Thermotoga neapolitana* β-glucosidase. The asparagine mutant (N291T) of this enzyme showed also altered regioselectivity. In particular TLC analysis of the transglycosylation products indicated that while the mutant retained its β-1,3 regioselectivity, β-1,4 and β-1,6 selectivities were lost when pNPG and arbutin were used as a donor and acceptor, respectively [46]. In another case, Feng *et al.* in 2005 described directed evolution applied to the β-glycosidase of *Thermus thermophilus* to increase the ability of this enzyme in transglycosylation reaction. The most efficient mutations of phenylalanine and asparagine (F401S and N282T), were located just in front of the acceptor subsite and the authors suggested that repositioning of the glycone in its subsite together with a better fit of the acceptor in the acceptor subsite, might favor the attack of a glycosyl acceptor in the mutant at the expense of water; this conclusion was based on molecular modeling techniques. They also concluded that their results suggest that directed evolution of the glycosidases in transglycosidases could be an alternative to the glycosynthase strategy; in fact, for certain mutants, synthesis by self-condensation of nitrophenyl glycosides became nearly quantitative [47]. However when the asparagine (N282T) mutant was analyzed with external added acceptor pNPGlcNAc using oNPGal as donor, the NMR study of kinetic formation of transglycosylation products showed that those due to self-condensation (Galβ1-3Gal-oNP and Galβ1-6Gal-oNP) are kinetically favored over the condensation product Galβ1-4GlcNAc-pNP, suggesting that oNPGal is a better acceptor than pNPGlcNAc; competition between self-condensation and condensation could be responsible for the moderate yield of the desired product Galβ1-4GlcNAc-pNP. Based on the analysis of catalytic context, the authors studied *in silico* the complex alanine mutant-acceptor (A221W)/pNPGlcNAc establishing that the docking of the acceptor is not perturbed by the mutation, compared to the WT case. Donor docking was also studied with A221W mutant complexed with oNPGal to analyze precisely the perturbation as compared with WT. In WT enzyme there is enough room for oNPGal in the acceptor sites allowing the self-condensation reaction. In the A221W mutant, a drastic steric conflict occurs between the galactose ring and the tryptophan substituting alanine and this effect was seen by the authors as the most promising for preventing the self-condensation reaction. The mutation A221W was thus introduced by directed mutagenesis. The analysis of products and relative yields of this double mutant N282T/A221W shows that these results are consistent with the MM calculations. The authors proposed their approach as convenient in that relatively stable activated sugars can be used [48]. While this work shows the value of a rational approach to eliminate the side effects of transglycosylation reactions, a thermophilic glycosynthase from *Sulfolobus solfataricus* was indeed shown to act in presence of external formate on stable activated sugars such as oNP-glyco donors [49].

A random mutagenic approach coupled to a screening procedure was applied in another thermophilic case in nature. The screening was based on the reduction of the hydrolysis of a potential transglycosylation product (lactosucrose, β-D-galactopyranosyl-(1-4)-D-sucrose) formed in presence of sucrose, thus providing mutant enzymes possessing improved synthetic properties for the transglycosylation [50]. The application of thermostable β-galactosidases such as the β-galactosidase BgaB from *Geobacillus stearothermophilus* is of interest for transgalactosylation because at higher temperatures higher lactose concentration can be used to favor the synthesis.

A complete change of product profile of the reaction was observed for one of the mutants obtained by site directed mutagenesis of α-amylase of *Bacillus amyloliquefaciens*. The Val289 residue substituted with tyrosine showed less than 15% activity compared to the wild-type, but it acquired transglycosylation activity, producing longer oligosaccharides. The analysis of all mutants produced led to the conclusion that changes in the hydrolytic property of α-amylase may be due to factors like the geometry and electrostatics of the environment around the active site [51].

It could be of interest in this paragraph to mention the development of methodologies for the screening of large libraries of mutant enzymes. In a paper of 2009 Konè *et al.* reported on digital screening methodology as a system dedicated to the screening for sugar-transfer activity [52], which gave great impetus to the study of glycosidases with efficient transglycosidases activity as an alternative to glycosyltransferases or glycosynthases. Recently, from genomic and metagenomic program results, protein sequences with unknown functions are increasing, thus effective screening methodologies have become an important aspect of this research. In the field of polysaccharide degrading enzymes, a profiling method reported in 2012 by the Helbert group is worth noting. Polysaccharidases are important enzymes that are used as specific tools to improve conversion of lignocellulosic biomass into sugar monomers prior to ethanol production, for: degradation of microbial polysaccharides, obtaining bioactive materials, for structural determination of unknown complex polysaccharides structures, *etc.* The profiling strategy is based on a series of filtrations necessary to eliminate any reducing sugars not directly generated by enzyme degradation. After enzymatic action, filtrates are assayed with a ferricyanide solution to reveal the reducing sugars produced by glycoside hydrolases or polysaccharide lyases; however matrix-assisted laser desorption ionization mass spectrometry (MALDI-MS), presenting several unique advantages for the structural characterization of degradation products of carbohydrates, is also considered as providing an effective methodology [53,54].

4. Conclusions

Although not exhaustively catalogued in this review, the scientific literature of the last decade, with regard to the process of enzymatic preparation of glycosides, enables the reader to conclude that research for uniqueness, in terms of enzymatic features, is still active. This is in contrast with the tempting generalization about enzymes as plastic biomolecules fully receptive to be engineered by appropriate changes, although this plasticity has been successfully demonstrated in some examples [1]. The list of naturally "unique" enzymes cited in this review includes examples found among the most promising subjects of biocatalytic applications: (i) hyaluronidases for producing and/or transferring hyaluronic oligosaccharides of pharmaceutical interest or (ii) disaccharidases such as endo-α-N-acetylgalactosaminidase (for the analysis of glycan chains) and (iii) the fungal rutinosidase used in the synthesis of related fluorogenic derivative of 4-methylumbelliferone, are of specific interest. Among engineered enzymes, even though glycosynthases have to be seen as the most successful approach to reach quantitative transfer yields in transglycosylations, engineered enzymes examples with mutations only in the active-site environ (and not at active site aminoacids), demonstrated that biocatalytic features are well manipulated also in this way, thus enlarging competence to regioselectivity and or specificity for heteroacceptors and not only to reaction yield as devised for glycosynthases.

This overview should increase the interest from biocatalyst practitioners towards both naturally unique enzymes and for infrequent mutations capable of inducing biocatalytically interesting features.

Acknowledgments

The support for bibliographic search facilities is provided by Consiglio Nazionale delle Ricerche.

References

1. Bornscheuer, U.T.; Huisman, G.W.; Kazlausaks, R.J.; Lutz, S.; Moore, J.C.; Robins, K. Engineering the third wave of biocatalysis. *Nature* **2012**, *485*, 185–194.
2. Riva, S. 1983–2013: The long wave of biocatalysis. *Trends Biotechnol.* **2013**, *31*, 120–121.
3. Trincone, A. Potential biocatalysts originating from sea environments. *J. Mol. Catal. B-Enzym.* **2010**, *66*, 241–256.
4. Dumon, C.; Songa, L.; Bozonneta, S.; Fauréa, R.; O'Donohue, M.J. Progress and future prospects for pentose-specific biocatalysts in biorefining. *Proc. Biochem.* **2012**, *47*, 346–357.
5. Eklunda, E.A.; Bodeb, L.; Freeze, H.H. Diseases Associated with Carbohydrates/Glycoconjugates. In *Comprehensive Glycoscience Volume 4: Cell Glycobiology and Development; Health and Disease in Glycomedicine*; Kamerling, J.P., Ed.; Elsevier: N.Y., USA, 2007; Volume 4, pp. 339–371.
6. Kren, V. Glycoside *vs.* Aglycon: The Role of Glycosidic Residue in Biological Activity. In *Glycoscience, Chemistry and Chemical Biology*; Fraser-Reid, B.O., Tatsuta, K., Thiem, J., Eds.; Springer-Verlag: Berlin, Heidelberg, 2008; pp. 2589–2644.
7. Sears, P.; Wong, C.-H. Intervention of carbohydrate recognition by proteins and nucleic acids. *Proc. Natl. Acad. Sci. USA* **1996**, *93*, 12086–12093.
8. Wang, Q.; Dordick, J.S.; Linhardt, R.J. Synthesis and application of carbohydrate-containing polymers. *Chem. Mater.* **2002**, *14*, 3232–3244.
9. Wildman, E.C.R. Classifying Nutraceuticals. In *Handbook of Nutraceuticals and Functional Foods*; Robert, E.C., Ed.; CRC Press: Boca Raton, London, New York Washington, D.C., USA, 2000; pp. 13–30.
10. Bojarova, P.; Rosencrantz, R.R.; Elling, L.; Kren, V. Enzymatic glycosylation of multivalent scaffolds. *Chem. Soc. Rev.* **2013**, *7*, 4774–4797.
11. De Roode, B.M.; Franseen, M.C.R.; van der Padt, A.; Boom, R.M. Perspectives for the industrial enzymatic production of glycosides. *Biotechnol. Prog.* **2003**, *19*, 1391–1402.
12. Trincone, A.; Giordano, A. Glycosyl hydrolases and glycosyltransferases in the synthesis of oligosaccharides. *Curr. Org. Chem.* **2006**, *10*, 1163–1193.
13. Desmet, T.; Soetaert, W.; Bojarova, P.; Kren, V.; Dijkhuizen, L.; Eastwick-Field, V.; Schiller, A. Enzymatic glycosylation of small molecules: Challenging substrates require tailored catalysts. *Chem. Eur. J.* **2012**, *18*, 10786–10801.
14. Hehre, E.J. Glycosyl transfer: A history of the concept's development and view of its major contributions to biochemistry. *Carbohydr. Res.* **2001**, *331*, 347–368.
15. Kobata, A. The history of glycobiology in Japan. *Glycobiology* **2001**, *11*, 99R–105R.

16. Koshland, D.E. Stereochemistry and the mechanism of enzymatic reactions. *Biol. Rev.* **1953**, *28*, 416–436.
17. Vivian, L.Y.; Withers, S.G. Breakdown of oligosaccharides by the process of elimination. *Curr. Opin. Chem. Biol.* **2006**, *10*, 147–155.
18. Vuong, T.V.; Wilson, D.B. Glycoside hydrolases: Catalytic base/nucleophile diversity. *Biotechnol. Bioeng.* **2010**, *107*, 195–205.
19. Thibodeaux, C.J.; Melancon, C.E.; Liu, H.-W. Unusual sugar biosynthesis and natural product glycodiversification. *Nature* **2007**, *446*, 1008–1016.
20. Hofinger, E.S.A.; Bernhardt, G.; Buschauer, A. Kinetics of Hyal-1 and PH-20 hyaluronidases: Comparison of minimal substrates and analysis of the transglycosylation reaction. *Glycobiology* **2007**, *17*, 963–971.
21. Takagaki, K.; Ishido, K.; Kakizaki, I.; Iwafune, M.; Endo, M. Carriers for enzymatic attachment of glycosaminoglycan chains to peptide. *Biochem. Biophys. Res. Comm.* **2002**, *293*, 220–224.
22. Madokoro, M.; Ueda, A.; Kiriake, A.; Shiomi, K. Properties and cDNA cloning of a hyaluronidase from the stonefish *Synanceia verrucosa* venom. *Toxicon* **2011**, *58*, 285–292.
23. Koutsioulis, D.; Landry, D.; Guthire, E.P. Novel endo-α-N-acetylgalactosaminidases with broader substrate specificity. *Glycobiology* **2008**, *18*, 799–805.
24. Fujita, M.; Shoda, S.; Haneda, K.; Inazu, T.; Takegawa, K.; Yamamoto, K. A novel disaccharide substrate having 1,2-oxazoline moiety for detection of transglycosylating activity of endoglycosidases. *Biochimi. Biophys. Acta* **2001**, *1528*, 9–14.
25. Huang, W.; Li, J.; Wang, L.-X. Unusual transglycosylation activity of *Flavobacterium meningosepticum* endoglycosidases enables convergent chemoenzymatic synthesis of core fucosylated complex N-glycopeptides. *Chembiochem* **2011**, *11*, 932–941.
26. Slámová, K.; Gažák, R.; Bojarová, P.; Kulik, N.; Ettrich, R.; Pelantová, H.; Sedmera, P.; Křen, V. 4-Deoxy-substrates for β-N-acetylhexosaminidases: How to make use of their loose specificity. *Glycobiology* **2010**, *20*, 1002–1009.
27. Hu, Y.; Luan, H.; Liu, H.; Ge, G.; Zhou, K.; Liu, Y.; Yang, L. Acceptor specificity and transfer efficiency of a β-glycosidase from the China white jade snail. *Biosci. Biotechnol. Biochem.* **2009**, *73*, 671–676.
28. Roccatagliata, A.J.; Maier, M.S.; Seldes, A.M.; Iorizzi, M.; Minale, L. Starifhs saponins. Part II. Steroidal oligoglycosides from the starfish *Cosmasterias lurida*. *J. Nat. Prod.* **1994**, *57*, 747–754.
29. Ye, M.; Yan, L.-Q.; Li, N.; Zong, M.-H. Facile and regioselective enzymatic 5β-galactosylation of pyrimidine 2β-deoxynucleosides catalyzed by β-glycosidase from bovine liver. *J. Mol. Cat. B Enzym.* **2012**, *79*, 35–40.
30. Binder, W.H.; Kahlig, H.; Schmid, W. Galactosylation by use of β-galactosidase: Enzymatic syntheses of disaccharide nucleosides. *Tetrahedron Asymm.* **1995**, *6*, 1703–1710.
31. Andreotti, G.; Trincone, A.; Giordano, A. Convenient synthesis of β-galactosyl nucleosides using the marine β-galactosidase from *Aplysia fasciata*. *J. Mol. Catal. B-Enzym.* **2007**, *47*, 28–32.
32. Zeng, Q.M.; Li, N.; Zong, M.H. Highly regioselective galactosylation of floxuridine catalyzed by β-galactosidase from bovine liver. *Biotechnol. Lett.* **2010**, *32*, 1251–1254.

33. Andreotti, G.; Giordano, A.; Tramice, A.; Mollo, E.; Trincone, A. Hydrolyses and transglycosylations performed by purified α-D-glucosidase of the marine mollusc *Aplysia fasciata*. *J. Biotechnol.* **2006**, *122*, 274–284.
34. Manzo, E.; Tramice, A.; Pagano, D.; Trincone, A.; Fontana, A. Chemo-enzymatic preparation of α-6-sulfoquinovosyl-1,2-O-diacylglycerols. *Tetrahedron* **2012**, *68*, 10169–10175.
35. Remond, C.; Plantier-Royon, R.; Aubry, N.; Maes, E.; Bliard, C.; O'Donohuea, M.J. Synthesis of pentose-containing disaccharides using a thermostable α-L-arabinofuranosidase. *Carbohyd. Res.* **2004**, *339*, 2019–2025.
36. Remond, C.; Plantier-Royon, R.; Aubry, N.; O'Donohuea, M.J. An original chemoenzymatic route for the synthesis of β-D-galactofuranosides using an α-L-arabinofuranosidase. *Carbohydr. Res.* **2005**, *340*, 637–644.
37. Sandoval, M.; Civera, C.; Berenguer, J.; García-Blanco, F.; Hernaiz, M.J. Optimised N-acetyl-D-lactosamine synthesis using *Thermus thermophilus* β-galactosidase in bio-solvents. *Tetrahedron* **2013**, *69*, 1148–1152.
38. Mazzaferro, L.S.; Piñuel, L.; Erra-Balsells, R.; Giudicessi, S.L.; Breccia, J.D. Transglycosylation specificity of *Acremonium* sp. α-rhamnosyl-β-glucosidase and its application to the synthesis of the new fluorogenic substrate 4-methylumbelliferyl-rutinoside. *Carbohydr. Res.* **2012**, *347*, 69–75.
39. Perugino, G.; Trincone, A.; Rossi, M.; Moracci, M. Oligosaccharide synthesis by glycosynthases. *Trends Biotechnol*. **2004**, *22*, 31–37.
40. Cobucci-Ponzano, B.; Moracci, M. Glycosynthases as tools for the production of glycan analogs of natural products. *Nat. Prod. Rep.* **2012**, *29*, 697–709.
41. Schmaltz, R.M.; Hanson, S.R.; Wong, C.-H. Enzymes in the synthesis of glycoconjugates. *Chem. Rev*. **2011**, *111*, 4259–4307.
42. Mocchetti, I. Exogenous gangliosides, neuronal plasticity and repair, and the neurotrophins. *Cell. Mol. Life Sci*. **2005**, *62*, 2283–2294.
43. Vaughan, M.D.; Johnson, K.; DeFrees, S.; Tang, X.; Warren, R.A.J.; Withers, S.G. Glycosynthase-mediated synthesis of glycosphingolipids. *J. Am. Chem. Soc.* **2006**, *128*, 6300–6301.
44. Rich, J.R.; Cunningham, A-M.; Gilbert, M.; Withers, S.G. Glycosphingolipid synthesis employing a combination of recombinant glycosyltransferases and an endoglycoceramidase glycosynthase. *Chem. Commun.* **2011**, *47*, 10806–10808.
45. Okada, M.; Nakayama, T.; Noguchi, A.; Yano, M.; Hemmi, H.; Nishino, T.; Ueda, T. Site-specific mutagenesis at positions 272 and 273 of the Bacillus sp. SAM1606 α-glucosidase to screen mutants with altered specificity for oligosaccharide production by transglucosylation. *J. Mol. Catal. B-Enzym.* **2002**, *16*, 265–274.
46. Ki-Won, C.; Park, K.-M.; Jun, S.-Y.; Park, C.-S.; Park, K.-H.; Cha, J. Modulation of the regioselectivity of a *Thermotoga neapolitana* β-glucosidase by site-directed mutagenesis. *J. Microbiol. Biotechnol.* **2008**, *18*, 901–907.
47. Feng, H-Y.; Drone, J.; Hoffmann, L.; Tran, V.; Tellier, C.; Rabiller, C.; Michel Dion, M. Converting a β-glycosidase into a β-transglycosidase by directed evolution. *J. Biol. Chem.* **2005**, *280*, 37088–37097.

48. Tran, V.; Hoffmann, L.; Rabiller, C.; Tellier, C.; Dion, M. Rational design of a GH1 β-glycosidase to prevent self-condensation during the transglycosylation reaction. *Protein Eng. Des. Sel.* **2010**, *23*, 43–49.
49. Trincone, A.; Giordano, A.; Perugino, G.; Rossi, M.; Moracci, M. Highly productive autocondensation and transglycosylation reactions with *Sulfolobus solfataricus* glycosynthase. *ChemBioChem* **2005**, *6*, 1431–1437.
50. Placier, G.; Watzlawick, H.; Rabiller, C.; Mattes, R. Evolved β-galactosidases from *Geobacillus stearothermophilus* with improved transgalactosylation yield for galacto-oligosaccharide production. *App. Environ. Microbiol.* **2009**, *75*, 6312–6321.
51. Priyadharshini, R.; Hemalatha, D.; Gunasekaran, P. Role of Val289 residue in the α-amylase of *Bacillus amyloliquefaciens* MTCC 610: An analysis by site directed mutagenesis. *J. Microbiol. Biotechnol.* **2010**, *20*, 563–568.
52. Koné, F.M.T.; Le Bechec, M.; Sine, J.P.; Dion, M.; Tellier, C. Digital screening methodology for the directed evolution of transglycosidases. *Prot. Engin. Des. Sel.* **2009**, *22*, 37–44.
53. Fer, M.; Préchoux, A.; Leroy, A.; Sassi, J.F.; Lahaye, M.; Boisset, C.; Nyvall-Collén, P.; Helbert, W. Medium-throughput profiling method for screening polysaccharide-degrading enzymes in complex bacterial extracts. *J. Microbiol. Meth.* **2012**, *89*, 222–229.
54. Ropartz, D.; Bodet, P.E.; Przybylski, C.; Gonnet, F.; Daniel, R.; Fer, M.; Helbert, W.; Bertrand, D.; Rogniaux, H. Performance evaluation on a wide set of matrix-assisted laser desorption ionization matrices for the detection of oligosaccharides in a high-throughput mass spectrometric screening of carbohydrate depolymerizing enzymes. *Rapid Commun. Mass Spectrom.* **2011**, *25*, 2059–2070.

6

Production of Fungal Glucoamylase for Glucose Production from Food Waste

Wan Chi Lam, Daniel Pleissner and Carol Sze Ki Lin *

School of Energy and Environment, City University of Hong Kong, Tat Chee Avenue, Kowloon, Hong Kong, China; E-Mails: wanclam@cityu.edu.hk (W.C.L.); dpleissn@cityu.edu.hk (D.P.); carollin@cityu.edu.hk (C.S.K.L).

* Author to whom correspondence should be addressed; E-Mail: carollin@cityu.edu.hk

---

Abstract: The feasibility of using pastry waste as resource for glucoamylase (GA) production via solid state fermentation (SSF) was studied. The crude GA extract obtained was used for glucose production from mixed food waste. Our results showed that pastry waste could be used as a sole substrate for GA production. A maximal GA activity of 76.1 ± 6.1 U/mL was obtained at Day 10. The optimal pH and reaction temperature for the crude GA extract for hydrolysis were pH 5.5 and 55 °C, respectively. Under this condition, the half-life of the GA extract was 315.0 minutes with a deactivation constant (k_d) 2.20 × 10^{-3}minutes^{-1}. The application of the crude GA extract for mixed food waste hydrolysis and glucose production was successfully demonstrated. Approximately 53 g glucose was recovered from 100 g of mixed food waste in 1 h under the optimal digestion conditions, highlighting the potential of this approach as an alternative strategy for waste management and sustainable production of glucose applicable as carbon source in many biotechnological processes.

Keywords: glucoamylase production; *Aspergillus awamori*; food waste hydrolysis; glucose production

---

1. Introduction

Food waste is a serious global problem, especially in many developed countries. In Hong Kong, over 3500 tons of food wastes are generated every day [1]. Currently, landfilling and incineration are the major practices for managing these wastes in many countries. These practices, however, may cause severe environmental pollutions and adds burden to the economy. Due to its high contents of carbohydrates and proteins, food wastes may serve as feedstock in biorefineries for production of fungal enzymes, e.g., glucoamylase (GA) and offers an innovative approach to waste management.

GA is a family of amylolytic enzymes that catalyze the cleavage of α-(1,4) glycosidic bonds in starch and release glucose as end product [2,3]. Glucose is the principle carbon source in many biotechnological processes and of great importance for fermentative chemicals and fuel production such as succinic acid, bio-plastic and ethanol. Starch is usually the major component of mixed food waste from restaurants [4–6], application of GA for food waste hydrolysis to recover glucose from starch, therefore, may not only offer a solution for managing food waste but also help to save precious resources.

Aspergillus awamori is a known secretor of GA with beneficial properties for industrial bioprocesses such as high productivity and enzyme activity at high temperatures [3,7]. Therefore, *Aspergillus awamori* is employed in this study for GA production through solid state fermentation (SSF). SSF is a fermentation process conducted in the absence of free water, thus it is desirable for industrial enzymes production since the enzymes produced at the end are not diluted by the amount water added in comparison to submerge fermentation. Consequently, the enzymes produced are at a much higher concentration. Pastry waste collected from local Starbucks contains 44.6% of starch and 7% of protein [8]. Starch has been shown as an inducer for GA synthesis by some *Aspergillus* producers [9–11]; thus, the significant amount of starch in pastry waste could be desirable for GA production. The smaller amount of protein, 7%, on the other hand could provide a source of nitrogen to promote the fungal growth and facilitate GA production. Therefore, pastry waste was selected in this study for GA production. The crude GA extract obtained without further purification was characterized in terms of optimal pH and reaction temperature, as well as thermo-stability. Additionally, application of the crude GA extract was studied for hydrolysis of mixed food waste collected from a local restaurant to produce high glucose solution.

2. Results and Discussion

2.1. Glucoamylase Production from Pastry Waste

To demonstrate the feasibility of pastry waste as the sole substrate for GA production, SSF was conducted with *Aspergillus awamori* without addition of any other nitrogen or carbon sources. GA production from pastry waste over time is shown in Figure 1. Production of GA reached maximal activity at Day 10. Approximately, 253.7 ± 20.4 U of GA was produced from one gram of pastry waste on dry basis (d.b.); the GA activity of the crude GA extract was 76.1 ± 6.1 U/mL. The result demonstrates that pastry waste can be used solely for GA production.

Figure 1. Glucoamylase (GA) production from pastry waste with *Aspergillus awamori* during solid state fermentation (SSF) for 11 days at 30 °C. Experiments were duplicated. The mean values are plotted and the standard errors are reported.

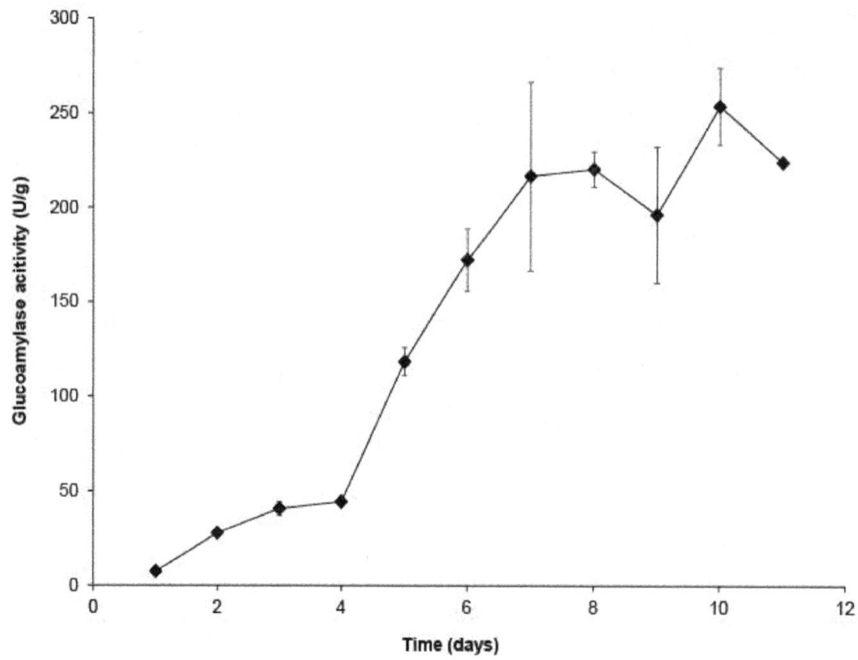

GA production with different wastes as substrates have been studied and the results are summarized in Table 1. In some studies, nitrogen in form of ammonium, urea and yeast extract was supplemented to substrates to facilitate fungal growth and GA production [4,12–14]. In contrast, nitrogen supplement was not involved in this study, but the GA activity of enzyme extract appears higher than in earlier studies with nitrogen supplement most likely due to a good balance of carbon (C) to nitrogen (N) and phosphorus (P) ratios. The data suggests that pastry waste is a promising substrate for GA production.

Table 1. GA production and yields from different studied substrates with or without nitrogen supplement through solid state fermentation.

Substrate	Crude GA concentration (U/mL)	Yield (U/g)	Fungus	Nitrogen supplement	References
Rice powder	N/A	71.3 ± 2.34 [a]	*Aspergillus niger*	+	[12]
Wheat bran	N/A	110 ± 1.32 [a]	*Aspergillus niger*	+	[12]
Mixed food waste	137	N/A	*Aspergillus niger*	+	[13]
Cowpea waste	970	N/A	*Aspergillus oryzae*	-	[14]
Wheat bran	4.4	48	*Aspergillus awamori*	-	[15]
Wheat pieces	3.32	81.3	*Aspergillus awamori*	-	[16]
Waste bread	3.94	78.4	*Aspergillus awamori*	-	[16]
Waste bread	N/A	114	*Aspergillus awamori*	-	[17]
Pastry waste	76.1 ± 6.1 [a]	253.7 ± 20.4 [a]	*Aspergillus awamori*	-	This study

[a] Values indicate means ± standard errors.

2.2. Characterization of Optimal Reaction Temperature and pH of the Crude Glucoamylase Extract

Optimal pH and reaction temperature of the crude GA extract were determined. The results are shown in Figure 2. In order to determine the optimal reaction pH of the crude GA extract, assays were carried out at various pHs from 3.5 to 7.5 and the results are indicated in Figure 2A. The maximal enzyme activity was obtained at pH 5.5, indicating it was the optimal pH for starch hydrolysis. Fungal GAs from *Aspergillus* strains are usually active at acidic pH, their enzyme activities vary from pH 3.5 to 7 depending on the strains and amino acid sequences (isoform) [3]. Similarly, GA assay was conducted at different temperatures as indicated in Figure 2B from 40–75 °C with 5 °C increment at pH 5.5 in order to determine the optimal reaction temperature for the crude GA extract and assuming that the pH factor is independent of the temperature factor. A significant increase in enzyme activity was observed as temperature increases from 40 to 55 °C. Maximal GA activity was observed from 55 to 65 °C suggesting the range for optimal reaction temperature of the crude GA extract and in accordance with optimal reaction temperatures in the range of 50 to 60 °C usually found for GAs from *Aspergillus* [3]. Further increase in reaction temperature greatly reduced the enzyme activity most likely due to enzyme denaturation.

Figure 2. Effect of (**A**) pH at 55 °C and (**B**) temperature at pH 5.5 on crude GA extract activity. Experiments were duplicated. The mean values are plotted and the standard errors are reported.

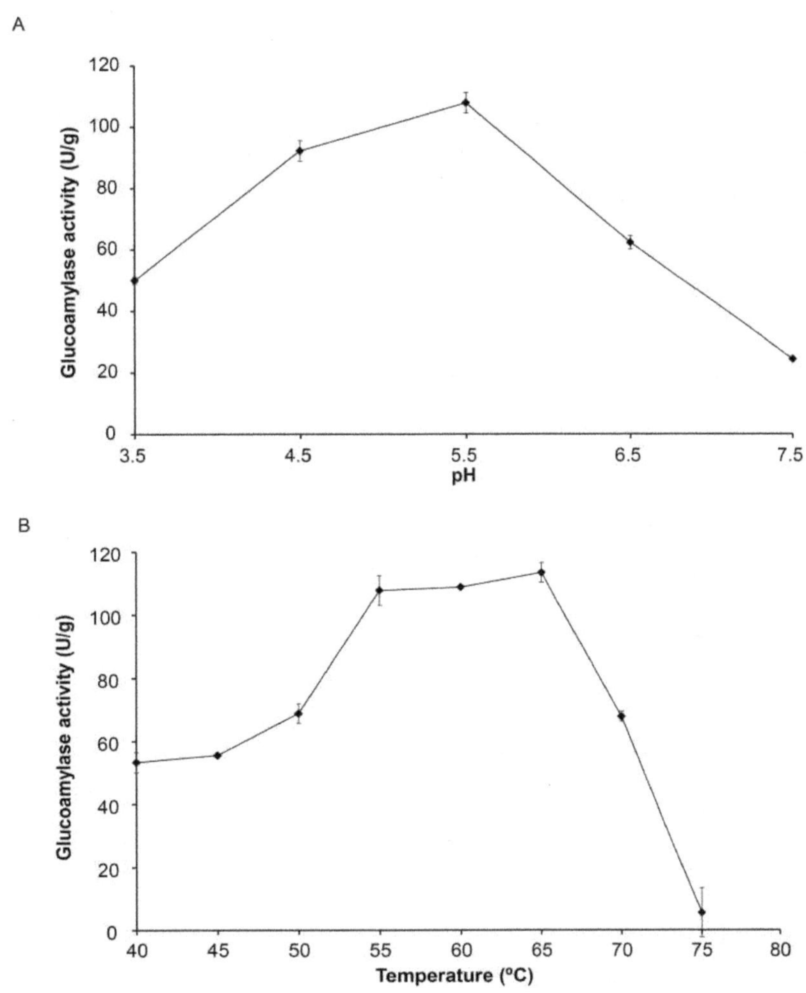

2.3. Thermo-Stability of the Crude Glucoamylase Extract at Optimal Reaction Temperatures

Since high GA activity was observed at 55, 60 and 65 °C, thermo-stability of the crude GA extract at these temperatures was further investigated in order to determine the optimal digestion temperature for the subsequent food waste hydrolysis experiment. The residual enzyme activity after heated at 55, 60 and 65 °C for over 90 min is shown in Figure 3. The rate constant (k_d, minutes^{-1}) for the first-order thermal deactivation was determined from the slope of the deactivation time course as shown in Figure 3 using Equation (1) [18], where E_t is the residual GA activity after heat treatment for time t. E_0 is the initial enzyme activity before heat treatment. The half-life of thermal deactivation ($t_{1/2}$) was determined according to Equation (2) [19]. The thermal deactivation of the crude GA extract exhibited a linear relationship showing that it followed first-order kinetics as reported [20]. The thermal deactivation rate constant k_d (minutes^{-1}) of the crude GA extract at 55 °C was found approximately 10 times slower than the rates at 60 and 65 °C, suggesting the crude GA is more thermo-stable at 55 °C. The k_d of the crude GA extract at 55 °C was 2.20×10^{-3} in comparison with 2.13×10^{-2} and 2.17×10^{-2} at 60 and 65 °C, respectively. The half-life ($t_{1/2}$) of the enzyme extract at 55 °C was 315 min in comparison with 32.5 and 31.9 min for 60 and 65 °C, respectively (Table 2). Since the enzymatic activity of the crude GA extract at 55 °C was close to the activity at 60 and 65 °C, but more stable, it was adopted as the optimal digestion temperature for the subsequent food hydrolysis experiment.

$$\ln (E_t/E_0) = -k_d t \quad (1)$$

$$t_{1/2} = \ln(2)/k_d \quad (2)$$

Figure 3. Thermal deactivation of crude GA extract at (◆) 55 °C, (■) 60 °C and (▲) 65 °C over 90 min at pH 5.5. Experiments were duplicated. The mean values are plotted and the standard errors were reported.

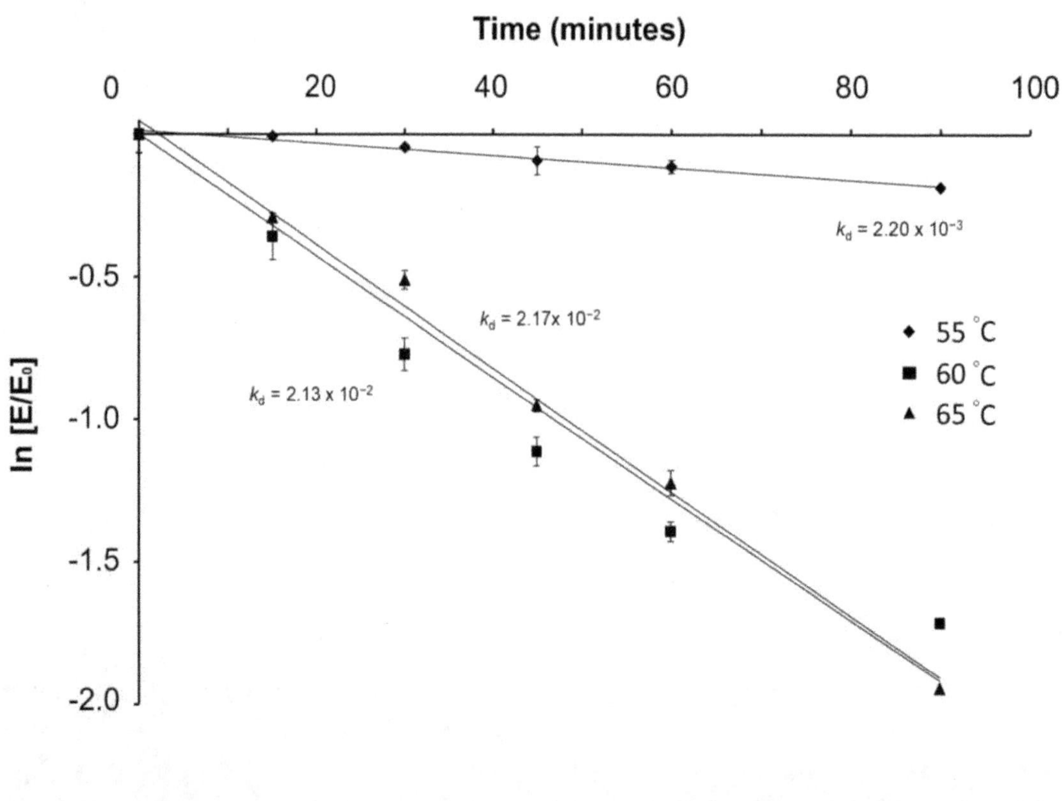

Table 2. Deactivation constant (k_d) and half-lives ($t_{1/2}$) of the crude GA extract at 55, 60 and 65 °C at pH 5.5.

Temperature (°C)	k_d (minutes^{-1})	$t_{1/2}$ (minutes)
55	2.20×10^{-3}	315.0
60	2.13×10^{-2}	32.5
65	2.17×10^{-2}	31.9

2.4. Application of Crude Glucoamylase Extract on Mixed Food Waste Hydrolysis for Glucose Production

In the reality, mixed food waste is rich in salt [21] that may inhibit the enzymatic hydrolysis. To verify if the crude GA extract produced was applicable to food waste digestion for glucose production, it was used to hydrolyse the food waste which was collected from a local restaurant, under its optimal digestion conditions (at pH 5.5 and 55 °C). Increasing concentration of enzyme was added to the food waste and the time required for hydrolysis to produce glucose was determined (Figure 4). Significant difference in glucose concentration was only observed for the first hour but not after. In all cases, the food waste hydrolysis by the enzyme extracts was completed in 1 h. At the end of the hydrolysis, approximately 12 g/L glucose was produced and that was corresponding to approximately 53 g glucose produced from 100 g of mixed food waste (d.b.), while no production of glucose occurred in a control without crude GA extract. The amount of glucose produced from 100 g of mixed food waste hydrolysis was consistent with our previous study using a different hydrolysis approach [5].

Figure 4. Hydrolysis of mixed food waste for glucose production in the presence of crude GA extract with (●) no enzyme, (◆) 7.1 U/mL, (■) 14.2 U/mL and (▲) 28.4 U/mL at pH 5.5 and 55 °C for 3 h. Experiments were duplicated. The mean values are plotted and the standard errors are reported.

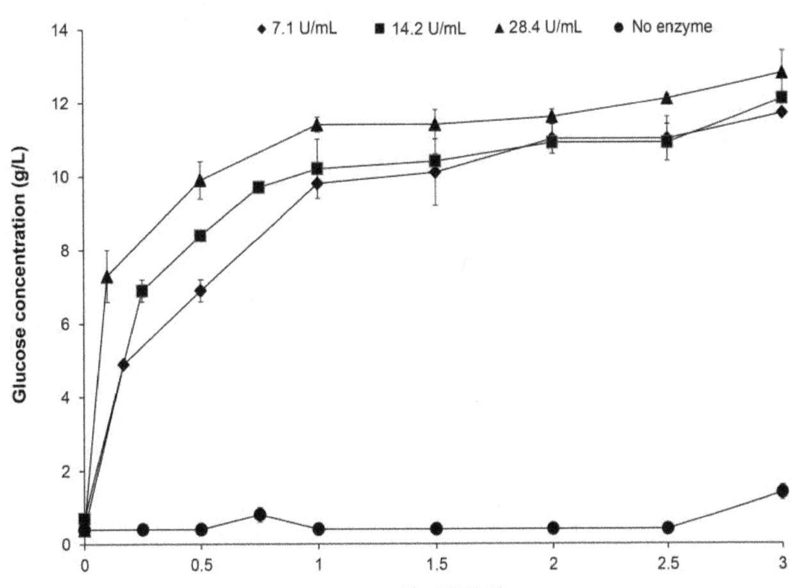

The two commonly used approaches for cereal-based waste and food waste hydrolysis to produce glucose rich solution, include simultaneous fungus culturing and hydrolysis with the enzymes actively secreted [5,8,16,22,23] or direct addition of enzyme solution to digest the substrate [24]. In the first case, food waste hydrolysis is usually completed after 24 h [5,8,16,22,23]. In this study, the latter approach was adopted. Food waste hydrolysis by the crude GA extract was completed in 1 h under optimal conditions found. Similar experiment has been reported by *Yan et al.* using commercial GA for food waste hydrolysis. In their studies, food waste hydrolysis was completed in 2.5 h when GA to substrate ratio reached 80–140 U/g food waste [24]. When the GA to substrate ratio in the solution is reduced to 7.4 U/g substrate, 24 h was needed for complete substrate hydrolysis [16]. The higher efficiency for food waste hydrolysis (in 1 h) in this study was likely due to the higher initial GA to substrate ratio.

2.5. Material Balance for Glucose Production from 1 kg Mixed Food Waste with Crude Glucoamylase Extract

Scheme I shows the material balance of the studied process for crude GA production from pastry waste and glucose recovery from 1 kg mixed food waste (d.b.). All the calculations are provided in the supplementary information. According to our study, hydrolysis of 1 kg mixed food waste (d.b.) could lead to 0.53 kg glucose production. Approximately, 1.4 kg pastry waste (d.b.) is required to produce sufficient amount of crude GA extract for 1 kg mixed food waste hydrolysis. Furthermore, the material balance only presents the theoretical values of the up-scaled process based on our laboratory-scale experimental data. However, upscale study is needed in order to demonstrate the process can be applied at industrial scale.

Scheme I. Material balance of the process in this study for processing 1 kg mixed food waste to produce glucose using crude GA extract produced from pastry waste based on the laboratory-scale experimental data.

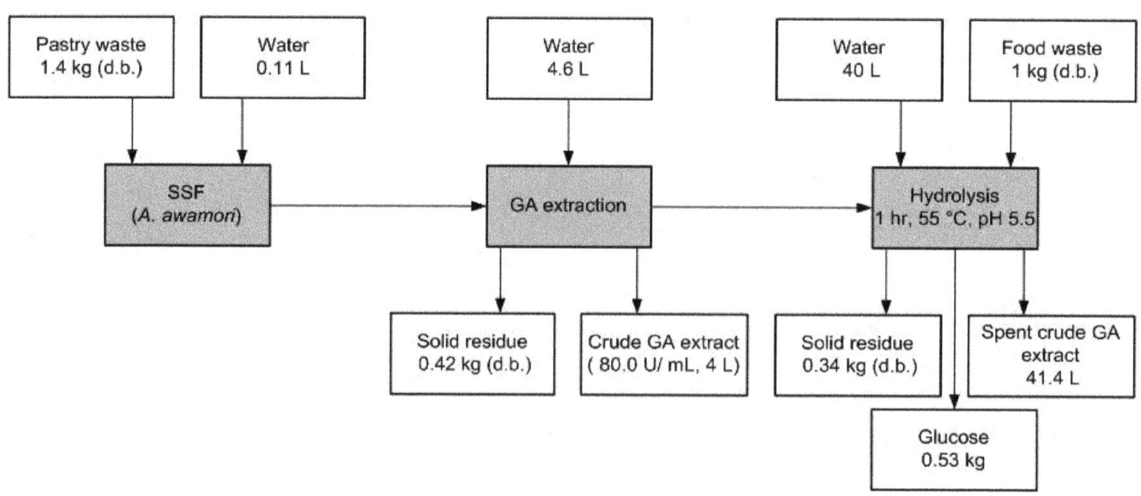

3. Experimental Section

3.1. Microorganism

Frozen spores of the fungus *Aspergillus awamori* (ATCC 14331), were used for SSF. Spores were suspended in demineralized water and loaded into conical flasks containing cornmeal agar and incubated at 30 °C for 7 days. Harvesting of fresh spores was carried out using 10% glycerol and the number of spores was counted using a haemocytometer. Fresh spore suspension of 1 mL was diluted to the required concentration with sterile demineralized water and used immediately for SSF.

3.2. Food Wastes Preparation

Pastry waste and mixed food waste were obtained from a local Starbucks store and canteen for SSFs and enzymatic hydrolysis experiments, respectively. Once the pastry waste was collected, it was homogenized with a kitchen blender and stored at −20 °C. Pastry waste was autoclaved before subjected to SSF and mixed food waste was lyophilized before enzymatic hydrolysis.

3.3. Solid State Fermentation for Glucoamylase Production

Sterilized pastry waste of 20 g was placed in a petri dish and inoculated with 1 mL of fresh spore suspension of *Aspergillus awamori* loaded onto the surface of the substrate. For each g of substrate, 5×10^5 spores were used. The plates were incubated under static condition at 30 °C for 11 days. Whole plate content was withdrawn regularly and analyzed for GA production.

3.4. Glucoamylase Extraction

Whole content of the fermented solid in the petri dish was mixed thoroughly in a kitchen blender containing 20 mL demineralized water and the mixture was transferred into a 500 mL Duran bottle. The kitchen blender was rinsed with another portion of 20 mL demineralized water and pooled with the previously obtained mixture. The suspension was then mixed for 30 min with a magnetic stirrer at 30 °C, followed by centrifugation at 22,000 × g for 10 min. The supernatant was collected and filtered with Whitman no. 1 filter paper. The solution obtained refers to crude GA extract.

3.5. Food Waste Hydrolysis for Glucose Production

Food waste hydrolysis was conducted by adding increasing amount of crude GA extract to test tubes containing 50 mg dried food waste in 2.2 mL of 0.2 M sodium acetate (pH 5.5). The reaction mixture was incubated at 55 °C for 3 h and mixed by pipetting for every 30 min. Aliquots of samples were withdrawn regularly from the reaction mixture and mixed with 10% (w/v) trichloroacetic acid prior to glucose determination.

3.6. Glucoamylase Activity

Activity of the crude GA extract was determined using the method described by Melikoglu *et al.* [17]. Wheat flour solution of 6% (w/v) was used as substrate and it was prepared in sodium acetate (pH 5.5) and gelatinized at 80 °C for 15 min before usage. The assay was conducted by mixing 0.25 mL of 5 times diluted crude GA extract and 0.5 mL of gelatinized wheat flour solution, and incubated at 55 °C. The reaction was terminated after 10 min by adding 0.25 mL 10% (w/v) trichloroacetic acid solution to the reaction mixture. The reaction mixture was centrifuged and the glucose concentration in the supernatant was analyzed using the Analox GL6 glucose analyzer. One unit (U) of GA activity is defined as the amount of enzyme that releases 1 μmol of glucose per minute under assay conditions and is expressed as U/g of dry substrate as described [12].

For determination of optimal reaction pH, the crude GA extract and the gelatinized wheat flour solutions were prepared in sodium acetate buffer from pH 3.5 to 7.5. Optimal reaction temperature determination of the crude GA extract was carried out at temperature range of 40–75 °C with 5 °C increment at optimal pH. Thermo-stability of the crude GA extract was investigated at different temperatures (55–65 °C) at optimal pH. Experiments were performed in duplicate and the standard errors were reported.

4. Conclusions

In this work, we have demonstrated the feasibility of pastry waste as feedstock for GA production. High GA yield (253.7 ± 20.4 U/g) of the crude enzyme extract was obtained without addition of nitrogen in comparison to other reported waste substrates, highlighting its potential as a feedstock for GA production in industrial scale. Under the optimal digestion conditions (pH 5.5 and 55 °C), the crude GA extract could hydrolyze mixed food waste in 1 h and generate around 53 g glucose from 100 g of mixed food waste. The work is of great significance as it shows sustainable GA production from food waste for potential municipal food waste treatment and sustainable chemicals production.

Acknowledgments

The authors acknowledge the Biomass funding from the Ability R&D Energy Research Centre (AERC) at the School of Energy and Environment in the City University of Hong Kong. We are also grateful to the donation from the Coffee Concept (Hong Kong) Ltd. For the 'Care for Our Planet' campaign, as well as a grant from the City University of Hong Kong (Project No. 7200248). Authors acknowledge the Industrial Technology Funding from the Innovation and Technology Commission (ITS/323/11) in Hong Kong.

References

1. Hong Kong SAR Environmental Protection Department. Monitoring of solid waste in Hong Kong – waste statistics for 2011, in Hong Kong 2011. Available online: https://www.wastereduction.gov.hk/en/materials/info/msw2011.pdf (accessed on 21 June 2013).
2. Sauer, J.; Sigurskjold, B.W.; Christensen, U.; Frandsen, T.P.; Mirgorodskaya, E.; Harrison, M.; Roepstorff, P.; Svensson, B. Glucoamylase: Structure/Function relationships, and protein engineering. *Biochim. Biophys. Acta* **2000**, *1543*, 275–293.
3. Norouzian, D.; Akbarzadeh, A.; Scharer, J.M.; Young, M.M. Fungal glucoamylases. *Biotechnol. Adv.* **2006**, *24*, 80–85.

4. Wang, X.Q.; Wang, Q.H.; Liu, Y.Y.; Ma, H.Z., On-Site production of crude glucoamylase for kitchen waste hydrolysis. *Waste Manag. Res.* **2010**, *28*, 539–544.
5. Pleissner, D.; Lam, W.C.; Sun, Z.; Lin, C.S.K. Food waste as nutrient source in heterotrophic microalgae cultivation. *Bioresour. Technol.* **2013**, *137*, 139–146.
6. Sayeki, M.; Kitagawa, T.; Matsumoto, M.; Nishiyama, A.; Miyoshi, K.; Mochizuki, M.; Takasu, A.; Abe, A. Chemical composition and energy value of dried meal from food waste as feedstuff in swine and cattle. *Anim. Sci. J.* **2001**, *72*, 34–40.
7. Koutinas, A.A.; Wang, R.; Webb, C. Estimation of fungal growth in complex, heterogeneous culture. *Biochem. Eng. J.* **2003**, *14*, 93–100.
8. Zhang, A.Y.-Z.; Sun, Z.; Leung, C.C.J.; Han, W.; Lau, K.Y.; Li, M.; Lin, C.S.K. Valorisation of bakery waste for succinic acid production. *Green Chem.* **2013**, *15*, 690–695.
9. Zambare, V. Solid state fermentation of *Aspergillus. oryzae* for glucoamylase production on agro residues. *Int. J. Life Sci.* **2010**, *4*, 16–25.
10. Ganzlin, M.; Rinas, U. In-Depth analysis of the *Aspergillus. niger* glucoamylase (glaA) promoter performance using high-throughput screening and controlled bioreactor cultivation techniques. *J. Biotechnol.* **2008**, *135*, 266–271.
11. Ventura, L.; González-Candelas, L.; Pérez-Gonzáez, J.A.; Ramón, D. Molecular cloning and transcriptional analysis of the *Aspergillus. terreus* gla1 gene encoding a glucoamylase. *Appl. Environ. Microbiol.* **1995**, *61*, 399–402.
12. Anto, H.; Trivedi, U.B.; Patel, K.C. Glucoamylase production by solid-state fermentation using rice flake manufacturing waste products as substrate. *Bioresour. Technol.* **2006**, *97*, 1161–1166.
13. Wang, Q.; Wang, X.; Wang, X.; Ma, H. Glucoamylase production from food waste by *Aspergillus. niger* under submerged fermentation. *Process. Biochem.* **2008**, *43*, 280–286.
14. Kareem, S.O.; Akpan, I.; Oduntan, S.B. Cowpea waste: A novel substrate for solid state production of amylase by *Aspergillus. oryzae*. *Afr. J. Microbiol. Res.* **2009**, *3*, 974–977.
15. Du, C.; Lin, S.K.C.; Koutinas, A.; Wang, R.; Dorado, P.; Webb, C. A wheat biorefining strategy based on solid-state fermentation for fermentative production of succinic acid. *Bioresour. Technol.* **2008**, *99*, 8310–8315.
16. Wang, R.; Godoy, L.C.; Shaarani, S.M.; Melikoglu, M.; Koutinas, A.; Webb, C. Improving wheat flour hydrolysis by an enzyme mixture from solid state fungal fermentation. *Enzyme Microb. Technol.* **2009**, *44*, 223–228.
17. Melikoglu, M.; Lin, C.S.K.; Webb, C. Stepwise optimisation of enzyme production in solid state fermentation of waste bread pieces. *Food Bioprod. Process.* **2013**, in press.
18. Lawton, J.M.; Doonan, S. Thermal inactivation and chaperonin-mediated renaturation of mitochondrial aspartate aminotransferase. *Biochem. J.* **1998**, *334*, 219–224.
19. Johannes, T.W.; Woodyer, R.D.; Zhao, H. Directed evolution of a thermostable phosphite dehydrogenase for NAD(P)H regeneration. *Appl. Environ. Microbiol.* **2005**, *71*, 5728–5734.
20. Allen, M.J.; Coutinho, P.M.; Ford, C.F. Stabilization of *Aspergillus. awamori* glucoamylase by proline substitution and combining stabilizing mutations. *Protein Eng.* **1998**, *11*, 783–788.
21. Myer, R.O.; Brendemuhl, J.H.; Johnson, D.D. Evaluation of dehydrated restaurant food waste products as feedstuffs for finishing pigs. *J. Anim. Sci.* **1999**, *77*, 685–692.

22. Leung, C.C.J.; Cheung, A.S.Y.; Zhang, A.Y.-Z.; Lam, K.F.; Lin, C.S.K. Utilisation of waste bread for fermentative succinic acid production. *Biochem. Eng. J.* **2012**, *65*, 10–15.
23. Dorado, M.P.; Lin, S.K.C.; Koutinas, A.; Du, C.; Wang, R.; Webb, C. Cereal-Based biorefinery development: Utilisation of wheat milling by-products for the production of succinic acid. *J. Biotechnol.* **2009**, *143*, 51–59.
24. Yan, S.; Yao, J.; Yao, L.; Zhi, Z.; Chen, X.; Wu, J. Fed batch enzymatic saccharification of food waste improves the sugar concentration in the hydrolysates and eventually the ethanol fermentation by *Saccharomyces. Cerevisiae* H058. *Braz. Arch. Biol. Technol.* **2012**, *55*, 183–192.

7

Enzyme-Catalyzed Synthesis of Unsaturated Aliphatic Polyesters based on Green Monomers from Renewable Resources

Yi Jiang [1,2], Albert J.J. Woortman [1], Gert O.R. Alberda van Ekenstein [1] and Katja Loos [1,2,*]

[1] Department of Polymer Chemistry, Zernike Institute for Advanced Materials, University of Groningen, Nijenborgh 4, 9747 AG Groningen, The Netherlands; E-Mails: Y.Jiang@rug.nl (Y.J.); A.J.J.Woortman@rug.nl (A.J.J.W.); G.O.R.Alberda.van.Ekenstein@rug.nl (G.O.R.A.E.)

[2] Dutch Polymer Institute (DPI), P.O. Box 902, 5600 AX Eindhoven, The Netherlands

* Author to whom correspondence should be addressed; E-Mail: K.U.Loos@rug.nl

---

Abstract: Bio-based commercially available succinate, itaconate and 1,4-butanediol are enzymatically co-polymerized in solution *via* a two-stage method, using *Candida antarctica* Lipase B (CALB, in immobilized form as Novozyme® 435) as the biocatalyst. The chemical structures of the obtained products, poly(butylene succinate) (PBS) and poly(butylene succinate-*co*-itaconate) (PBSI), are confirmed by ^{1}H- and ^{13}C-NMR. The effects of the reaction conditions on the CALB-catalyzed synthesis of PBSI are fully investigated, and the optimal polymerization conditions are obtained. With the established method, PBSI with tunable compositions and satisfying reaction yields is produced. The ^{1}H-NMR results confirm that carbon-carbon double bonds are well preserved in PBSI. The differential scanning calorimetry (DSC) and thermal gravimetric analysis (TGA) results indicate that the amount of itaconate in the co-polyesters has no obvious effects on the glass-transition temperature and the thermal stability of PBS and PBSI, but has significant effects on the melting temperature.

Keywords: enzymatic polymerizations; CALB; unsaturated aliphatic polyesters; bio-based monomers; succinate; itaconate; 1,4-butanediol; renewable resources

---

1. Introduction

Utilizing renewable resources for the replacement of depleting fossil stocks is an appealing research topic, both in the academic and industrial areas [1–5]. It is a promising approach to solve the severe environmental problems induced by the increasing petroleum consumptions nowadays and the plausible energy shortage in the future. As abundant carbon-neutral renewable resources, biomass stocks are generated directly from solar energy in a short cycle. A great number of monomers and macromonomers can be produced from biomass stocks by natural biological activities or chemical modifications [6,7]. These bio-based monomers provide numerous opportunities for the synthesis of green and novel polymers.

Unsaturated polyesters are widely used as thermosetting resins in various industrial areas [8–10]. They are usually produced by polycondensation of diacids and diols based on petroleum stocks, using titanium or tin alkoxides as catalyst [11]. The synthesis temperature is usually above 150 °C [11,12]. Many monomers with chemically or thermally unstable moieties are not suitable for polyester synthesis, due to uncontrollable side reactions induced by such a high temperature, like gelation, decomposition and discoloration [11–13]. Besides, the residual metals from the conventional catalysts are hard to remove, which may cause undesirable pollution upon disposal [12].

Candida antarctica Lipase B (CALB) is a very versatile biocatalyst for polyester synthesis, working with various monomers and organic solvents under mild conditions [12,14–27]. CALB-catalyzed synthesis of polyesters from green monomers has recently gained increasing popularity. Succinate [28–31], fatty acids from plant oils [11,32–37], isosorbide [29,30] and 1,4-butanediol [8,28,31,38–40] are extensively studied. The bio-based polyesters produced are eco-friendly, since the monomers and the catalysts are all generated from renewable resources, and the polymers are biodegradable [9,28,39].

Itaconate is a commercially available green bio-based monomer. Its acid derivative, itaconic acid, has been industrially fermented from carbohydrates using *Aspergillus terreus* since the 1960s [41]. This monomer has interesting photoactive and biocompatible properties [9]. It is an ideal building block for constructing unsaturated polyesters with potential biomedical and engineering applications [9,42,43]. However, up until now, itaconate has not been well studied for polyester synthesis, neither by conventional catalysts nor by biocatalysts. Limited kinds of itaconate-based polyesters were synthesized using conventional chemical catalysts [8–10,43–48]. Only two papers referred to the enzymatic polymerizations of itaconate with other monomers: Barrett *et al.* reported enzymatic co-polymerizations of dimethyl itaconate and adipic acid with 1,4-cyclohexanedimethanol and poly(ethylene glycol) [9]; and Rajkhowa *et al.* reported Lipase-catalyzed polymerization of diglycidyl ether of bisphenol A and itaconic anhydride [49]. To the best of our knowledge, the enzyme-catalyzed co-polymerization of succinate, itaconate and 1,4-butanediol has not yet been studied. This is probably due to the low enzyme polymerizability of itaconate, which is caused by its short chain length [28] and the stereo-hindrance effect of the carbon-carbon double bond suspended around the carbonyl group. We believe CALB is the perfect biocatalyst for the synthesis of itaconate-based polyesters, due to its wide monomer adaptability, rendering polyesters in which the thermal unstable carbon-carbon double bonds could be preserved, since the enzymatic polymerization will be performed at mild temperatures under 100 °C [12,17].

We present an environmental friendly approach towards unsaturated aliphatic polyesters. Bio-based succinate, itaconate and 1,4-butanediol are enzymatically co-polymerized in solution *via* a two-stage method, using CALB as the catalyst. The general synthesis strategy is illustrated in Figure 1. Monomers are oligomerized at 80 °C under nitrogen atmosphere during the first stage. Then, the oligomers are polycondensed at the same temperature under high vacuum during the second stage. To achieve the best polymerization results, the effects of different reaction conditions on CALB-catalyzed synthesis of poly(butylene succinate-*co*-itaconate) (PBSI) are extensively investigated, and the optimal polymerization conditions are obtained. With the method we established, poly(butylene succinate) (PBS) and a series of PBSI are synthesized. The chemical structures, molecular weight and thermal properties of the co-polyesters are characterized by different methods.

Figure 1. *Candida antarctica* Lipase B (CALB)-catalyzed co-polymerization of succinate, itaconate and 1,4-butanediol.

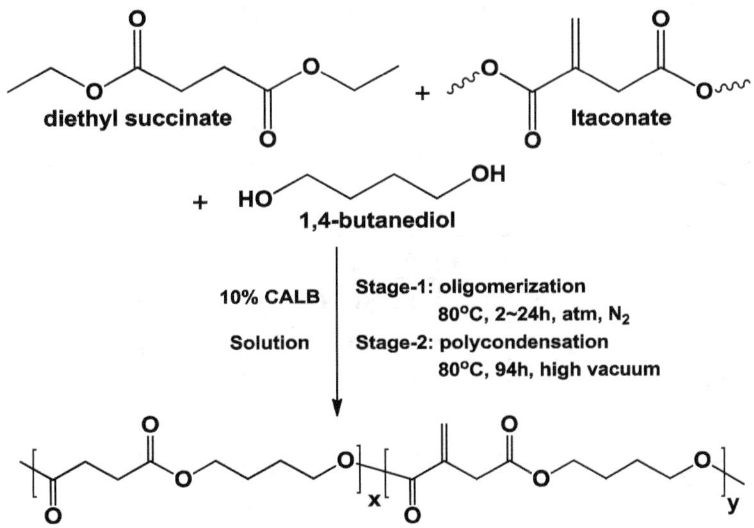

2. Results and Discussion

2.1. *Effects of Polymerization Conditions on* CALB-*Catalyzed Co-Polymerization of Diethyl Succinate, Dimethyl Itaconate and 1,4-Butanediol*

Succinic acid, itaconic acid and 1,4-butanediol are commercially available bio-based monomers· They are good building blocks for polyester synthesis. However, succinic acid has a rather low solubility in 1,4-butanediol under enzymatic polycondensation conditions, which leads to low polymerization efficiency [28]. Itaconic acid has the same problem, according to our observation[S] reported in Section 2.2. To avoid phase separation during polymerization, diethyl succinate and dimethyl itaconate were used as the acyl donors.

For the purpose of producing PBSI with the highest amount of itaconate, molecular weight and reaction yield, the effects of several polymerization conditions on CALB-catalyzed co-polymerization of diethyl succinate, dimethyl itaconate and 1,4-butanediol were investigated, including solvent, solvent dosage, oligomerization time during the first stage and vacuum during the second stage.

2.1.1. Effect of Solvent on CALB-Catalyzed Co-Polymerization of Diethyl Succinate, Dimethyl Itaconate and 1,4-Butanediol

Dodecane, diethylene glycol dimethyl ether (diglyme) and diphenyl ether were evaluated, as they have been proven to be suitable for enzymatic synthesis of polyesters [13,28,50]. As summarized in Table 1, PBSI with the highest mole percentage of itaconate (X_I) and number average molecular weight (M_n) was synthesized using diphenyl ether as solvent. However, the highest reaction yield was obtained in diglyme; the second highest reaction yield in diphenyl ether.

As presented in Table 1, when the molar feed ratio of dimethyl itaconate was 15%, the values of product X_I obtained in diphenyl ether, dodecane and diglyme were 10.5%, 3.5% and 0.3%, respectively. The corresponding M_n calculated from ^{1}H-NMR spectra were 1,304 g/mol, 963 g/mol and 854 g/mol. The reaction yields were 55.5%, 54.3% and 68.8%, respectively.

Meanwhile, when the feed ratio of dimethyl itaconate was increased to 25%, the values of product X_I achieved in diphenyl ether, dodecane and diglyme were 15.9%, 3.7% and 5.9%, respectively. The corresponding M_n were 1,403 g/mol, 730 g/mol and 795 g/mol. The reaction yields were 22.6%, 18.3% and 30.2%, respectively.

Table 1. The effect of solvent on CALB-catalyzed synthesis of poly(butylene succinate-*co*-itaconate) (PBSI).

		Molar composition						Molecular weight	
		Feed (%)			PBSI [a] (%)				
Solvent	log *P*	F_S	F_I	F_B	X_S	X_I	X_B	M_n (g/mol) [b]	Yield (%)
Diphenyl ether	4.1	35	15	50	40.5	10.5	49.0	1,304	55.5
Dodecane	6.8	35	15	50	46.5	3.5	50.0	963	54.3
Diglyme	−1.3	35	15	50	49.2	0.3	50.5	854	68.8
Diphenyl ether	4.1	25	25	50	35.3	15.9	48.8	1,403	22.6
Dodecane	6.8	25	25	50	46.5	3.7	49.8	730	18.3
Diglyme	−1.3	25	25	50	43.2	5.9	50.9	795	30.2

F_S, F_I, F_B: molar feed ratio of succinate, itaconate and 1,4-butanediol; X_S, X_I, X_B: mole percentage of succinate, itaconate and butylene units in PBSI; polymerization conditions: 200 wt% of solvent; Stage-1: 80 °C, 24 h, N_2; Stage-2: 80 °C, 94 h, vacuum 40 mmHg; [a] molar composition determined by integration of the ^{1}H-NMR spectra; [b] number average molecular weight (M_n) calculated from ^{1}H-NMR spectra; log *P*, logarithm of partition coefficient.

It has already been reported that diphenyl ether is the preferred solvent to achieve higher M_n for lipase-catalyzed synthesis of polyesters [13,28,50], which is in good accordance with our current results. Three factors could be attributed to our case, including the log *P* (logarithm of partition coefficient) value of the solvent, the accessibility of CALB and the miscibility of the intermediate and the final products in the solvent. The log *P* values of diphenyl ether and dodecane are higher than 1.9. They are more efficient for producing high molecular weight polyesters than diglyme. Furthermore, we found that CALB dispersed well in diphenyl ether and diglyme, but adhered tightly to the flasks in dodecane. Additionally, PBS with low molecular weight and PBSI are fully miscible in diphenyl ether at 80 °C, but precipitated fast in diglyme and dodecane. Therefore, PBSI with higher M_n is produced in diphenyl ether, since CALB is more accessible and the intermediate products are diffused better, which provide sufficient time and space for the chain growth and the transesterification of PBSI.

Moreover, we found that the enzymatic polymerizability of dimethyl itaconate was higher in diphenyl ether in comparison to the other two solvents. In addition, the molar composition of PBSI produced in diphenyl ether matched better with the feed composition of the monomers. The enzymatic polymerizability of dimethyl itaconate in dodecane and diglyme was quite low; only less than 6% of dimethyl itaconate was co-polymerized, although the reaction yield was considerably high. This is because PBS and PBSI are not soluble in dodecane and diglyme at 80 °C. In diglyme and dodecane, 1,4-butanediol prefers to react with diethyl succinate first, since the enzyme polymerizability of diethyl succinate is higher than that of dimethyl itaconate. Low molecular weight PBS and PBSI composed with a little amount of itaconate was produced and precipitated fast from the reaction.

Therefore, diphenyl ether is the most suitable solvent for CALB-catalyzed synthesis of PBSI.

2.1.2. Effect of Solvent Dosage on CALB-Catalyzed Co-Polymerization of Diethyl Succinate, Dimethyl Itaconate and 1,4-Butanediol

Figure 2 illustrates the values of product M_n, X_I and reaction yield as a function of diphenyl ether dosage. PBSI with the highest values of M_n, X_I and reaction yield was obtained from the reaction with 150 wt% of diphenyl ether (in relation to the total amount of monomers).

Figure 2. The effect of diphenyl ether dosage on CALB-catalyzed synthesis of PBSI.

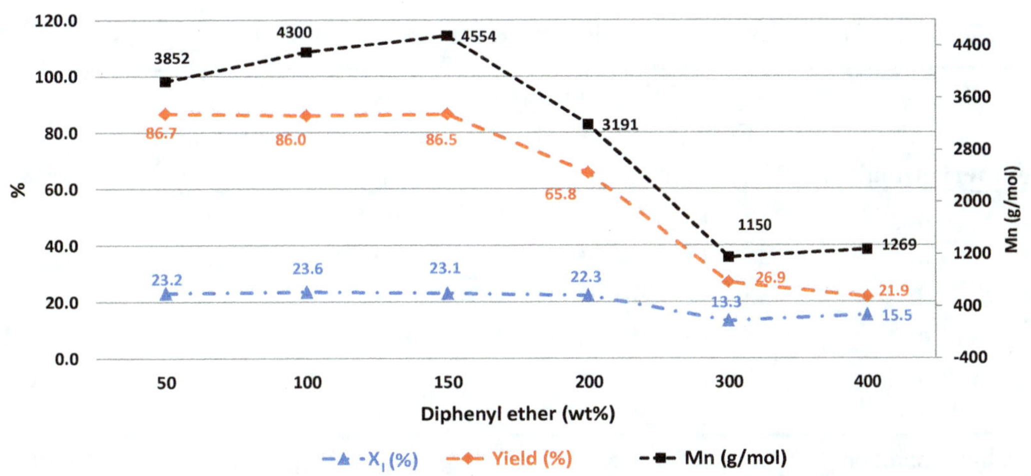

Polymerization conditions: 50%–400 wt% of diphenyl ether; $F_S/F_I/F_B$ = 25%/25%/50%; Stage-1: 80 °C, 2 h, N_2; Stage-2: 80 °C, 94 h, vacuum 2 mmHg; M_n was calculated from ^{1}H-NMR spectra.

As presented in Figure 2, the values of product X_I and reaction yield remained almost the same, when the dosage of diphenyl ether was increased from 50% to 150%. The product X_I was around 23%, and the reaction yield was close to 87%. The corresponding M_n increased from 3,852 g/mol to 4,554 g/mol. The X_I value calculated from ^{1}H-NMR was in good agreement with the feed ratio of itaconate. The synchronous increase of product M_n with the solvent dosage was due to the better diffusion of the reactants in dilute reaction.

By increasing the solvent dosage from 150% to 400%, the values of product M_n and reaction yield decreased significantly. The M_n reduced from 4,554 g/mol to 1,269 g/mol, while the reaction yield decreased from 86.5% to 21.9%. As for the X_I value, it remained similar, around 23%, when the solvent dosage was increased to 200%. On further dilution to 300%, it dropped to 13.3%, and on dilution to 400%, it remained more or less the same: 14.6%. We suspect that the increase of the residual alcohol amount and the decrease of the polymerization rate in dilute reaction could be the reasons. The alcohols were more difficult to remove by vacuum at a certain low concentration in dilute solution. The absolute amount of the residual alcohols was higher in the reactions with higher solvent dosage. At the same time, the polymerization rate was reduced, since the concentrations of CALB and the reactants were lower. The transesterification of oligomers with itaconate was hindered, and only low molecular weight co-polyesters were produced. The low molecular weight PBSI composed with a higher amount of itaconate is soluble in methanol. In this case, only PBSI with a lower amount of itaconate was obtained after purification, which led to the lower value of the reaction yield.

As a result, the best diphenyl ether dosage for *in vitro* synthesis of PBSI is 150 wt% (in relation to the total amount of monomers).

2.1.3. Effect of Oligomerization Time during the First Stage on CALB-Catalyzed Co-Polymerization of Diethyl Succinate, Dimethyl Itaconate and 1,4-Butanediol

As shown in Table 2, the selected oligomerization time has no obvious effects on the composition and the reaction yield of PBSI. PBSI with the highest M_n was synthesized from the reaction conducting 2 h oligomerization during the first stage.

Table 2. The effect of oligomerization time on CALB-catalyzed synthesis of PBSI.

	Molar composition							
	Feed (%)			**PBSI [a] (%)**			**Molecular weight**	
Oligomerization time	**F_S**	**F_I**	**F_B**	**X_S**	**X_I**	**X_B**	**M_n (g/mol) [b]**	**Yield (%)**
2 h	35	15	50	35.7	14.1	50.2	3,936	87.7
6 h	35	15	50	35.2	14.9	49.9	3,212	87.6
12 h	35	15	50	36.3	14.4	49.3	2,960	84.1
2 h	25	25	50	27.0	23.1	49.9	2,935	73.1
6 h	25	25	50	28.9	21.5	49.6	2,459	68.5
12 h	25	25	50	27.5	22.8	49.7	2,609	71.6

Polymerization conditions: 150 wt% of diphenyl ether; Stage-1: 80 °C, 2–12 h, N_2; Stage-2: 80 °C, 94 h, vacuum 2 mmHg; [a] molar composition determined by integration of the ^{1}H-NMR spectra; [b] number average molecular weight calculated from ^{1}H-NMR spectra.

When the feed ratio of dimethyl itaconate was 15%, the composition of PBSI agreed well with the feed composition of the monomers, which was independent of the time of oligomerization. The reaction yield was quite satisfying (higher than 84%). However, the product M_n decreased from 3,936 g/mol to 2,960 g/mol, with increasing the oligomerization time from 2 h to 12 h.

Similar trends were obtained when the feed ratio of dimethyl itaconate was 25%. In spite of increasing the oligomerization time from 2 h, 6 h and up to 12 h, the composition and reaction yield of PBSI remained almost the same. The product M_n, however, decreased a little from 2,935 g/mol (2 h oligomerization) to 2,609 g/mol (12 h oligomerization).

The enzymatic oligomerization process of diethyl succinate, dimethyl itaconate and 1,4-butanediol was monitored by *in situ* ^{1}H-NMR, as shown in Figure 3. The ^{1}H-NMR spectra of the intermediate products reacted for more than 30 min were almost identical, except the signals belonging to the active hydroxyl groups at 1.7–2.2 ppm. It is suspected that the monomers were fully transformed to oligomers after 30 min during the first stage.

Therefore, it can be concluded that 2 h of oligomerization during the first stage will be sufficient enough for CALB-catalyzed synthesis of PBSI.

Figure 3. *In situ* ^{1}H-NMR investigation of the oligomerization process of: (**a**) diethyl succinate (35%), dimethyl itaconate (15%) and 1,4-butanediol (50%); (**b**) diethyl succinate (25%), dimethyl itaconate (25%) and 1,4-butanediol (50%).

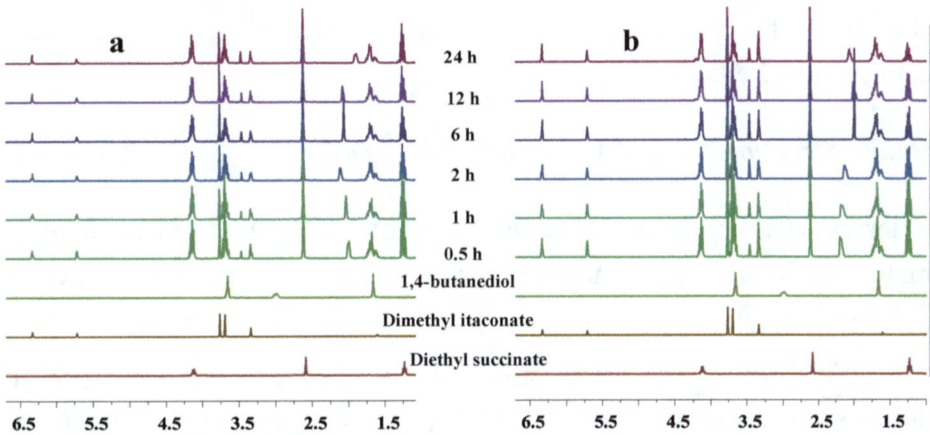

2.1.4. Effect of Vacuum during the Second Stage on CALB-Catalyzed Co-Polymerization of Diethyl Succinate, Dimethyl Itaconate and 1,4-Butanediol

As presented in Table 3, the vacuum during the second stage has a significant impact on the CALB-catalyzed co-polymerization. PBSI with the highest values of X_I, M_n and reaction yield was produced by the co-polymerizations with the highest vacuum applied (2 mmHg). Furthermore, the composition of PBSI produced under higher vacuum matched better with the feed composition of the monomers.

Table 3. The effect of vacuum on CALB-catalyzed synthesis of PBSI.

	Molar composition						**Molecular weight**	
	Feed (%)			**PBSI [a] (%)**				
Vacuum	**F_S**	**F_I**	**F_B**	**X_S**	**X_I**	**X_B**	**M_n (g/mol) [b]**	**Yield (%)**
2 mmHg	35	15	50	35.7	14.1	50.2	3,936	87.7
10–20 mmHg	35	15	50	37.1	13.1	49.8	2,552	56.2
40 mmHg	35	15	50	40.5	10.5	49.0	1,304	55.5
2 mmHg	25	25	50	27.0	23.1	49.9	2,935	73.1
10–20 mmHg	25	25	50	31.7	19.1	49.2	2,328	19.5
40 mmHg	25	25	50	35.2	15.9	48.9	1,203	22.6

Polymerization conditions: 150 wt% of diphenyl ether; Stage-1: 80 °C, 2 h, N_2; Stage-2: 80 °C, 94 h, vacuum at 2–40 mmHg; [a] molar composition determined by integration of the ^{1}H-NMR spectra; [b] number average molecular weight calculated from ^{1}H-NMR spectra.

When the feed ratio of dimethyl itaconate was 15%, the values of product X_I, M_n and reaction yield increased significantly with an increase of vacuum. The values of product X_I from the reactions under vacuum of 40 mmHg, 10–20 mmHg and 2 mmHg were 10.5%, 13.1% and 14.1%, respectively. The corresponding product M_n was 1,304 g/mol, 2,552 g/mol and 3,936 g/mol. The reaction yield was 55.5%, 56.2% and 87.7%, respectively.

The same trend of product X_I, M_n and reaction yield as a function of vacuum was identified, when the feed ratio of dimethyl itaconate was 25%. By lowering the pressure from 40 mmHg to 2 mmHg, the values of product X_I increased from 15.9% to 23.1%. The product M_n increased from 1,203 g/mol to 2,935 g/mol. Additionally, the reaction yield increased from 22.6% to 73.1%.

The effect of vacuum is quite reasonable, since the residual alcohols and water can be further eliminated from the reaction under higher vacuum, which facilitates the chain growth of co-polyesters. It is obvious that the reduced pressure shall be regulated to 2 mmHg during the second stage for *in vitro* synthesis of PBSI in diphenyl ether.

2.2. Effect of Itaconate Structure on CALB-*Catalyzed Synthesis of PBSI in Diphenyl Ether*

Four itaconate derivatives were studied, namely itaconic acid, dimethyl itaconate, diethyl itaconate and dibutyl itaconate. The byproducts generated during polycondensation are water, methanol, ethanol and n-butyl alcohol, respectively.

Phase separation of itaconic acid in the presence of the other reactants was observed. This is due to the fact that itaconic acid has a rather low solubility in the mixture of diphenyl ether, diethyl succinate and 1,4-butanediol.

Dimethyl itaconate dissolved in the reaction medium when at temperatures above 60 °C. The co-polymerization with dimethyl itaconate, diethyl itaconate or dibutyl itaconate was homogeneous.

PBSI could not be obtained in a sufficient amount when using itaconic acid as the unsaturated monomer. The reaction yield was extremely low, less than 4%, as shown in Figure 4a.

Figure 4. The effect of itaconate structure on CALB-catalyzed synthesis of PBSI: (**a**) the reaction yield of PBSI as a function of the feed ratio of itaconate; (**b**) the mole percentage of itaconate in PBSI (X_I) as a function of the feed ratio of itaconate.

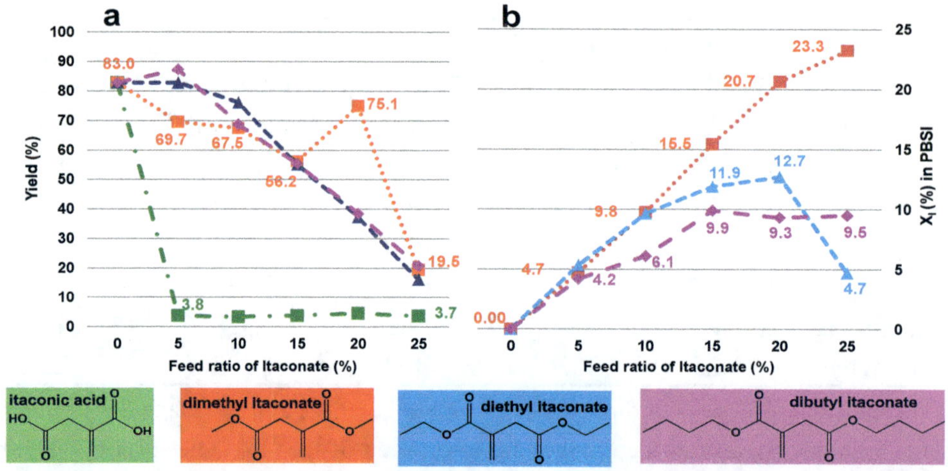

Polymerization conditions: 150 wt% of diphenyl ether; Stage-1: 80 °C, 2 h, N_2; Stage-2: 80 °C, 94 h, vacuum 10–20 mmHg.

For the co-polymerizations with dialkyl itaconate esters, the reaction yield of PBSI decreased with increasing the feed ratio of itaconate, as displayed in Figure 4a, e.g., the reaction yield decreased from 83.0% to 19.5% when the feed ratio of dimethyl itaconate increased from 0% to 25%. The same trend of the reaction yield as a function of the feed ratio of itaconate was also observed in other co-polymerizations with diethyl itaconate or dibutyl itaconate.

Moreover, for reactions with dimethyl itaconate, the value of product X_I was in good accordance with the corresponding feed ratio of itaconate, as presented in Figure 4b. While the feed ratio of itaconate was 0%, 5%, 10%, 15%, 20% and 25%, the value of product X_I was 0.0%, 4.7%, 9.8%, 15.5%, 20.7% and 23.3%, respectively. The deviation between the X_I value calculated from ^{1}H-NMR spectra and the corresponding feed ratio of itaconate was quite small.

However, for co-polymerizations with diethyl itaconate, the value of product X_I deviated obviously from the corresponding feed ratio of itaconate, especially when the feed ratio of diethyl itaconate was higher than 10%. While the feed ratio of diethyl itaconate was 0%, 5%, 10%, 15%, 20% and 25%, the corresponding product X_I was 0.0%, 5.4%, 9.7%, 11.9%, 12.7% and 4.7%, respectively. The value of product X_I only matched with the feed ratio of diethyl itaconate when less than or equal to 10% of diethyl itaconate was added. The highest X_I value was only 12.7%, which was produced from the reaction with 20% of diethyl itaconate.

For co-polymerizations with dibutyl itaconate, the highest X_I value achieved was only 9.9%. The value of product X_I increased from 0.0% to 9.9% as the corresponding feed ratio of dibutyl itaconate ascended from 0% to 15%. A plateau value was reached of around 10% when the feed ratio of dibutyl itaconate was further increased from 15% to 25%.

As a result, the enzyme polymerizability sequence of itaconate derivatives in diphenyl ether is dimethyl itaconate, diethyl itaconate, dibutyl itaconate and itaconic acid, from high to low. Itaconic acid is not favored by CALB, since it is immiscible in solution, which makes it inaccessible to the biocatalyst. Dimethyl itaconate is the most polymerizable derivative by CALB, since the reaction byproduct, methanol, can be removed easily under vacuum, because of the low boiling temperature of 64.7 °C, compared with ethanol (bp = 78.4 °C) and n-butyl alcohol (bp = 117.7 °C). Besides, the chemical reactivity of the alkyl esters is another reason. It is well known that methyl esters are chemically much more reactive than ethyl and butyl esters.

Therefore, dimethyl itaconate is the most preferred unsaturated monomer for *in vitro* synthesis of PBSI in diphenyl ether.

In conclusion, combining the results from Sections 2.1 to 2.2, we have established the optimal polymerization conditions for CALB-catalyzed co-polymerization of diethyl succinate, itaconate and 1,4-butanediol, which are as follows:

(1) using diphenyl ether as solvent;

(2) the dosage of diphenyl ether is 150 wt% (in relation to the total amount of monomers);

(3) oligomerization for 2 h during the first stage;

(4) regulating vacuum to 2 mmHg during the second stage;

(5) applying dimethyl itaconate as the unsaturated monomer.

2.3. CALB-Catalyzed Synthesis of PBSI Using Optimal Polymerization Conditions

Diethyl succinate, dimethyl itaconate and 1,4-butanediol were co-polymerized in the presence of CALB, using the optimal conditions we established. A series of co-polyesters was produced by alternating the feed ratio of dimethyl itaconate from 0% to 30%. No products were obtained if the feed ratio of dimethyl itaconate was increased to 35% and 50%. For the two control reactions without CALB, also no polymers were obtained after purification. The molar composition, reaction yield, molecular weight and the thermal properties of PBS and PBSI are summarized in Table 4.

As shown in Table 4 and plotted in Figure 5a, co-polyesters with tunable compositions and satisfying reaction yields were achieved, by adjusting the feed ratio of dimethyl itaconate from 0% to 25%. The mole percentage of itaconate in PBSI can be controlled from 0% to 23.5%. The corresponding reaction yield was quite good, higher than 75%.

Table 4. Summary of the results for the CALB-catalyzed synthesis of PBS and PBSI.

	Molar composition										Molecular Weight			
	Feed (%)			PBSI [a] (%)							NMR	GPC		
Co-polyester	F_S	F_I	F_B	X_S	X_I	X_B	Yield (%)	T_g [b] (°C)	T_m [c] (°C)	T_d [d] (°C)	M_n [e]	M_n	M_w	M_w/M_n
PBS	50	0	50	50.5	0.0	49.5	85.7	−35.9	112.9	406.1	4,463	6,017	11,520	1.91
$PB_{50}S_{40}I_{10}$	40	10	50	39.7	10.4	49.9	84.6	−38.8	94.6	402.0	4,696	10,128	19,236	1.90
$PB_{50}S_{35}I_{15}$	35	15	50	35.0	15.7	49.3	90.0	−36.7	80.3	410.9	6,494	13,288	22,642	1.70
$PB_{50}S_{30}I_{20}$	30	20	50	30.9	20.1	49.0	75.1	−37.5	71.2	404.7	5,670	8,394	11,892	1.42
$PB_{50}S_{25}I_{25}$	25	25	50	27.2	23.5	49.3	86.5	−38.4	57.4	407.9	4,554	11,096	16,587	1.49
PBSI	20	30	50	36.4	15.6	48.0	20.9	−49.6	69.8	396.9	1,004	1,948	2,203	1.13
NA [f]	15	35	50	-	-	-	-	-	-	-	-	-	-	-
NA [f]	0	50	50	-	-	-	-	-	-	-	-	-	-	-
Control-1 [g]	50	0	50	-	-	-	-	-	-	-	-	-	-	-
Control-2 [g]	25	25	50	-	-	-	-	-	-	-	-	-	-	-

[a] Molar composition determined by integration of the ^{1}H-NMR spectra; [b] glass-transition temperature (T_g) from differential scanning calorimetry (DSC) by a quench method; [c] melting temperature (T_m) from DSC measurements, first scan; [d] the temperature of the maximal rate of decomposition (T_d) from thermal gravimetric analysis (TGA) measurement; [e] number average molecular weight calculated from ^{1}H-NMR spectra; [f] no polymer was obtained after precipitation; [g] control reactions without CALB. No polymer was obtained after precipitation.

The carbon-carbon double bonds were well preserved in the unsaturated co-polyesters, as confirmed by ^{1}H-NMR spectra in Figure 6. No resonances can be assigned to the deterioration of carbon-carbon double bonds. As displayed in the ^{1}H-NMR spectra of $PB_{50}S_{35}I_{15}$ and $PB_{50}S_{25}I_{25}$, the ratio between the integration values of the two separated peaks assigned to the protons of the carbon-carbon double bonds (6.32 ppm and 5.72 ppm, respectively) and the peak assigned to the protons of the methylene group of itaconate (3.33 ppm) was close to 1:1:2, which was exactly the same as that of dimethyl itaconate. Besides, the final products were white semicrystalline powders without discoloration. It is clear that no side reaction occurred during the enzymatic co-polymerizations.

Figure 5. **(a)** The mole percentage of itaconate (X_I) and the reaction yield of PBSI as a function of the feed ratio of dimethyl itaconate (F_I); **(b)** Overlay of gel permeation chromatography (GPC) elution curves in chloroform; PBSI was synthesized *via* the optimal conditions.

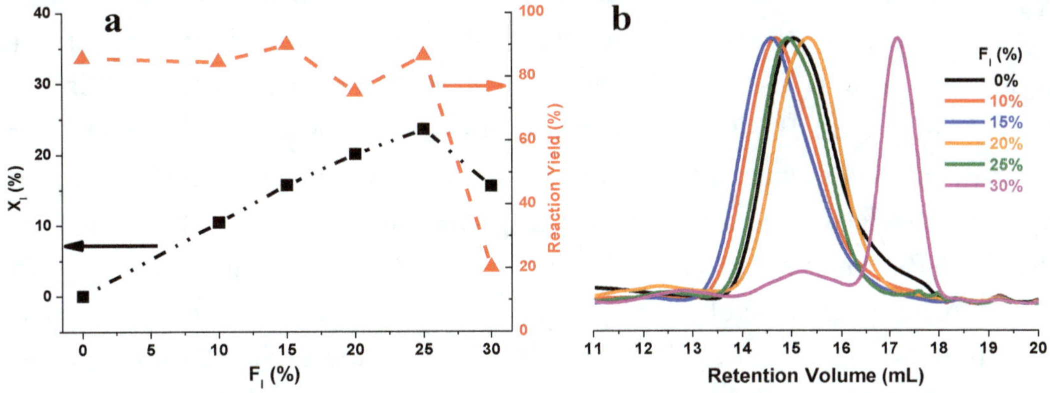

Figure 6. ^{1}H-NMR spectra of dimethyl itaconate, $PB_{50}S_{35}I_{15}$ and $PB_{50}S_{25}I_{25}$ in $CDCl_3$.

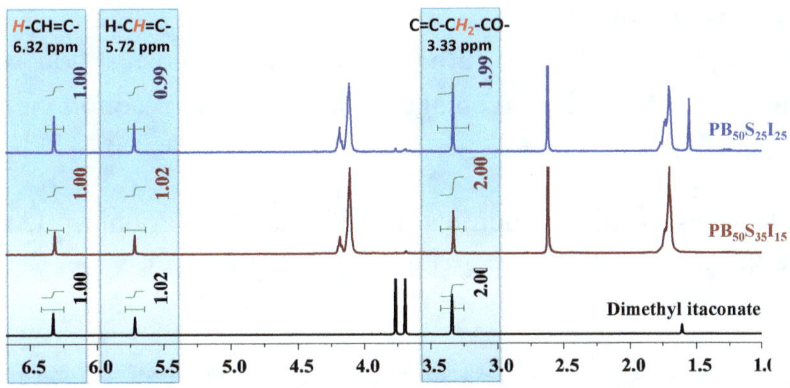

The product M_n calculated from ^{1}H-NMR was around 4,000–6,500 g/mol, while the feed ratio of dimethyl itaconate was less than or equal to 25%. GPC results indicate that the corresponding M_n was around 6,000–14,000 g/mol. The polydispersity index (PDI, M_w/M_n) was between 1.42 and 1.91. The molecular weight of the saturated polyesters was lower than that of the unsaturated counterparts, and the PDI was higher. This is due to the higher melting temperature of PBS, which is around 112 °C. During the polycondensation step, the reaction with PBS solidified after 19 h at 80 °C. The chain growth and transesterification of PBS stopped in the solid state. In contrast to this, the reaction with PBSI proceeded continuously in the liquid state.

The highest value of X_I achieved in PBSI was 23.5%, from the reaction with 25% of dimethyl itaconate. If the feed ratio of dimethyl itaconate was increased to 30%, the product X_I reduced to 15.6%. The corresponding product M_n and the reaction yield decreased significantly to 1,948 g/mol and 20.9%, as shown in Figure 5a. If the feed ratio of dimethyl itaconate was further increased to 35% and 50%, no polymer was obtained after precipitation in cold methanol.

As shown in Figure 5b, PBSI synthesized from the reaction with 30% of dimethyl itaconate had two peaks. The first peak was at a retention volume between 13.6 to 16.2 mL, similar to that of the co-polyesters with high molecular weight; however, the concentration was much lower. The major peak was at a retention volume between 16.2 to 18.2 mL, which means that the molecular weight of the most PBSI produced was quite low. It is clear that co-polyesters with two different molecular weight distributions were produced from reaction with 30% of dimethyl itaconate.

The limited incorporation of itaconate in PBSI could be attributed to the relatively low enzyme polymerizability of dimethyl itaconate compared to diethyl succinate. As shown in Figure 7, there are six possible microstructures in PBSI. It is easier for CALB to produce the microstructure of succinate-butylene-succinate (S-B-S) than that of succinate-butylene-itaconate (S-B-I) and itaconate-butylene-itaconate (I-B-I). The formation sequence from easy to difficult is S-B-S, S-B-I-1, S-B-I-2, I-B-I-1, I-B-I-2 and I-B-I-3. While the feed ratio of itaconate is 25%, the molar ratio between succinate, itaconate and 1,4-butanediol is 1:1:2. The possible dominant microstructures in PBSI could be $-(\text{S-B-I-B})_n-$. If the feed ratio of itaconate is higher than 25%, the possible structures could be $-\text{I-B-}(\text{S-B-I-B})_n\text{-I-B-}$, while some extra I-B-I structures are formed. However, the formation of the I-B-I microstructure is quite difficulty. Oligomers with itaconate ends will be terminated in chain growth, and the transesterification of oligomers with I-B-I structures will be hindered. Consequently, the molecular weight of PBSI will be lower. Low molecular weight PBSI is soluble in methanol and will be washed away during the purification steps. This is why we obtain only PBSI with the highest mole percentage of itaconate around 25%.

This hypothesis also explains well why PBSI obtained from the reaction with 30% of dimethyl itaconate had lower values for M_n, mole percentage of itaconate and reaction yield, as well as, had two retention volume peaks, as plotted in Figure 5.

Figure 7. Possible types of microstructures in PBSI, the formation sequence from easy to difficult: S-B-S < S-B-I-1 < S-B-I-2 < I-B-I-1 < I-B-I-2 < I-B-I-3.

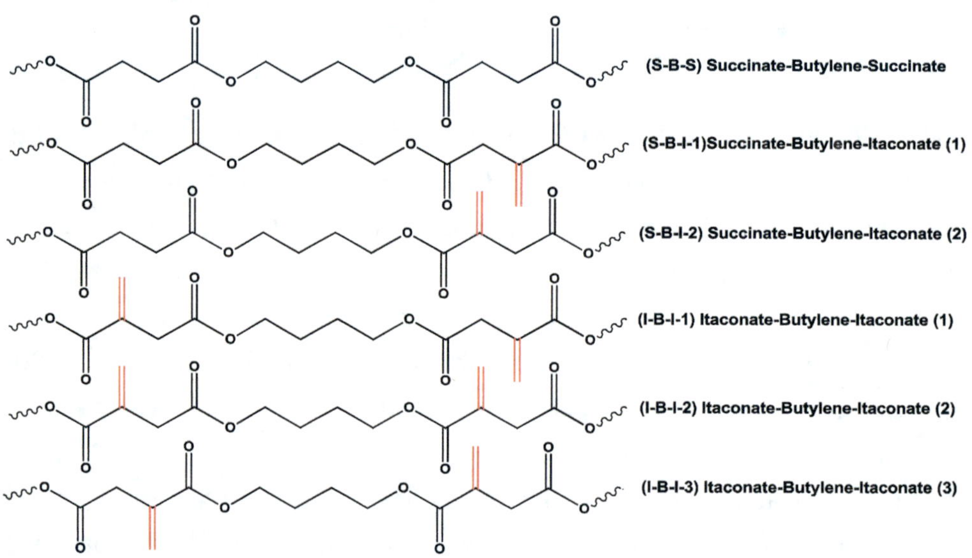

2.4. Thermal Properties of PBS and PBSI

The glass transition temperature (T_g), the melting temperature (T_m) and the temperature of the maximal rate of decomposition (T_d) of PBS and PBSI are plotted as a function of the mole percentage of itaconate in Figure 8. We found that the amount of itaconate composed in the co-polyesters has no obvious effects on the T_g and the thermal stability of PBSI. The T_g and T_d of PBS and PBSI were similar, as shown in Figure 8a. The T_g of the saturated PBS was −35.9 °C. It was slightly higher than that of the rest of the unsaturated PBSI, which was around −36–−38 °C. The T_d of all polyesters was around 400 °C.

However, the T_m of the co-polyesters is significantly affected by the amount of itaconate composed in PBSI. The T_m of the saturated PBS was 112.9 °C, which was 55.5 °C higher than that of the unsaturated counterpart with 23.5% of itaconate. The T_m of PBSI decreased almost linearly as a function of the mole percentage of itaconate, as shown in Figure 8b.

Figure 8. (**a**) the T_g and T_d of PBSI as a function of the mole percentage of itaconate (X_I); (**b**) the T_m of PBSI as a function of the mole percentage of itaconate (X_I); PBSI was synthesized *via* the optimal conditions.

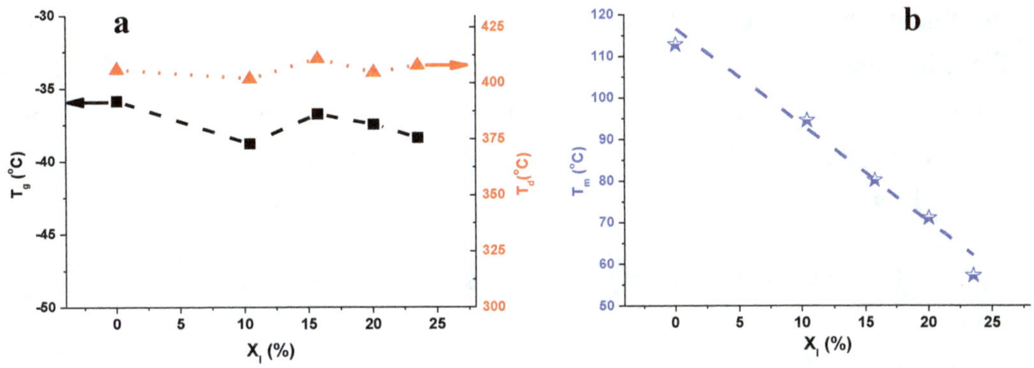

3. Experimental

3.1. Materials

Lipase acrylic resin from *Candida antarctica* Lipase B (CALB, in immobilized form as Novozyme® 435, 5,000+ U/g), diethyl succinate (99%), itaconic acid (99%+), dimethyl itaconate (99%), 1,4-butanediol (99%+), dodecane (99%+, anhydrous), diethylene glycol dimethyl ether (99.5%, anhydrous) and diphenyl ether (99%) were purchased from Aldrich. Diethyl itaconate (98%+) and dibutyl itaconate (95%+) were purchased from TCI EUROPE. Diphenyl ether was vacuum distilled and stored with 4Å molecular sieves before use. The other chemicals were used as received.

3.2. General Procedure for CALB-*Catalyzed Co-Polymerization of Succinate, Itaconate and 1,4-Butanediol in Solution*

CALB (10 wt% in relation to the total amount of monomers, for all reactions) was first fed into a 25 mL flask and stored in a desiccator with phosphorus pentoxide at room temperature under high vacuum for 16 h. Then diethyl succinate (5.55 mmol), itaconate (5.55 mmol), 1,4-butanediol (11.10 mmol) and the solvent (usually 150 wt% in relation to the total amount of monomers) were added. With the protection of nitrogen, the monomers were oligomerized at 80 °C for 2–24 h under atmospheric pressure during the first stage. The oligomers were further polymerized for another 94 h at the same temperature under the reduced pressure of 2–40 mmHg during the second stage.

After polymerization, chloroform (25 mL) was added into the reaction flask. CALB was filtered out and washed with chloroform (35 mL) three times. The collected solution was condensed by rotary evaporation at 40 °C under a vacuum of 300 mbar. The products were precipitated in cold methanol

(−20 °C) and washed with methanol three times. Finally, the polyesters were dried in a vacuum oven at 40 °C for 2–3 days before analysis. The final polymers obtained were white semicrystalline powders.

^{1}H-NMR (400 MHz, $CDCl_3$, ppm) analysis of PBS and PBSI: 6.25–6.35 (1H, s, ***H***-CH=C-CO-, from itaconate), 5.65–5.75 (1H, s, H-C***H***=C-CO-, from itaconate), 4.05–4.25 (4H, m, -CO-O-C$\boldsymbol{H}_2$-, from 1,4-butanediol), 3.30–3.40 (2H, s, -C=C-C$\boldsymbol{H}_2$-CO-, from itaconate), 2.58–2.68 (4H, s, -OC-C$\boldsymbol{H}_2$-, from succinate), 1.60–1.78 (4H, m, -O-CH_2-C$\boldsymbol{H}_2$-C$\boldsymbol{H}_2$-CH_2-O-, from 1,4-butanediol); low intensity signals due to end group-groups were observed at 3.75–3.79 (3H, s, -O-C$\boldsymbol{H}_3$, from itaconate), 3.63–3.73 (2H, m, -C$\boldsymbol{H}_2$-OH, from 1,4-butanediol) and 1.15–1.32 (3H, m, -OCH_2C$\boldsymbol{H}_3$, from succinate). The spectra are shown in Figure 9a.

^{13}C-NMR (100 MHz, $CDCl_3$, ppm) analysis of PBS and PBSI: 172.2 (-***C***O-, from succinate), 170.6 (-***C***O-, from itaconate), 166.0 (-C=C-***C***O-O-, from itaconate), 133.8 (-***C***=C-, from itaconate), 128.4 (-C=***C***-, from itaconate), 62.5–65.0 (-CO-***C***H_2-, from 1,4-butanediol), 37.6 (-***C***H_2-, from itaconate), 29.0 (-***C***H_2-, from succinate), 25.2 (-***C***H_2-, from 1,4-butanediol); low intensity signals assigned to end-groups were determined at 62.2 (-***C***H_2-, from 1,4-butanediol) and 51.8 (-***C***H_3, from itaconate). The spectra are shown in Figure 9b.

Figure 9. **(a)** ^{1}H-; **(b)** ^{13}C-NMR spectra of PBS and PBSI.

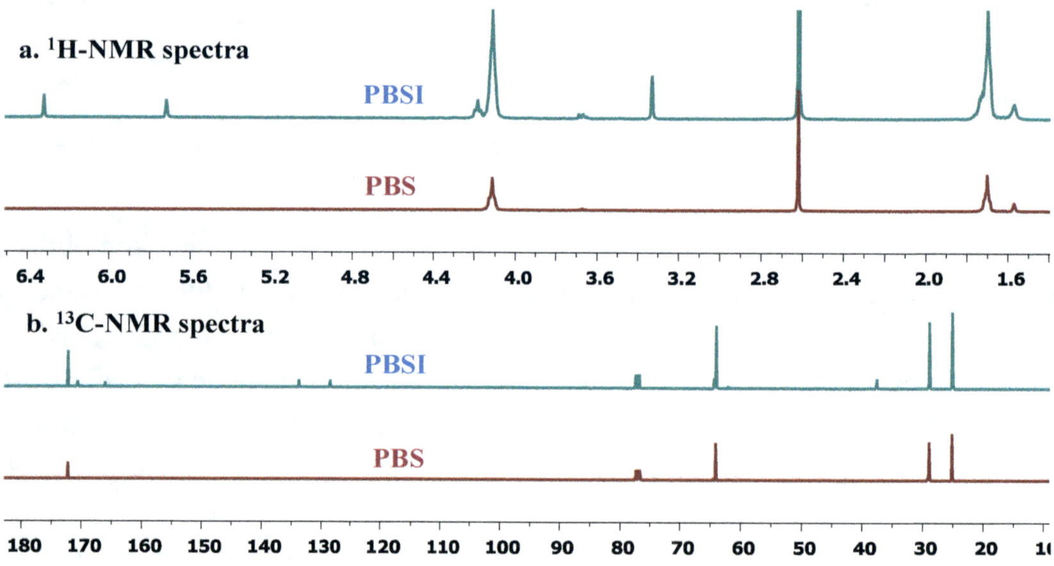

3.3. Control Reactions for Enzymatic Polymerization

Diethyl succinate, dimethyl itaconate and 1,4-butanediol were co-polymerized without CALB using the same method as we established. No polymer was obtained after precipitation.

3.4. In situ *^{1}H-NMR Investigation of the Oligomerization Process of Diethyl Succinate, Dimethyl Itaconate and 1,4-Butanediol in Diphenyl Ether*

CALB (10 wt% in relation to the total amount of monomers) was pre-dried according to the general procedure described in Section 3.2. Diphenyl ether (150 wt% in relation to the total amount of monomers) was used in this study. Diethyl succinate, dimethyl itaconate and 1,4-butanediol were enzymatically oligomerized at 80 °C under atmospheric pressure with the protection of nitrogen for 24 h.

At pre-selected time intervals, about 20 mg of the solution was withdrawn from the reaction. They were added directly into a NMR tube containing 1 g of $CDCl_3$ for ^{1}H-NMR analysis.

3.5. Instrumental Methods

^{1}H- and ^{13}C-NMR spectra of PBS and PBSI were characterized on a Varian VXR spectrometer (400 MHz for ^{1}H-NMR analysis and 100 MHz for ^{13}C-NMR analysis), using $CDCl_3$ as solvent. The chemical shifts reported were referenced to the resonances of tetramethylsilane (TMS) or the solvent.

The molecular weight (M_n and M_w) and the polydispersity index (PDI) were measured by gel permeation chromatography (GPC) using a Viscotek GPC equipped with three detectors (LS detector: Viscotek Ralls detector; VS detector: Viscotek Viscometer Model H502; RI detector: Shodex RI-71 Refractive Index detector), using a guard column (PLgel 5 μm Guard, 50 mm) and two columns (PLgel 5 μm MIXED-C, 300 mm, from Agilent Technologies) at 30 °C. Chloroform of HPLC grade was used as the eluent at a flow rate of 1.0 mL/min. The molecular weight calculations were performed based on the universal calibration. Narrow polydispersity polystyrene standards (Agilent and Polymer Laboratories), with a weight-average molecular weight from 645 to 3,001,000 g/mol, were used to generate the universal calibration curves.

The glass transition temperature and the melting temperature of the co-polyesters were measured by differential scanning calorimetry (DSC) using a TA-Instruments Q1000 DSC. A quench method was conducted for the T_g measurement. The heating rate was 20 °C/min.

Thermal gravimetric analysis (TGA) was performed on a Perkin Elmer Thermo Gravimetric Analyzer TGA7. Samples were measured at a scan rate of 10 °C/min under nitrogen environment.

3.6. Calculation of the Molar Composition of PBSI from ^{1}H-NMR Spectra

Figure 10. Calculation of the molar composition of PBSI from $_1$H-NMR spectra.

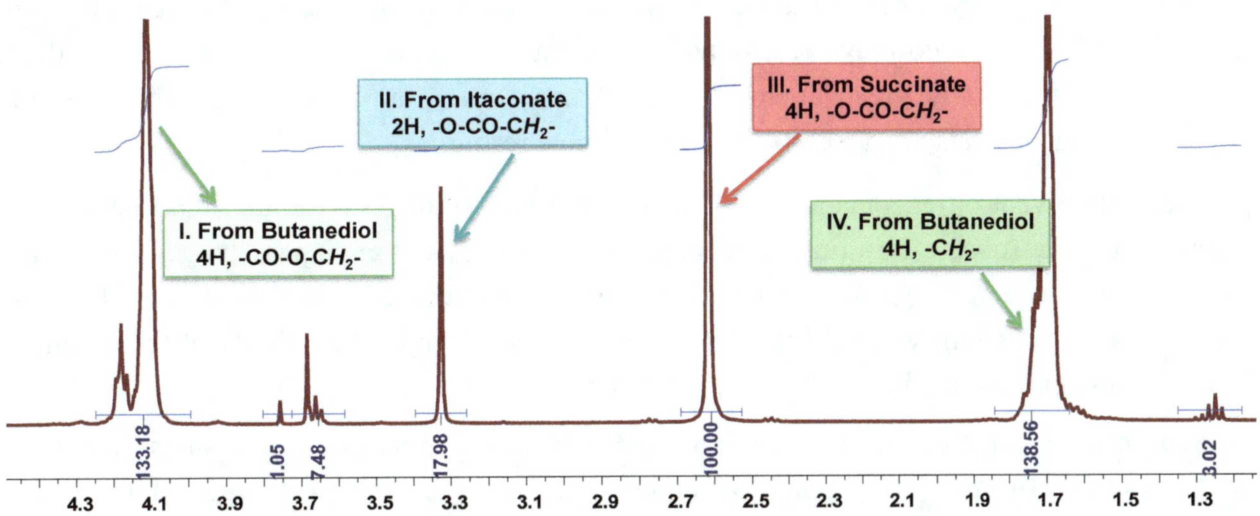

$$X_S(\%) = \frac{A_{III}/4}{A_I/4 + A_{II}/2 + A_{III}/4} * 100 \quad X_I(\%) = \frac{A_{II}/2}{A_I/4 + A_{II}/2 + A_{III}/4} * 100 \quad X_B(\%) = \frac{A_I/4}{A_I/4 + A_{II}/2 + A_{III}/4} * 100 \qquad (1)$$

The molar composition of PBSI was calculated from ^{1}H-NMR spectra, as presented in Figure 10 and Equation (1). A_I, A_{II} and A_{III} are the integration values of Peak I, Peak II and Peak III, respectively, as displayed in Figure 10. X_S, X_I and X_B represent the mole percentage of succinate, itaconate and butylene unit in PBSI.

3.7. Calculation of the Number Average Molecular Weight of PBSI from ^{1}H-NMR Spectra

The number average molecular weight (M_n) was characterized from ^{1}H-NMR spectra, as shown in Figure 11 and Equation (2). I_1, I_4, and I_5 are the integration values of the peaks assigned to the backbones of PBSI, which are originated from 1,4-butanediol, itaconate and succinate, respectively, as shown in Figure 11. I_2, I_3 and I_6 are the integration values of the peaks assigned to the end groups.

Figure 11. Calculation of the M_n from ^{1}H-NMR spectra.

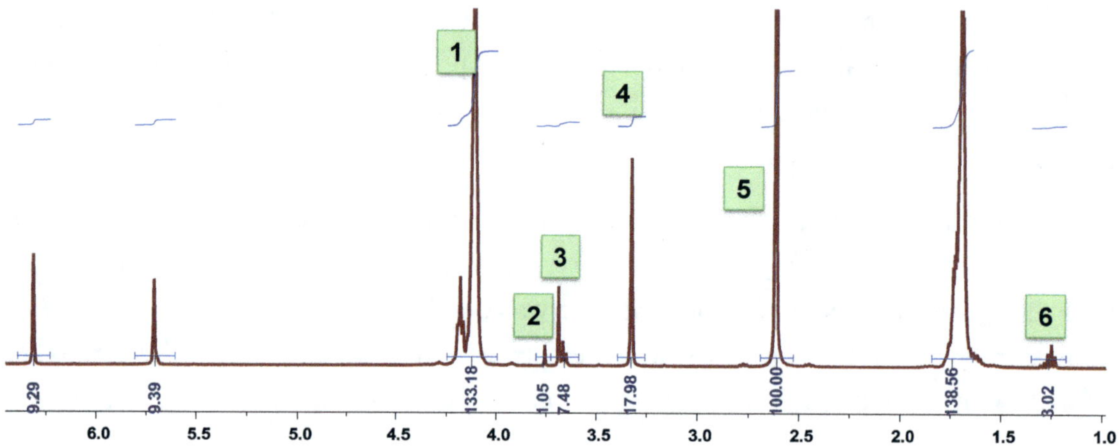

$$\overline{M_n} = \frac{I_1/4 * 88.12 + I_4/2 * 96.10 + I_5/4 * 84.09 + I_2/3 * 31.03 + I_3/2 * 83.12 + I_6/3 * 45.03}{0.5 * (I_2/3 + I_3/2 + I_6/3)} \quad (2)$$

4. Conclusions

Fully bio-based poly(butylene succinate) and poly(butylene succinate-*co*-itaconate) were synthesized by CALB-catalyzed enzymatic co-polymerizations of succinate, itaconate and 1,4-butanediol *via* the two-stage method. It is a totally green approach toward unsaturated aliphatic polyesters, since all the building blocks and catalysts are generated from renewable resources.

The effect of solvent on enzymatic co-polymerization was studied. We found that diphenyl ether was the most suitable solvent, resulting in the highest product M_n and mole percentage of itaconate, as well as satisfying reaction yields. We believe that such an effect could be attributed to the log *P* value of the solvent, the accessibility of CALB in the reaction medium and the solubility of the intermediate and final products in the solvent.

The study on the effect of diphenyl ether dosage indicated that the preferred solvent amount was 150 wt% (in relation to the total amount of monomers). The M_n, mole percentage of itaconate and reaction yield of PBSI increased with the increase of diphenyl ether dosage first, and then, decreased. The diffusion of the reactants, the residual byproducts amount in the reaction and the polymerization rate are the three factors contributing to such an effect.

The selected oligomerization time during the first stage had no obvious effects on CALB-catalyzed synthesis of PBSI. The *in situ* NMR investigations indicated that the oligomerization process was completed after 30 min.

We found that the vacuum during the second stage had a significant influence on enzymatic polycondensation. The best results were achieved under the highest vacuum, since the equilibrium of polycondensation was further shifted to the final products by removing the residual water and alcohols.

The enzymatic polymerizability sequence of itaconate derivatives in diphenyl ether was identified, which was dimethyl itaconate, diethyl itaconate, dibutyl itaconate and itaconic acid, from high to low. The boiling temperature of the reaction byproducts and the accessibility of itaconate to the biocatalyst are two major reasons attributed to this sequence.

With the method here, PBS and a series of PBSI were successfully synthesized with good reaction yields. The amount of itaconate in PBSI was tunable, by adjusting the feed ratio of dimethyl itaconate. The carbon-carbon double bonds of itaconate were well preserved in PBSI. The molecular weight of PBS was lower than that of the unsaturated counterparts, but the PDI was a bit higher.

The values of the product M_n, mole percentage of itaconate and reaction yield decreased significantly with the increasing of the feed ratio of dimethyl itaconate to 30%. If it was further increased to 35%, or even higher, no product was obtained.

We found that the amount of itaconate in PBSI has no obvious effects on the glass-transition temperature and the thermal stability of the co-polyesters, but has significant effects on the melting temperature. The glass transition temperature of PBS and PBSI was similar, around −35–−38 °C. The temperature of the maximal rate of decomposition remained similar, around 400 °C. However, the melting temperature of the obtained co-polyesters decreased almost linearly as the mole percentage of itaconate in PBSI increased.

Acknowledgments

This work forms part of the research program of the Dutch Polymer Institute (DPI), #727c Polymers Go Even Greener. The authors thank Jelena Ćirić and Joop Vorenkamp for the GPC characterizations.

References

1. Gandini, A. Monomers and macromonomers from renewable resources. In *Biocatalysis in Polymer Chemistry*; Loos, K., Ed.; Wiley-VCH: Weinheim, Germany, 2010; pp. 1–34.
2. Dove, A. Polymer science tries to make it easy to be green. *Science* **2012**, *335*, 1382–1384.
3. Mathers, R.T. How well can renewable resources mimic commodity monomers and polymers? *J. Polym. Sci. Pol. Chem.* **2012**, *50*, 1–15.
4. Mülhaupt, R. Green polymer chemistry and bio-based plastics: Dreams and reality. *Macromol. Chem. Phys.* **2013**, *214*, 159–174.

5. Robert, C.; de Montigny, F.; Thomas, C.M. Tandem synthesis of alternating polyesters from renewable resources. *Nat. Commun.* **2011**, *2*, e586.
6. Werpy, T.; Petersen, G.R. *Top Value Added Chemicals from Biomass Volume I—Results of Screening for Potential Candidates from Sugars and Synthesis Gas*; US Department of Energy: Oak Ridge, TN, USA, 2004; pp. 1–67.
7. Bozell, J.J.; Petersen, G.R. Technology development for the production of biobased products from biorefinery carbohydrates—The US Department of Energy's "top 10" revisited. *Green Chem.* **2010**, *12*, 539–554.
8. Teramoto, N.; Ozeki, M.; Fujiwara, I.; Shibata, M. Crosslinking and biodegradation of poly(butylene succinate) prepolymers containing itaconic or maleic acid units in the main chain. *J. Appl. Polym. Sci.* **2005**, *95*, 1473–1480.
9. Barrett, D.G.; Merkel, T.J.; Luft, J.C.; Yousaf, M.N. One-step syntheses of photocurable polyesters based on a renewable resource. *Macromolecules* **2010**, *43*, 9660–9667.
10. Jasinska, L.; Koning, C.E. Unsaturated, biobased polyesters and their cross-linking *via* radical copolymerization. *J. Polym. Sci. Pol. Chem.* **2010**, *48*, 2885–2895.
11. Zhang, Y.-R.; Spinella, S.; Xie, W.; Cai, J.; Yang, Y.; Wang, Y.-Z.; Gross, R.A. Polymeric triglyceride analogs prepared by enzyme-catalyzed condensation polymerization. *Eur. Polym. J.* **2013**, *49*, 793–803.
12. Gross, R.A.; Ganesh, M.; Lu, W. Enzyme-catalysis breathes new life into polyester condensation polymerizations. *Trends Biotechnol.* **2010**, *28*, 435–443.
13. Mahapatro, A.; Kalra, B.; Kumar, A.; Gross, R.A. Lipase-catalyzed polycondensations: Effect of substrates and solvent on chain formation, dispersity, and end-group structure. *Biomacromolecules* **2003**, *4*, 544–551.
14. Albertsson, A.C.; Srivastava, R.K. Recent developments in enzyme-catalyzed ring-opening polymerization. *Adv. Drug Deliv. Rev.* **2008**, *60*, 1077–1093.
15. Kobayashi, S. Recent developments in lipase-catalyzed synthesis of polyesters. *Macromol. Rapid Commun.* **2009**, *30*, 237–266.
16. Uyama, H.; Kobayashi, S. Enzymatic synthesis of polyesters *via* polycondensation. In *Enzyme-Catalyzed Synthesis of Polymers*; Kobayashi, S., Ritter, H., Kaplan, D., Eds.; Springer-Verlag Berlin: Berlin, Germany, 2006; Volume 194, pp. 133–158.
17. Miletic, N.; Loos, K.; Gross, R.A. Enzymatic polymerization of polyester. In *Biocatalysis in Polymer Chemistry*; Loos, K., Ed.; Wiley-VCH: Weinheim, Germany, 2010; pp. 83–130.
18. Gross, R.A.; Kumar, A.; Kalra, B. Polymer synthesis by *in vitro* enzyme catalysis. *Chem. Rev.* **2001**, *101*, 2097–2124.
19. Miletic, N.; Abetz, V.; Ebert, K.; Loos, K. Immobilization of *Candida antarctica* lipase B on polystyrene nanoparticles. *Macromol. Rapid Commun.* **2010**, *31*, 71–74.
20. Matsumura, S. Enzymatic synthesis of polyesters *via* ring-opening polymerization. In *Enzyme-Catalyzed Synthesis of Polymers*; Kobayashi, S., Ritter, H., Kaplan, D., Eds.; Springer-Verlag Berlin: Berlin, Germany, 2006; Volume 194, pp. 95–132.
21. Stavila, E.; Alberda van Ekenstein, G.O.R.; Loos, K. Enzyme-catalyzed synthesis of aliphatic-aromatic oligoamides. *Biomacromolecules* **2013**, *14*, 1600–1606.

22. Miletic, N.; Fahriansyah; Nguyen, L.T.T.; Loos, K. Formation, topography and reactivity of *Candida antarctica* lipase B immobilized on silicon surface. *Biocatal. Biotransform.* **2010**, *28*, 357–369.
23. Miletic, N.; Nastasovic, A.; Loos, K. Immobilization of biocatalysts for enzymatic polymerizations: Possibilities, advantages, applications. *Bioresour. Technol.* **2012**, *115*, 126–135.
24. Stavila, E.; Arsyi, R.Z.; Petrovic, D.M.; Loos, K. Fusarium solani pisi cutinase-catalyzed synthesis of polyamides. *Eur. Polym. J.* **2013**, *49*, 834–842.
25. Stavila, E.; Loos, K. Synthesis of lactams using enzyme-catalyzed aminolysis. *Tetrahedron Lett.* **2013**, *54*, 370–372.
26. Binns, F.; Harffey, P.; Roberts, S.M.; Taylor, A. Studies of lipase-catalyzed polyesterification of an unactivated diacid/diol system. *J. Polym. Sci. Pol. Chem.* **1998**, *36*, 2069–2079.
27. Binns, F.; Harffey, P.; Roberts, S.M.; Taylor, A. Studies leading to the large scale synthesis of polyesters using enzymes. *J. Chem. Soc. Perkin Trans. 1* **1999**, 2671–2676.
28. Azim, H.; Dekhterman, A.; Jiang, Z.; Gross, R.A. *Candida antarctica* lipase B-catalyzed synthesis of poly(butylene succinate): Shorter chain building blocks also work. *Biomacromolecules* **2006**, *7*, 3093–3097.
29. Habeych, D.I.; Juhl, P.B.; Pleiss, J.; Vanegas, D.; Eggink, G.; Boeriu, C.G. Biocatalytic synthesis of polyesters from sugar-based building blocks using immobilized *Candida antarctica* lipase B. *J. Mol. Catal. B-Enzym.* **2011**, *71*, 1–9.
30. Juais, D.; Naves, A.F.; Li, C.; Gross, R.A.; Catalani, L.H. Isosorbide polyesters from enzymatic catalysis. *Macromolecules* **2010**, *43*, 10315–10319.
31. Jiang, Z. Lipase-catalyzed synthesis of aliphatic polyesters *via* copolymerization of lactone, dialkyl diester, and diol. *Biomacromolecules* **2008**, *9*, 3246–3251.
32. Ebata, H.; Toshima, K.; Matsumura, S. Lipase-catalyzed synthesis and properties of poly[(12-hydroxydodecanoate)-co-(12-hydroxystearate)] directed towards novel green and sustainable elastomers. *Macromol. Biosci.* **2008**, *8*, 38–45.
33. Kobayashi, T.; Matsumura, S. Enzymatic synthesis and properties of novel biodegradable and biobased thermoplastic elastomers. *Polym. Degrad. Stab.* **2011**, *96*, 2071–2079.
34. Tsujimoto, T.; Uyama, H.; Kobayashi, S. Enzymatic synthesis of cross-linkable polyesters from renewable resources. *Biomacromolecules* **2001**, *2*, 29–31.
35. Tsujimoto, T.; Uyama, H.; Kobayashi, S. Enzymatic synthesis and curing of biodegradable crosslinkable polyesters. *Macromol. Biosci.* **2002**, *2*, 329–335.
36. Uyama, H.; Kuwabara, M.; Tsujimoto, T.; Kobayashi, S. Enzymatic synthesis and curing of biodegradable epoxide-containing polyesters from renewable resources. *Biomacromolecules* **2003**, *4*, 211–215.
37. Jiang, Z.; Azim, H.; Gross, R.A.; Focarete, M.L.; Scandola, M. Lipase-catalyzed copolymerization of ω-pentadecalactone with p-dioxanone and characterization of copolymer thermal and crystalline properties. *Biomacromolecules* **2007**, *8*, 2262–2269.
38. Mazzocchetti, L.; Scandola, M.; Jiang, Z. Enzymatic synthesis and structural and thermal properties of poly(ω-pentadecalactone-co-butylene-co-succinate). *Macromolecules* **2009**, *42*, 7811–7819.

39. Liu, W.; Wang, F.; Tan, T.; Chen, B. Lipase-catalyzed synthesis and characterization of polymers by cyclodextrin as support architecture. *Carbohydr. Polym.* **2013**, *92*, 633–640.
40. Jiang, Z. Lipase-catalyzed copolymerization of dialkyl carbonate with 1,4-butanediol and omega-pentadecalactone: Synthesis of poly-(omega-pentadecalactone-co-butylene-co-carbonate). *Biomacromolecules* **2011**, *12*, 1912–1919.
41. MuralidharaRao, D.; Hussain, S.M.D.J.; Rangadu, V.P.; Subramanyam, K.; Krishna, G.S.; Swamy, A.V.N. Fermentatative production of itaconic acid by aspergillus terreus using *Jatropha* seed cake. *Afr. J. Biotechnol.* **2007**, *6*, 2140–2142.
42. Guo, B.; Chen, Y.; Lei, Y.; Zhang, L.; Zhou, W.Y.; Rabie, A.B.M.; Zhao, J. Biobased poly(propylene sebacate) as shape memory polymer with tunable switching temperature for potential biomedical applications. *Biomacromolecules* **2011**, *12*, 1312–1321.
43. Wei, T.; Lei, L.; Kang, H.; Qiao, B.; Wang, Z.; Zhang, L.; Coates, P.; Hua, K.-C.; Kulig, J. Tough bio-based elastomer nanocomposites with high performance for engineering applications. *Adv. Eng. Mater.* **2012**, *14*, 112–118.
44. Retuert, J.; Yazdanipedram, M.; Martinez, F.; Jeria, M. Soluble itaconic acid ethylene-glycol polyesters. *Bull. Chem. Soc. Jpn.* **1993**, *66*, 1707–1708.
45. Ramo, V.; Anghelescu-Hakala, A.; Nurmi, L.; Mehtio, T.; Salomaki, E.; Harkonen, M.; Harlin, A. Preparation of aqueous crosslinked dispersions of functionalized poly(d,l-lactic acid) with a thermomechanical method. *Eur. Polym. J.* **2012**, *48*, 1495–1503.
46. Sakuma, T.; Kumagai, A.; Teramoto, N.; Shibata, M. Thermal and dynamic mechanical properties of organic-inorganic hybrid composites of itaconate-containing poly(butylene succinate) and methacrylate-substituted polysilsesquioxane. *J. Appl. Polym. Sci.* **2008**, *107*, 2159–2164.
47. Brugel, W.; Demmler, K. Synthesis and properties of itaconic acid-containing polyester resins. *J. Polym. Sci. Pol. Sym.* **1969**, *22*, 1117–1137.
48. Sepulchre, M.O.; Sepulchre, M. Synthesis of unsaturated polyesters from dipotassium salts of cis-aconitic, itaconic and mesaconic acids and 1,4-dibromobutane. *Macromol. Symp.* **1997**, *122*, 291–296.
49. Rajkhowa, R.; Varma, I.K.; Albertsson, A.C.; Edlund, W. Enzyme-catalyzed copolymerization of oxiranes with dicarboxylic acid anhydrides. *J. Appl. Polym. Sci.* **2005**, *97*, 697–704.
50. Linko, Y.Y.; Wang, Z.L.; Seppala, J. Lipase-catalyzed linear aliphatic polyester synthesis in organic-solvent. *Enzyme Microb. Technol.* **1995**, *17*, 506–511.

8

Biocatalysis for Biobased Chemicals

Rubén de Regil and Georgina Sandoval *

Unidad de Biotecnología Industrial, CIATEJ, A.C. Av. Normalistas 800, Col. Colinas de la Normal, Guadalajara, Jal, C.P. 44270, Mexico; E-Mail: ruben_de_regil@yahoo.com

* Author to whom correspondence should be addressed; E-Mails: gsandoval@ciatej.mx; georgina@confluencia.net.

Abstract: The design and development of greener processes that are safe and friendly is an irreversible trend that is driven by sustainable and economic issues. The use of Biocatalysis as part of a manufacturing process fits well in this trend as enzymes are themselves biodegradable, require mild conditions to work and are highly specific and well suited to carry out complex reactions in a simple way. The growth of computational capabilities in the last decades has allowed Biocatalysis to develop sophisticated tools to understand better enzymatic phenomena and to have the power to control not only process conditions but also the enzyme's own nature. Nowadays, Biocatalysis is behind some important products in the pharmaceutical, cosmetic, food and bulk chemicals industry. In this review we want to present some of the most representative examples of industrial chemicals produced *in vitro* through enzymatic catalysis.

Keywords: biocatalysis; hydrolases; transferases; polymers; biodiesel; fine chemicals; prebiotics

1. Introduction

It is becoming evident that in order to sustain the standard way of life of the developed and in-development world, it will be necessary to make some adjustments either to our consumption habits or to our sources of supplies of energy and materials. In the latter case, Biotechnology, as a diversified discipline in which chemistry, physics, biology, optics, electro-magnetism and thermodynamics converge, possess the knowledge tools to play a relevant and vital role to address the challenges of growth, ageing, employment, limited sources of raw materials, energy and water supply and living standard. Industrial biobased processes will increasingly become the *de facto* alternative to build a sustainable economy. It is estimated that, by 2030, the products of white biotechnology and bioenergy will account for 30% of industrial production worth €300 bn [1].

Conversion of biomass by Biocatalysis is likely to become standard technology which will contribute significantly to open up access to large feedstock supplies for bioprocesses and the production of transport fuels. On the health and ageing field, personalized nutrition, tailor-made medicine will become a reality in the coming years thanks to novel, specific biotech drugs and regenerative medicine obtained by more efficient and greener processes.

Biocatalysis refers to the transformation of substances of chemical or biological origin through the use of the enzymes produced by diverse living organisms. Enzymes may carry out reactions in a free, or immobilized form, or within the living cell in which they naturally live.

Isolated enzymes obtained as purified and concentrated extracts, whether used immobilized onto a support or in free form, may produce neater products than when the whole microorganism cell is used for the biotransformation. This is because of the absence of other kinds of enzymes from the internal cell machinery, which may subsequently modify the product, and the absence of internal cellular components during the lysis-extraction step. This advantage of isolated enzymes has a positive impact on yield and purification costs. However, isolated enzymes are known to be less stable in a purified form than within the cell. The lower stability of free enzymes has been improved to some extent by immobilization and crosslinking techniques, the results of which vary depending on the type of enzyme, pH memory, extent of crosslinking, immobilization support and the procedure itself.

Whole microorganism cells, obtained in sufficient amount by fermentation, are used instead of isolated enzymes when their enzymes become highly unstable or non-functional outside the inner environment of the cell, so catalysis is confined within the cell and products are later extracted from it.

Chemical synthesis in the food, feed, industrial and pharma sectors is the field to which Biocatalysis has contributed the most, but it has also contributed to the bioremediation sector, another field in which enzymes are having a significant role.

The relative success of Biocatalysis applied to Synthetic Chemistry has been due mainly to the high enantio- and regioselectivities that enzymes exhibit towards their substrates, because enzymes speed up chemical processes that would otherwise run very slowly or without selectivity, and because enzymes function under mild reaction conditions. These advantages have allowed chemists to avoid the burden of group-protecting procedures, saving time, materials (including the harsh, dangerous or toxic ones) and energy costs. Other advantages of enzymes are that they are easy to control and biodegradable. Thus, Biocatalysis has proved in many cases to be a superior pathway than conventional chemical synthesis pathways, not only in the simplicity of accomplishing the reactions, but also from an economical and environmental point of view.

An important landmark in Biocatalysis was the use of organic solvents as reaction media, as it was thought that enzymes may only work in aqueous medium, naturally. Organic solvents became an important aid in dissolving organic, hydrophobic molecules, which account for much of the library of organic chemicals used in Synthetic Chemistry. The possibility to carry out reactions in organic media favored the reactions rate precisely because of the better solubility of reactants. Purification steps became shorter and easier because the avoidance of surfactants needed in aqueous media and the use of low boiling point solvents easy to evaporate. Moreover, hydrolytic, secondary, water-dependent reactions responsible for lower product yields, and growth of microbes in containers and pipes, is avoided because of the absence of water and the harsh effects of solvents to living cells.

However, Biocatalysis still needs to overcome some limitations such as the enzyme costs of production and the relative narrow stability window of many enzymes under diverse reaction conditions, which restricts their applications onto broader industrial processes. Chemicals used in industrial settings are mostly derived from fossil oil, whereas active ingredients in pharmaceutical processes may be new and/or structurally complex. In general, a vast number of chemicals used in industry are artificial, new man-made chemicals for which enzymes did not evolve to work with and thus are not suitable for the synthetic needs in industry. So there are high discrepancies between the enzyme´s function in nature and the functions needed in the transformation industry. High throughput enzyme screening and protein engineering techniques, particularly Directed Evolution, are some of the tools being employed to advance faster in this aim to find enzymes for artificial substrates, and to create enzymes with new capabilities.

In this review we want to present some of the most representative examples of industrial chemicals produced *in vitro* through enzymatic catalysis, and hence, *in vivo* fermentation products are not included.

2. Enzymes in Food Industry

Applications of enzymes in the food industry are found in almost every sector of this industry: confectionary and sweeteners, fruit and vegetables, dairy, brewery and beverages, meat, dietary and the nutraceuticals niche [2]. However, most of these applications are of hydrolytic nature focused in debranching, improving solubility, clarification, and hydrolysis aimed at diverse goals depending on the food being treated.

2.1. Prebiotics and Sweeteners

There are some examples of food chemicals being synthesized by enzymes. One of them is Disaccharide Difructose Anhydride III (DFA III), which is a non-reducing, non-cariogenic sweetener with probiotic properties [3]. DFA III can be synthesized by enzyme inulase II (EC 2.4.1.93), (Figure 1), which was first identified in *Arthrobacter ureafaciens* [4] and later in other bacteria [5–7]. However, native enzymes are not thermotolerant enough to resist high temperatures required for a large scale process. Thus, the *ift*-gene encoding for the inulase II from Buo141, a thermotolerant strain, was cloned and expressed into *E. coli* XL1-blue and further modified and immobilized to obtain the biocatalyst which was reported to have 1.7×10^6 U/L [3].

Prebiotics is another field in which enzymes are being used. Oligosaccharides are non-digestible saccharide polymers containing 3–10 monomeric sugar units that are found in low amounts in human milk [8,9], onion, garlic, banana, soybean and chicory [10]. They are classified as prebiotics because they selectively promote the growth of bifidobacteria and lactobacilli in the intestine, which are regarded to have a beneficial effect to human health [10–14].

Figure 1. Enzymatic synthesis of Difructose Anhydride III (DFA III) from Inulin using enzyme inulase II, an inulin fructo-transferase.

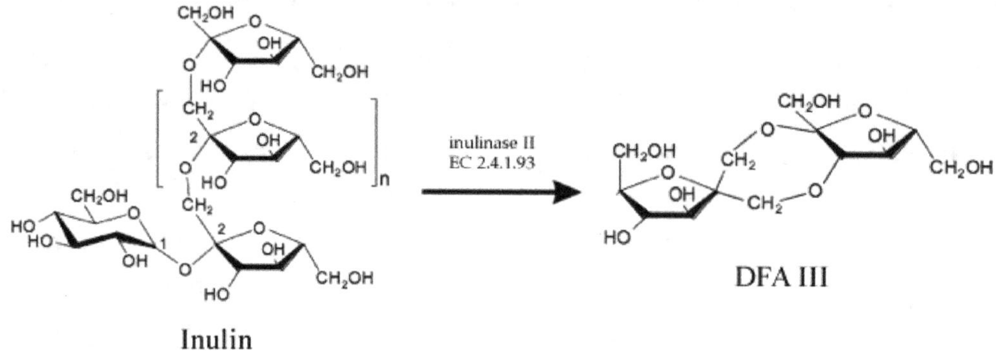

Oligosaccharides composed either of fructose or galactose units are named fructo-oligosaccharides (FOS) and galacto-oligosaccharides (GOS) respectively.

Galacto-oligosaccharides (GOS) can be synthesized from lactose when this sugar acts as the acceptor and transgalactosylation is catalyzed by β-galactosidase (Figure 2). In addition to the prebiotic activity, GOS have also been reported to contribute to (i) reduction of serum cholesterol and lipid level; (ii) synthesis of B-complex vitamins; (iii) enhance absorption of dietary calcium [14–16]; (iv) protection from infection and decrease of pathogenic bacteria; (v) stimulate absorption of some minerals [17]. However, these health promoting properties vary depending on the chemical composition, structure and degree of polymerization and these features depend, in turn, on the origin of the β-galactosidase [18].

Figure 2. Simplified enzymatic reaction to produce Galacto-oligosaccharides (GOS) from lactose using β-galactosidase. Subscript *n* may range from 0–6 though most native enzymes produce GOS with *n* between 1–2.

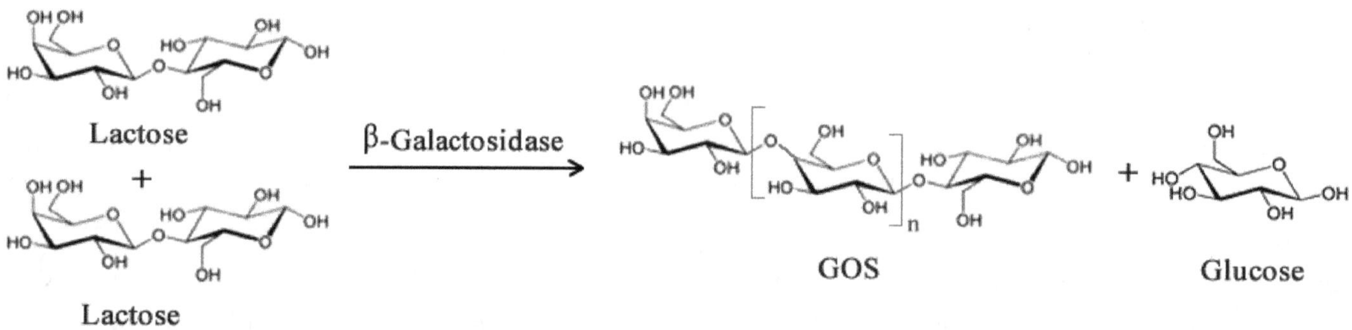

The most recurring means of the synthesis of GOS is by enzymatic catalysis from lactose using glycosyltransferases (EC 2.4) or glycoside hydrolases (EC 3.2.1) [19]. Glycosyltransferases are not used in the synthesis of GOS due to their low availability, the need for nucleotide sugar substrates and prohibitive prices. Instead, glycoside hydrolases are used, despite the fact that they are less stereoselective than the former [20]. Glycoside hydrolases from archeas (*Sulfolobus solfactaricum* and *Pyrococcus furiosus*), bacteria (*Saccharopolyspora* sp., *Bifidobacterium* sp., *Thermotoga* sp., *Thermus* sp., *Bacillus* sp., *Geobacillus* sp., *Caldicellulosiruptor sp.*, *Lactobacillus* sp., *Streptococcus* sp., *Enterobacter* sp., *Escherichia* sp.) and yeasts (*Aspergillus* sp., *Penicillinum* sp., *Talaromyces* sp.,

Trichoderma sp., *Kluyveromyces* sp., *Sirobasidium* sp., *Sterigmatomyces* sp., *Rhodotorula* sp., *Sporobolomyces* sp., *Rhizopus* sp.) have been tested for the synthesis of GOS [19,21].

Milk whey is a lactose-rich source which has been regarded in the past as a waste by-product from cheese production highly contaminant. So β-galactosidases have become a valued natural tool to produce a beneficial product out from a contaminant waste in a single step.

As a general finding, reports show that GOS yields average 30%–35% of total sugars [22,23] and are directly proportional to the initial lactose concentration. The major product is the trisaccharide accounting for as much as 80% of total GOS synthesized [12,24]. It was estimated that at least 3,500 tons of GOS are enzymatically synthesized from whey lactose, out of 30,000 tons of total world annual production [25]. They are commercialized to enrich mainly infant formulas.

Just like GOS, Fructo-oligosaccharides (FOS) have low caloric value and have much the same prebiotic properties due to their indigestibility in the upper gastrointestinal tract: they are non-cariogenic and have ability to promote the activity of beneficial colonic lactic acid bacteria and to modulate the intestinal immune response [26–28]. However, unlike GOS, FOS are naturally occurring sugars biosynthesized by numerous plants such as asparagus, sugar beet, onion, artichoke, *etc.* [29,30]. Two main types of FOS can be found in nature, the inulin type, which are fructose polymers with linkages β(2–1) biosynthesized by inulosucrases (E.C. 2.4.1.9), and the levan type, which are fructose polymers with linkages β(2–6) biosynthesized by levansucrases (E.C. 2.4.1.10) [31–33].

Two different classes of FOS mixtures are produced commercially, one is based on inulin by controlled enzymatic hydrolysis, and the other is based on sucrose by a transfructosylation processes with fructosyltransferases, as shown in Figure 3 (β-fructofuranosidase, E.C. 3.2.1.26 or β-d-fructosyltransferase, EC 2.4.1.9) [34,35]. FOS from inulin hydrolysis contain longer fructose chains (Degree of Polymerization, DP 2–9) and may have either a fructose or glucose unit at the end of the chain. On the other hand, FOS from sucrose transfructosylation contain shorter polymer chains (DP 2–4) and always end with a glucose unit [35].

Figure 3. Simplified enzymatic reactions used commercially to obtain Fructo-oligosaccharides (FOS).

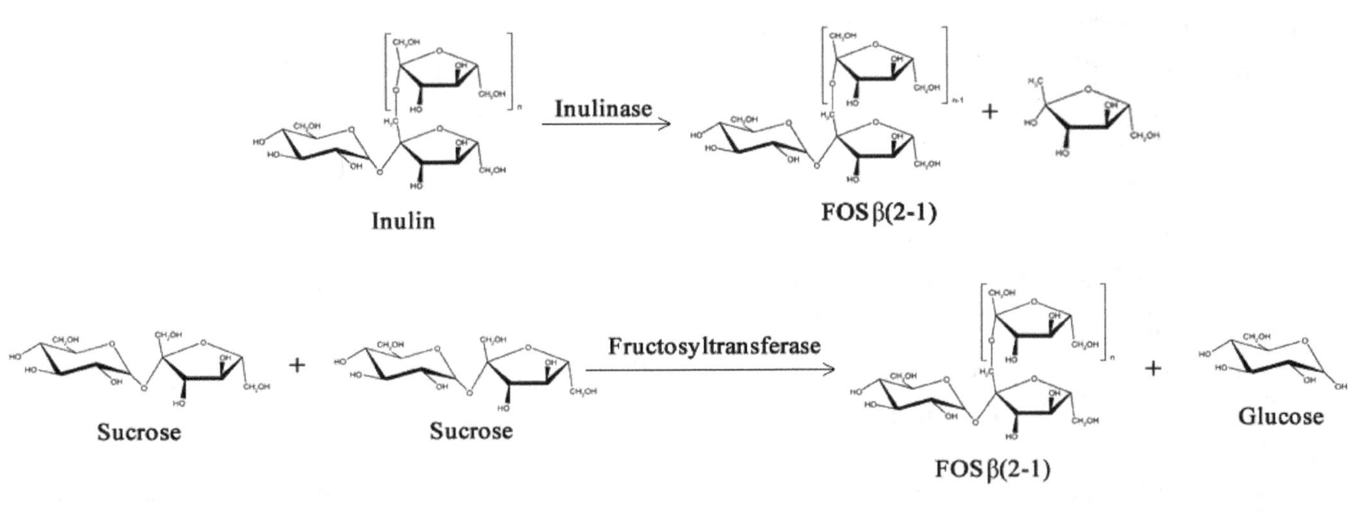

A commercial enzyme Pectinex Ultra SP-L was used immobilized onto Eupergit C to synthesize FOS from a 600 g/L of sucrose solution. Reactions were carried out at 65 °C in pH 6.5 during 24 h with a yield of 57% FOS of total sugars [36].

Despite the big production numbers that prebiotics have, the overall European market for prebiotics is still at the beginning stage, with the $87-million fructan (inulin and fructo-oligosaccharide) segment as the most developed part. In US the prebiotics market is estimated to reach US$198.3 million in 2014 [37].

2.2. Structured Lipids

Specialty fats and oils are lipids with special functional or nutritional effects in the human body. Among these, structured lipids (SLs) have a predominant importance within this field [38,39]. SLs are tailor-made fats and oils which incorporate specific new fatty acids to triacylglycerols, or have different composition and positional distribution of existing fatty acids, within its glycerol backbone. Among the various types of SLs, the MLM (medium, long, medium fatty acid chain length esterified at the *sn*-1, *sn*-2 and *sn*-3 positions of the glycerol backbone respectively) have been a subject of great interest because of their ability to provide quick energy through the pancreatic lipase hydrolytic release of the medium chain fatty acids which become rapidly oxidized by the liver [40]. The remaining long chain *sn*-2-monoacylglyceride is left to be absorbed through the lymphatic system [41]. This long chain monoacylglyceride can be an omega-6 fatty acid or any other poly unsaturated fatty acid (PUFA) which will provide the already known health benefits to the body [42].

SLs are claimed to have a modulation effect on the immune system, to improve the lipid clearance in the blood [43], and as a fat for special nutritional feeding [44]. Fatty acids at the *sn*-2 position are more readily absorbed *in vivo* than those at the *sn*-1 or 3 positions, thus, SLs of the type MLM that include an essential fatty acid in the *sn*-2 position are desirable nutrition sources for people who suffer from malabsorption or a pancreatic condition [44,45].

SLs can be obtained either chemically or enzymatically using lipases [42]. The high regioselectivity of lipases makes Biocatalysis particularly well suited for the synthesis of SLs, avoiding the synthesis of unwanted by-products, which lower the yields [46]. However, enzymatic synthesis of SLs has yet to be optimized in order to avoid the high costs of large-scale purification of unreacted substrates (Free fatty acids, Triacyglicerides, Diacylgricerides, Esters) which must be removed upon completion of reaction [47].

Two basic approaches have been followed to carry out the enzymatic synthesis of MLM SLs: (1) Transesterification and (2) Acidolysis [48]. Lipase transesterifications for the synthesis of MLM SLs have been performed using different vegetal and fish oils as source of essential long PUFAs, and coconut and palm kernel oil as a source of medium chain fatty acids (C6:0–C12:0) to be incorporated in the *sn*-1 and *sn*-3 positions [42,49]. Thus, to synthesize MLM SLs, *sn*-1 and *sn*-3, regiospecific lipases are favored. In Figure 4 these two approaches are depicted using a *sn*-1,3-specific lipase.

Casas-Godoy *et al.* [50] incorporated caprilyc and capric acid into olive oil using *Yarrowia lipolytica* immobilized on Accurel. Their optimized reaction allowed them to obtain an incorporation rate of 25% and 21% for capylic and capric acid respectively within 2 days at 40 °C and 5% (w/w) enzyme load.

Figure 4. The main two approaches to synthesize MLM Structured Lipids enzymatically: (**A**) Transesterification and (**B**) Acidolysis. Subproducts like LML Structured lipids may also be obtained, but only the product of interest is shown. L = Long Chain; M = Medium Chain; MCFA = Medium Chain Fatty Acid; MCFAEt = Medium Chain Fatty Acid Ester (see also [48]).

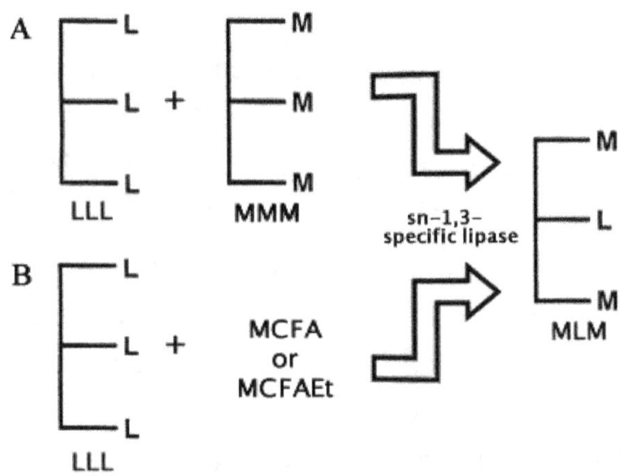

In another work, castor oil was dehydrated, isomerized and then transesterified with methyl laurate using *Thermomyces lanuginosa* and *Carica papaya*. The incorporation and transesterification yields were 59% and 88% for *T. lanuginosa* and 44% and 67% for *C. papaya*, respectively. This was achieved in 72 h at 60 °C with a 10% (w/w) enzyme load. Transesterification in the *sn*-2 position was observed and accounted for less than 8% [51].

Palm oil trolein has been subject to enrichment in its *sn*-2 position with DHA and ARA through acidolysis using Novozym 435. The research Group achieved an incorporation of 11% of DHA at *sn*-2 position (17% total) and 5.5% of ARA at *sn*-2 position (7% total) in 24 h at 60 °C and 10% enzyme load [52]. In a similar work, Khodadadi *et al.* [49] transesterified palm oil with tricaprylin in order to modify the former. Their results showed that Linolenic acid chains in palm oil are more easily transesterified than Linoleic or Oleic. Their best results were around 20% incorporation of caprylic chains into *sn*-1, *sn*-2 and *sn*-3 positions of which around a half corresponded to *sn*-2 position. The experiments were carried out with Novozym 435, Lipozyme TL-IM, Lipozyme RM-IM and Amano DF, though the best results were with Novozym 435, 24 h of reaction time, 50 °C and 1% enzyme load. Tecelão *et al.* [53] recently achieved 21% and 8% of incorporation of oleic acid and DHA respectively, to tripalmitin using *Carica papaya* latex at 60 °C, in a solvent-free system during 24 h.

Nagao *et al.* [54] achieved almost 100% of fatty acid incorporation with a DHA-rich by-product waste from a tuna oil industrial, hydrolysis process employed to extract DHA. Their reaction conditions were 30 °C, 120 h, vacuum at 2 kPa and 10% of Novozym 435 load. The degree of esterification with DHA was 51% and 17% at *sn*-1,3 and *sn*-2 position respectively.

3. Enzymes in the Bulk Chemistry Industry

Although industrial biotransformations are mainly used for the production of fine chemicals, there are a few examples where Biocatalysis is also used to produce commodity chemicals such as acrylamide, and biodiesel. The use of biocatalysts employed to assist in synthetic routes to complex molecules of industrial interest is growing steadily [55–57].

3.1. Biodiesel

The lipase-catalyzed synthesis of biodiesel offers advantages over conventional methods for its production: lipases are able to work in gentle conditions and with a variety of triglyceride substrates, including waste oils and fats, corrosion problems present in chemical synthesis are avoided in the lipase-catalyzed synthesis, easy recovery of biocatalyst and glycerol, high levels of Free Fatty Acids (FFA) may also be esterified by lipases, and low environmental impact [58,59]. Furthermore, the separation and purification step of biodiesel is easier [60,61], resulting in a more attractive technical and environmental process. Numerous lipases have been studied for biodiesel production, with diverse triglyceride substrates and acyl acceptors. The use of edible oils is today not an option due to the food *vs.* fuel issue, which also makes them a more expensive raw material. Research studies so far have converted a series of non-edible fats and oils replacing the traditional alkaline transesterification. Process flow, reactor design and cost performance have been some of the main elements studied. It has been shown that high productivity, enzyme reuse and low reaction times can be achieved [62–64]. Nevertheless, enzymatic biodiesel still faces technical and economic challenges and has room for improvement to make enzymatic biodiesel a more attractive option for industrial production.

The most used method to produce enzymatic biodiesel is by transesterification of oil with an alcohol as acyl acceptor (Figure 5). Acyl acceptors, pose a challenge because the most available and cheap ones for industrial production and those who meet international specifications for biodiesel, are methanol and ethanol, which exert a strong denaturating action towards lipases [65]. Nevertheless, relatively successful efforts to overcome this inconvenience have emerged, like a pioneering work on stepwise alcohol addition [66], acyl acceptor alterations [67,68] and solvent engineering [69,70] which have sorted out this problem with relatively efficacy. Isopropanol [71], Isobutanol [72], 1-butanol [71], 2-butanol [72], 2-ethyl-1-hexanol [73], methyl acetate [74] and ethyl acetate [68] have been reported to be used as acyl acceptors. It was later documented that glycerol, a by-product of biodiesel synthesis, exerted lipase inhibition as well, which is more likely due to mass transfer limitation in the immobilized lipase [75].

Figure 5. Enzymatic reaction of transesterification of oils with alcohol to produce fatty acid esters (biodiesel).

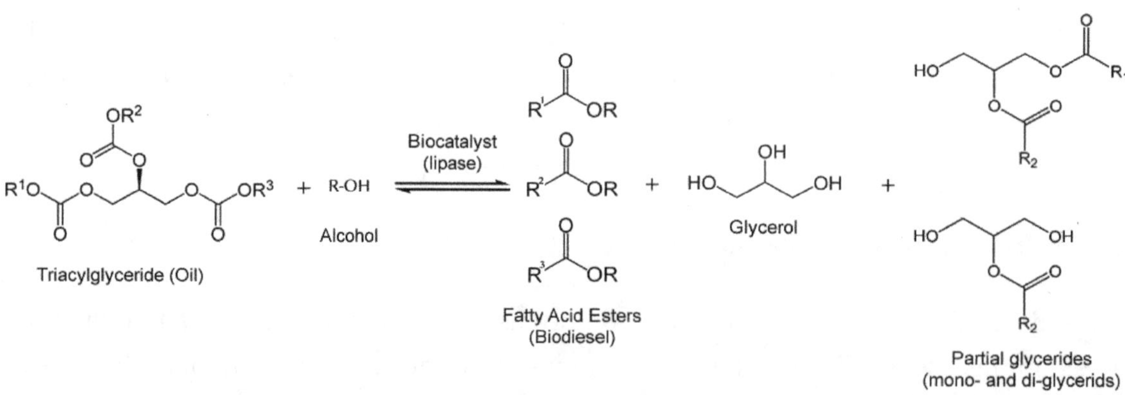

Free and immobilized lipases of fungal and yeast origin such as *Mucor* sp., *Rhizopus* sp., *Candida* sp., *Aspergillus* sp., *Thermomyces* sp., *Penicillinum* sp. *and Pseudomonas* sp. have shown to be good biocatalysts for biodiesel synthesis as yields for most of them are reported to be 80%–95% [58]. Enzyme immobilization is a cost which greatly influences the economic viability when scaling-up industrial biodiesel projects; thus, alternative methods of preparing the enzyme is a current active topic in the field [76].

Organic solvents have been used to solve problems like substrate solubility and mass transfer [63]. Hydrophobic organic solvents have been used because they improve trygliceride and fatty substrates solubility [77], but leave viscous glycerol insoluble, which poses mass transfer limitations as stated above. Despite the denaturing effects of hydrophilic organic solvents over enzymes [78,79], some solvents like 1,4-dioxane and *tert*-butanol have been proven to generate high transesterification yields [69,70,80]. Incubation of lipases in alcohols with carbon number ≥3 has been documented to improve enzymatic activity in the synthesis of methyl esters [81]. Acetone has also been tested; however, the maximum yields of biodiesel have reached 40% [82].

The use of waste animal fat from slaughterhouses as raw material for the production of biodiesel by transesterification with ethanol was recently evaluated by Rivera *et al.* [69]. Enzymes N435 and Lipozyme RM IM at 45 °C at a 2% (w/w) enzyme load, in an organic (*tert*-butanol) and solvent-free systems were used. The biodiesel yields were 80% for the solvent-free system after 2 days of reaction, and 65% for the organic system after 1 day of reaction.

Recently, solvents like supercritical carbon dioxide have been used and biodiesel yields accounted for as much as 81% at 60 °C with a 15% (w/w) enzyme load during 4 h [83]. Ionic liquids have also been used for biodiesel synthesis [84,85] and one of the best results was achieved after 12 h at 50 °C in 1-ethyl-3-methylimidazolium trifluoromethane-sulfonate ([C2mim][OTf]) with an 80 % yield [86].

Intracellular lipases in the form of whole-cell biocatalysts, present a feasible alternative for biodiesel production as well. Their main advantage is the lack of laborious and costly extraction and purification procedures needed when an isolated, free or immobilized, enzyme is used [87]. In order to use whole-cell Biocatalysis, immobilization is advocated. One of the most accepted whole-cell biocatalyst systems for industrial applications is the use of filamentous fungi immobilized onto biomass support particles (BSPs) [88–90]. A*spergillus* sp. and *Rhizopus* sp., have been identified as robust for whole-cell immobilization, and thus have become the most widely studied fungi for biodiesel production [58]. Oils from yeasts and fungi are an alternative source of oils for biodiesel, as they can accumulate long-chain triacylglycerols with 16 and 18 carbons. The oil contents obtained from several yeast strains such as *Cryptococcus*, *Lipomyces*, *and Rhodotorula* species have reached 60%–70% of their dry weight [91].

3.2. Industrial Polymers

Biocatalysis has opened new synthetic strategies for organic chemists because it has been applied to synthetic chemistry in order perform stereochemical complex reactions in a relatively simple, straightforward way, which otherwise would require laborious protection-deprotection steps, toxic reactants and/or high-temperature/pressure procedures [92,93]. Thus, these advantages of using enzymes have been adopted in the field of polymer synthesis, in which most polymers are difficult to produce or to control by conventional chemical methods. Polymers with specific structures can be prepared enzymatically. In contrast, attempts to attain similar levels of polymer structural control by conventional chemical methods may prove to be a true challenge. From an environmental stand point, polymers derived from enzyme-mediated catalysis, whether polyesters, polyphenols, polysaccharides, proteins, or other polymers, are in most cases biodegradable, a feature highly sought after these days.

The most common type of enzymes which are capable to catalyze polymerization reactions are: Transferases, Oxidoreductases and Hydrolases.

3.2.1. Transferases

Transferases transfer a glycosyl group from a sugar nucleotide donor to specific nucleophilic acceptors. Despite they impose a high degree of regio- and stereochemical control to the glycosidic bond they form, and thus having a great synthetic potential for interesting polymeric materials, their use for synthesis *in vitro* is limited and even prohibitive by both, the availability of enzymes due to expression and solubility, and by the high costs of expensive activated donor sugars such as uridine diphosphate. They are very sensitive biocatalysts so their isolation and use on a larger scale is not practical, and they are mostly reserved for specialty research endeavors [94].

3.2.2. Oxidoreductases

On the other hand, Oxidoreductases and particularly Hydrolases are more robust enzymes, less sensitive and easier to obtain and use *in vitro*. Polymers typically produced by these three types of enzymes are shown in Table 1.

Table 1. Common enzymes and the typical polymers synthesized by them.

Enzymes	Polymers synthesized
Oxidoreductases	
Peroxidases	Polyphenols, polyanilines, vinyl polymers
Laccases	
Tyrosinases	
Glucose oxidases	
Transferases	
Glycosyl transferases	Polysaccharides, cyclic oligosaccharides, polyesters
Acyl transferases	
Hydrolases	
Glycosidases	Polysaccharides, polyesters, polycarbonates, polyamides, polythioesters, polyphosphates
Lipases	
Peptidases	
Proteases	

3.2.2.1. Polyaromatics

Polyaromatics are widely found in nature such as lignin, and flavonoid compounds. Much of their role in nature is as structural component conferring structural strength to living systems, in the case of lignin, and as a bioactive molecule in the case of flavonoids. Phenol is the single aromatic most important compound for industrial applications. Current polymeric materials produced commercially from phenolic compounds as industrial plastics include Bakelite and poly-2,6-dimethyl-1,4-phenylene oxide) (PPO) which bear good toughness and temperature-resistant properties. However, the oxidative polymerization of phenol with a conventional catalyst usually gives insoluble products with uncontrolled structure [95]. Moreover, as Bakelite and PPO are phenol-formaldehyde resins, the toxicity of formaldehyde has brought limitations on their industrial production.

On the other hand, an enzyme-catalyzed polymerization, offers not only the typical benefits of Biocatalysis, but the possibility to have a better control over polymerization [96].

In a recent work, Salvachúa *et al.* [97] used Versatil peroxidase (VP) to crosslink several monomeric lignans and peptides. Results show lignan oligomers of a maximum DP of 9 with a Number Average Molecular Weight Mn = 3,300. As per the reactions with peptides, peptide oligomers with a maximum DP of 11 and a Mn = 6,500 were obtained. Hetero-oligomers between lignans and peptides were also synthesized by VP. Oligomers of a DP = 6 and a Mn = 2,300 were obtained where the ratio peptide: lignan was 1:5. The reactions were done at room temperature with 1.5 U mL^{-1} of VP, 0.1 mM H_2O_2 and 0.1 mM Mn^{2+}, during 24 h.

A peroxidase-catalyzed polymerization was performed under mild reaction conditions, using an aqueous buffer alcohol resulting in a soluble polyphenol with M_n of 3,000–6,000 [98,99]. The catalysts included horseradish peroxidase (HRP) and soybean peroxidase (SBP). The polymer solubility increased with increasing the oxyphenylene unit content (32%–59%) which was controlled by varying the methanol amount [100], and the solution pH [101]. Monodisperse polymer particles (250 nm diameter) were synthesized from phenol and subsituted phenols like m-cresol, p-cresol and polyphenylphenol, with HRP in a dispersion system 1,4-dioxane/buffer [102,103]. The mechanism of a peroxidase-catalyzed polymerization of phenols comprises three stages: Radical formation, Radical transfer and Radical coupling, shown in Figure 6.

Figure 6. Mechanism of phenol polymer formation (see also [104]).

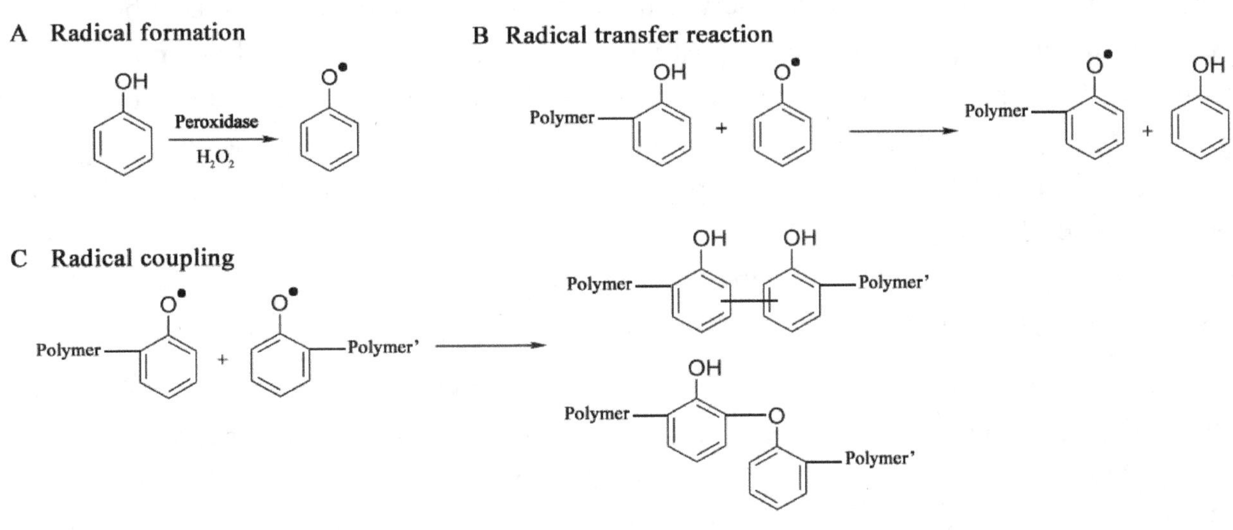

It was revealed that meta-substituted phenol polymerizations catalyzed by HRP and SBP give rise both to higher yields when small and large-bulky substituents are used, respectively [105]. Cardanol, a natural phenol derived from cashew nut shell liquid (CNSL) with a C15 unsaturated alkyl meta-substituent, has good perspectives for industrial utilization such as resins, friction lining materials, and surface coatings. Therefore, an enzymatic approach to develop derivatives from this compound is attractive. Kobayashi *et al.* [106] obtained an artificial polymer structurally similar to "urushi", a Japanese lacquer, through an SBP-polymerization of Cardanol. The polymer had a molecular weight between 2,000–4,000 Da., soluble in polar organic solvents and possessed a tough and hard property as a film with a glossy surface finish. Kim *et al.* [107] also synthesized polycardanol with SBP and found it had better anti-biofouling activity to *Pseudomonas fluorescens* compared to polypropylene.

Various other m-substituted phenols have been polymerized by HRP and SBP, for example meta-alkylphenols, meta-halogenated phenols, and meta-phenylphenol [108]. These enzymatically synthesized m-substituted polymers were applied to positive-type photoresists for printed wire boards, because of their high solubility toward alkaline solution and high thermal stability [108,109].

Ortho-substituted phenols have been polymerized with SBP affording a variety of oligophenols (dimers to pentamers) and some of their oxidation products, including quinones and demethylated quinones. Some of these are considered to serve as biologically important compounds with therapeutic potentials [110]. In separate reaction systems, it was possible to synthesize poly(phenylene oxide) (PPO) from two different *o*-substituted phenols, 2,6-dimethylphenol and 3,5-dimethoxy-4-hydroxybenzoic acid (syringic acid), using HRP, SBP and a laccase [111–113]. The polymers were formed exclusively of 2,6-dimethyl- or 2,6-dimethoxy-1,4-oxyphenylene units respectively, which are similar in structure to a widely used high-performance engineering plastic PPO.

Para-alkylphenols were polymerized by HRP in an aqueous 1,4-dioxane solution to synthesize polymers with Mn of several thousands, whereby the polymer yield increased as the n-alkyl chain length increased from 1 to 5 [102,114]. Oxidative polymerization of natural *para*-substituted phenols like tyrosine ethyl ester or methyl ester or phenol derivative hydroquinone-β-D-glucopyranoside (arbutin) [115] were carried out using HRP catalyst in an aqueous buffer solution. A mixture of phenylene and oxyphenylene units was observed in the tyrosine polymer, whereas the arbutin polymer, after degycosilation, consisted of C-C *o*-position hydroquinone units. Resulting polymers had a molecular weight from 1,500–4,000. Reactions catalyzed by HRP proceeded, involving radical species, yet took place chemoselectively. *Para*-phenylphenol is one well-studied *para*-substituted phenols in peroxidase-catalyzed polymerizations. Polymers up to 26,000 Da have been synthesized with HRP [99]. However, the polymerization of *para*-phenylphenol lacks precise structure control due to the existence of both stabilized *ortho*- and *para*-single electron radical species, causing different kinds of linkages and leading to a complex structural polymer [116].

Other numerous reports manage different aspects of the enzymatic polymerization of phenols: use of mixtures of aqueous-organic solvents with 1,4-dioxane, methanol, DMF [117], or total organic systems (isooctane) [118], ionic liquids [119], using cyclodextrins as a phenol solubilizing agent [120], incorporating PEG as a template aid [121], or carrying out the reaction in micelles and reverse micelles [122] or in capsules [123].

An important group of natural phenols widely distributed throughout the plant kingdom are flavonoids. Flavonoids are benzo-γ-pyrone compounds consisting of phenolic and pyrane rings. Their biological and pharmacological effects as antioxidant, antimutagenic, anticarcinogenic, antiviral and antiinflammatory have been extensively reviewed [124–127].

Flavanols or commonly known as catechins are the major group of polyphenols present in green tea. The main catechins in green tea are (+)-catechin, (−)-epicatechin (EC), (−)-epigallocatechin (EGC), (−)-epicatechin gallate (ECG), and (−)-epigal-locatechin gallate (EGCG). Poly(cathechin) has been synthesized by HRP in aqueous-organic solvents having yields of DMF-soluble polymers around 30% and molecular weights of 3,000 [128]. It has also been polymerized by a laccase derived from *Myceliophthore* (ML) in a mixture of acetone-buffer pH 5 rendering an 8,000 Da polymer [129]. In this case, as in other phenol polymerizations [100], the organic solvent/buffer ratio greatly affected the solubility of the synthesized polymer. The antioxidant properties of catechins and ECGC, not only remained, but increased in a concentration dependent manner in poly(catechin) and poly(ECGC) synthesized by HRP and ML [129,130]. Superoxide anion scavenging activity of poly(ECGC) was much superior than monomeric ECGC and poly(catechin) [130]. Inhibitory activity of Xanthine oxidase, which was negligible in monomeric catechin or ECGC, was markedly observed in poly(catechin) and in poly(ECGC), inclusive was higher than allopurinol, a commercial inhibitor frequently used for gout treatment [129,131]. Recently, a laccase was used to polymerize catechin in order to attach it to biomedical catheters in order to reduce bacterial biofilm formation [132].

Rutin has also been polymerized in organic-aqueous systems by Laccase ML-mediated catalysis. Yields superior to 70% were mostly observed and molecular weights between 7–9 kDa were registered [133].

Flavonols (quercetin) and isoflavones were also subject to an enzymatic oxidative polymerization by HRP and SBP in aqueous-organic solvent systems. Weight Average Molecular Weight Mw and Number Average Molecular Weight Mn for poly(quercetin) was 10,000 and 2,500 respectively, with a yield around 50%. Poly(rutin), poly(catechin), poly(daidzein), poly(formononetin) and others were also obtained in similar numbers [128].

Functionalization of polymers with flavonoids has also been investigated. Quercetin-functionalized chitosan has been the most studied system to yield antimicrobial products [134–137]. However, naringin-PVDF, morin-resin, catechin-inulin systems have also been reported [138–140].

3.2.3. Hydrolases

Hydrolases are the enzyme class with most widespread use and applications in the world. Around 75% of all commercialized industrial enzymes are hydrolytic in action. Despite proteases and carbohydrases are the most demanded subclass of enzymes accounting for around 40% of world's enzyme sales, lipases remain as the biocatalyst subclass most widely used in organic chemistry [141,142].

Lipases have been used to catalyze reactions in polycondensations of dicarboxylic acids with diols [143], ring-opening polymerization (ROP) of lactones [144], cyclic carbonates [145] and polycondensations of hydroxycarboxylic acids [146]. More examples are described in the following sections.

3.2.4. Polyesters

Polyesters occupy the 4th place among the most important biomacromolecules in living systems. They are macromolecules with good biodegradability, biocompatibility and permeability, and thus are highly suitable for biomedical applications, such as implant biosorbable materials, for tissue engineering and delivery vehicles for drug or genes [147]. Polyesters are very valuable industrial materials that are broadly used like poly(ethylene terephthalate) (PET), poly(butylene succinate) (PBS), poly(ε-caprolactone) (PCL), and poly(lactic acid) (PLA).

Two major approaches can be taken to perform enzymatic polyester synthesis: (1) polycondensation and (2) ring-opening polymerization (ROP). PET and PBS are produced in large-scale by polycondensation whereas PCL and PLA are produced via ROP. Classical cationic, anionic or metallic chemical synthesis of polyesters may render them unsuitable for biomedical applications due to the toxicity of these reagents. Enzyme catalyzed synthesis of esters has circumvented this and other problems since enzymatic reactions do not require high energy input (temperature or pressure), most enzymes can carry out regio-, chemo- and enantiospecific reactions with no need of protecting groups, exhibit high catalytic activity, they may be reused and as a consequence of all this, reactions are environmentally and health safer than the conventional chemical route [148].

3.2.5. Polycondensations

3.2.5.1. Lipase-Catalyzed Polycondensations

In lipase-catalyzed polycondensations dicarboxylic acids are esterified with diols to produce polyesters. Reactions have been realized mostly in organic solvents, in solvent-free media and in aqueous systems. However, there have been studies in ionic liquids [149] and supercritical fluids as well [150]. In the early years of exploration of these systems, Okumura *et al.* [151] set a series of reactions with dicarboxylic acids and diols in aqueous systems using a lipase from *Aspergillus niger* and he obtained oligomers of 3–7 units. Extracting water molecules from the system, either with molecular sieves or applying vacuum, seemed to improve the polymer size up to a DP of 20 in an adipic acid-1,4-butanediol reaction [152], and up to a Mn = 77 kDa in an adipic acid-1,6-hexanediol reaction [153]. Organic solvents with high boiling points were identified to favor the polyester synthesis. In an aim to study natural hydroxy-acids, Gómez-Patiño *et al.* [154] obtained oligomers from tomato cuticle monomer, 10,16-dihydroxyhexadecanoic acid and its methyl ester. *Candida antarctica* B was the best performer among the 5 lipases tested in organic media reactions obtaining the largest oligomer with a DP = 3 and Mw = 1,206.

Reactions between dicarboxylic acids and diols conducted in water with *Candida antarctica* (N435), but with a continuous dehydrating process, are claimed to give rise to good yields [155].

Polycondensations with structurally different monomers have been realized, for example siloxane-containing diacids with diols (Mw = 20,000) [156], diacids with glycerol (Mw = 2,000–6,000) [157,158], thio-glycerol with diacids (Mw = 170,000) [159], diacids with a variety of sugars, sorbitol, erythritol, xylitol, ribitol, mannitol, glucitol, galactitol yielded polyesters with Mw = 10 kDa for galactitol and 75 kDa for Mannitol. The regioselecitivy of the sugar esterifications was mostly consistent in all the sugars to be on the primary hydroxyl groups [160].

Activation of dicarboxylic acids has been employed by means of esterification the carboxylic groups with methanol or ethanol [161], by esterification [162] or by vinyl activation [163]. Azim *et al.* [164], in an interesting strategy, firstly polymerized succinic acid with 1,4-butanediol to obtain a cyclic oligomer, which later was repolymerized by ring-opening to obtain a high molecular weight polymer (Mw = 130 kDa). Activation of dicarboxylic acids by means of esterification the carboxylic groups with halogenated alcohols like 2-chloroethanol, 2,2,2-trifluroethanol and 2,2,2-trichloroethanol lead to an increase of Mw by a factor of 5 [162]. However, the halogenated alcohols produced as a consequence of the reaction may damage the biocatalyst. Activation of carboxylic groups by vinyl esterification produced in most cases polyesters with Mw around 25 kDa [103,165,166].

Through reaction optimizations, Yao *et al.* [167] were able to achieve a polyester Mw = 16.6 kDa using 1,8-octanediol, adipic acid and L-malic acid as monomers. The incorporation of a sugar to be used as a third monomer is a likely means to modify the properties of conventional polymers, and an interesting strategy to prepare a variety of functional materials.

Recently it was reported the polymerization of β-alanine by a lipase catalyzed polycondensation reaction between β-alanine esters. Lipase B from *Candida antarctica* aggregated in CLEAs was the biocatalyst that converted β-alanine esters into polymers in high yield (up to 90%) and with high DP (up to 53 units). The best results were obtained by using methyl esters and methyl-tert-butyl ether (MTBE) as the solvent medium at 50 °C for 16 h [168].

3.2.5.2. Protease-Catalyzed Polycondensations

Proteases, like some other hydrolases, catalyze not only hydrolytic reactions (peptide hydrolysis in this case) but also peptide bond formation under appropriate conditions to give polypeptides. There exist several oligopeptides with varied biologically active properties that have been synthesized widely using proteases as catalysts via condensation reactions [169]. Some examples include, aspartame synthesized by thermolysin [170], oxytocin by papain, thermolysin, and chymotrypsin, and somatostatin by thermolysin, and chymotrypsin [171]. Various attempts to synthesize long-chain polyaminoamides were done by a protease-catalyzed polycondensation using chymotrypsin, trypsin, subtilisin and papain. However, only lower molecular weight polymers were obtained [172].

3.2.6. Polyamides

It was recently reported the synthesis of polyamides by lipase Novozym 435 in high yields (93%) and a short lapse of time (30 min). Authors attribute this high productivity to the use of “designed” monomers of the type: ω-amino-α-alkoxy acetic acid ethyl ester, which bears an ω-amino group and an oxygen atom in β position. The reactions were performed in bulk at 80 °C with an enzyme load of 45% (v/v) and the resultant polyamides had a DP = 15 (PDI = 1.6) and a Mw between 3,000–4,000 [173].

3.2.6. Vinyl Polymers

One field that has expanded tremendously in the past two decades is the enzyme-initiated radical polymerization of aromatic and vinyl monomers [149]. Vinyl monomers studied for enzymatic polymerization are basically classified into (meth)acrylic type and styryl type, which both involve the formation of a radical species. HRP and Laccase peroxidases have been reported to perform well in the meth(acrylic)-type polymerization of acrylamide. In a HRP-mediated free radical polymerization of vinyl monomers, the HRP-mediator system has three components: enzyme, oxidant (H_2O_2) and an initiatior, such as β-diketones which are commonly used. On the other hand, when several oxidoreductases, such as laccase, lipoxidase, and sarcosine oxidase, were used as catalysts, the acrylamide polymerization proceeded even without H_2O_2 or β-diketones. This was the case for laccase from *Pycnoporus coccineus* which induced the acrylamide polymerization in water at 50–80 °C under argon to produce poly(acrylamide) with Mn = 1×10^6 , Mw/Mn ≈ 2 for 4–24 h and up to 81% yields [174]. Since the laccase-mediator system (LMS) may induce the radical polymerization of vinyl monomer with no H_2O_2 and is environmentally benign, different conditions for polymerization with commercial laccase *Myceliophthora thermophila* were further examined and optimal conditions were found as slightly acidic reaction media at around 50 °C using β-diketones and O_2 as the initiator and oxidant respectively. The molecular weight of poly(acrylamide) was Mn 6–28 × 10^4; Mw/Mn = 2.5–3.2 and could be controlled through the ratio of monomer to enzyme [175].

Poly(acrylamide) polymers of Mn of 1.5–4.6 × 10^5 with Mw/Mn 2.0–2.4, and acrylamide conversion between 70%–90% were synthesized by HRP in water at room temperature with 2,4-pentanedione as an initiator intermediate [176].

Enzyme-catalyzed vinyl polymerizations have been demonstrated in recent years with significant control of polymer molecular weight and yield depending on reaction conditions [149]. Resins of poly(sodium acrylate) are used as water-absorbent materials for cleaning surfaces, in water and oil conditioning, personal care products, and disposable materials for medical applications [177].

3.2.7. Nylons

Polyamides, also called nylons, display improved physical properties compared with polyolefins and polyesters due to the directionally specific inter-chain hydrogen bonds and significantly enhanced melting points [178]. Nylons can be synthesized either by polycondensation or by ROP. Ragupathy *et al.* [179] presented two and three step polycondensation and ROP methods to enzymatically produce nylon-8,10, nylon-8,13, nylon-6,13, and nylon-12,13. He used 1,8-diaminooctane (DAO) and diethyl sebacate (DES) for the polycondensations and DAO and lactone ethyleneridecanoate (ETD) for ROP. Novozym 435 in diphenyl ether at 150 °C carried out the polymerizations. Monomer conversion reached 97% and 90% and polyamides synthesized had an Mw of 5,000 and 8,000 for polycondensation and ROP reactions respectively.

3.2.8. Acrylate and Styrene Polymerization

Methyl-methacrylate (MMA) was polymerized by en enzyme-catalyzed reaction using the ternary system (enzyme, oxidant, and initiator) in water and water-miscible organic cosolvents such as DMF, acetone, dioxane, and THF. Soybean peroxidase in aqueous solution afforded a 48% yield of poly(methyl methacrylate) (PMMA) with Mn = 9.3×10^5 and Mw/Mn = 6.8. When HRP II (type II) was used, a 45% yield of PMMA was obtained with Mn = 6.3×10^5 and Mw/Mn = 3.0. It was found that yields of PMMA increased when cosolvents with low dielectric constant like dioxane and THF were used [180].

Respect to styrenic polymerizations, HRP in water was used to polymerize styrene and some derivatives. The polymerization was investigated using several β-diketones as initiators, and proceeded in a mixed solution of H_2O/THF = 3 (v/v). Polystyrene synthesized using 2,5-cyclopentanedione as initiator resulted in 60% yield with Mn and Mw/Mn of 6.7×10^4 and 1.9 respectively. Polymer yield, Mn and Mw/Mn of the resulting polystyrene depended on the type of initiator used. Synthesis of polymers from styrene derivatives, 4-methylstyrene and 2-vinylnaphthalene were also investigated and afforded polystyrenes in a >90% yield of polymer with a Mn of 1.15×10^5 and Mw/Mn = 2.28 [181].

3.2.9. Ring-Opening Polymerizations (ROP)

Among the two major synthetic polymerization approaches, ring-opening polymerization (ROP) has been most extensively studied to polyester synthesis.

Polymerization of lactones via ring-opening catalysis carried out by lipases has been studied since the early 1990s. This catalytic activity for ROP of lactone monomers has been the research field on diverse lipases such as *Pseudomonas fluorescens*, *Pseudomonas cepacia*, *Candida rugosa*, *Candida cylindracea*, *Candida antarctica*, *Aspergillus niger*, *Mucor miehei*, *Penicillium roqueforti*, *Rhizopus japanicus*. *Yarrowia lipolytica*, *Carica papaya* and porcine pancreas lipase [182–189]. Cutinase from *Humicola insolens* (HiC) has also been reported to carry out ROP [190].

Poly(1,4-dioxan-2-one) (polyDO) is a desirable biocompatible polymer with good flexibility and tensile strength which might be considered for biomedical applications. This polyester was synthesized by ROP of 1,4-dioxan-2-one by lipase B from *Candida antarctica* at 60 °C, resulting in a polymer of Mw = 41,000 [191]. In another study, Jiang *et al.* [192] claimed to synthesize by ring-opening copolymerization a polyester from 1,4-dioxan-2-one (DO) with pentadecalactone (PDL) to give a copolyester of poly(DO-co-PDL) with Mw > 30,000. They used *Candida antarctica* B lipase in toluene or diphenyl ether at 70 °C for 26 h.

Lipase 2 from *Yarrowia lipolytica* and *Carica papaya* latex were recently employed to polymerize ε-CL in bulk. Lipase 2 from *Y. lipolytica* was found to be very efficient as catalyst for several reactions [193] and latex from *C. papaya* is a low-cost auto-immobilized biocatalyst. The polymerization yields were 74% and 40% with an Mw of 1,350 and 1,100 respectively. Reactions were realized at 150 °C and 6 h for *Y. lipolytica* and 70 °C and 24 h for *C. papaya* [182].

As to macrolides, lipase *Pseudomonas fluorescens* was able to polymerize 11-undecanolide (12-membered, UDL), 12-dodecanolide (13-membered, DDL), 15-pentadecanolide (16-membered, PDL), and 16-hexadecanolide (17-membered, HDL) [114,186–189,194–196]. Lipases from *Candida cylindracea* and *Pseudomonas fluorescens* catalyzed the ROP of UDL in bulk with quantitative yields affording polyesters of Mn of 23,000 and 25,000 [197,198]. In another study, ROP by lipases CA, lipase CC, lipase PC, lipase PF, or PPL in bulk at 45–75 °C for 5 days was done over HDL, the largest unsubstituted lactone monomer studied so far. The study showed the synthesis of polyHDL with Mn reaching to 5,800 in high yields [199]. It was discovered that a larger ring monomer, like PDL showed a greater reactivity towards polymerization than a smaller monomers like 1,4-polydioxan-2-one (DO) [192].

Ebata *et al.* obtained a poly(ε-CL) from a cyclic dimer of ε-CL (14-memberd) that was polymerized at 70 °C in toluene by lipase B from *Candida antarctica*, affording quantitatively a poly(ε-CL) with Mn of 18,000 [200]. This research Group, under the same strategy used to produce oligomers which are then subject to subsequent polymerization, claimed that a poly(butylene-succinate) (PBS) polyester of Mw 130,000 was obtained from cyclic dimers of butylene-succinate of Mn = 390. The reaction was carried out in toluene at 120 °C using 40% (w/w) of lipase *Candida antarctica* B for 24 h [201].

It was found that among ε-CL substituted monomers, ω-methyl ε-CL showed the least reactivity towards the lipase-catalytic ROP [202]. On the other hand, α-methyl or δ-methyl ε-CL afforded polymers of Mw = 8,400–11,000 with yields of 74% and 93% respectively [203].

An equilibrium phenomenon was observed between the synthesis of cyclic and linear polymers in a lipase-catalyzed ROP reaction of lactones. Cyclic oligomers in addition to major linear polyesters coexist in equilibrium [204,205]. This cyclic-linear equilibrium is solvent-dependent, for example Novozym 435 was used to catalyze the ROP of ε-CL at 60 °C in bulk or in an organic solvent. In bulk, the polymer products coexisted as a cyclic dimer of ε-CL and linear poly(ε-CL) in 2% and in 98% abundance, respectively. When the reaction was carried out in in acetonitrile, THF or 1,4-dioxane the same proportion was observed, as for example the equilibrium yields in acetonitrile were 70% for the cyclic oligomers (17% of dilactone and 53% of cyclic oligomers) and 30% for linear poly(ε-CL). When the reaction was carried out in isooctane, the proportion inverted to 3% and 97%, respectively, as in the bulk reaction [204].

Another hydrolase enzyme, a depolymerase, was found to catalyze ROP reactions *in vitro*. This depolymerase is a poly(hydroxybutyrate) (PHB) (EC 3.1.1.75), and was extracted from *Pseudomonas stutzeri*, *Alcaligenes faecalis* and *Pseudomonas lemoignei*. *A. faecalis* PHB depolymerase was assayed with a series of small, medium and large lactones and the best polimerization activity (93% yield) was obtained with β-butyrolactone, resulting in the formation of polyesters with a Mw = 16,000. On the other hand, medium and large lactones (ε-caprolactone, 11-undecanolide, and 12-dodecanolide), which are readily polymerized by lipases, were scarcely polymerized by PHB depolymerase [206].

3.3. Commodity Chemicals

One of the most noteworthy cases of biocatalytic production of a commodity chemical is the bioconversion of acrylonitrile to acrylamide [207]. This reaction, schematically simplified in Figure 7, is carried out by a nitrile hydratase whose catalytic activity transforms a cyanide group into an amide. Mitsubishi Rayon Co., Ltd. (Tokyo, Japan) currently produces over 20,000 tons annually of acrylamide using a third-generation biocatalyst, *Rhodococcus rhodochrous* J1, developed for commercial use by Nitto Chemical Industries. Acrylamide is produced from acrylonitrile in a continuous system of fixed-bed reactors at 10 °C with polyacrylamide-immobilized J1 cells. The process achieves conversion of acrylonitrile to acrylamide in ~99.9% yield, and the catalyst productivity is >7,000 g of acrylamide per g dry cell weight [208].

Figure 7. Enzymatic conversion of acrylonitrile to acrylamide by means of a nitrile hydratase.

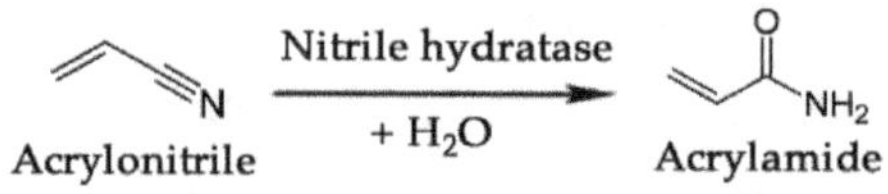

Lonza Guangzhou Fine Chemicals produces nicotinamide through a chemoenzymatic process shown in Figure 8 [209,210]. They start from 2-methyl-1,5-diaminopentane, a byproduct from production of nylon-6,6, which is converted to 3-methylpyridine, which in turn is ammoxidized. The resulting 3-cyanopyridine is hydrolyzed to nicotinamide using immobilized *Rhodococcus rhodochrous* J1 cells [211,212]. The Plant's capacity is 3,400 tons per year of nicotinamide. The enzymatic process affords the desired amide at >99.3% selectivity at 100% conversion, whereas the chemical process requires caustic hydrolysis of 3-cyanopyridine but also produces 4% nicotinic acid as byproduct.

Figure 8. Chemoenzymatic route to obtain nicotinamide developed by Lonza [209,210].

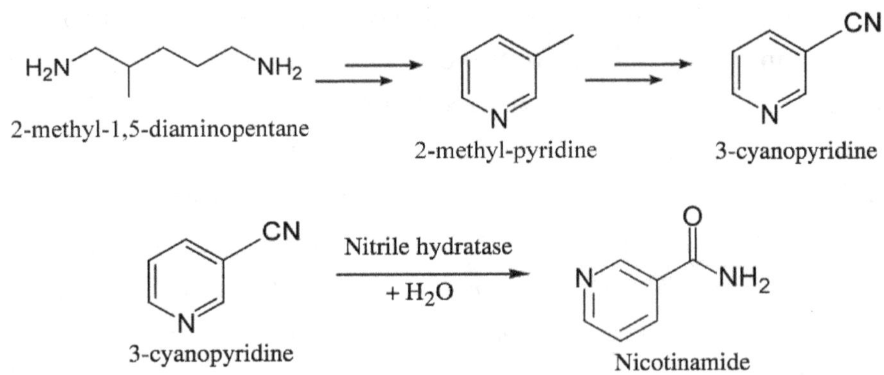

A commodity chemical which is aimed to substitute to some extent the universally oil-derived terephthalic acid is 2,5-furandicarboxylic acid (FDC) [213]. Terephthalic acid is one of the main raw materials to produce polyesters. FDC relies in the supply of 5-hydroxymethylfurfural (HMF), which is the product of the dehydration of hexoses. Oxidation of HMF yields FDC and may be realized by using heterogeneous or electrochemical catalysis [214,215]. However, two biocatalytic processes have been devised to help in having a reliable and sustainable supply for HMF: a whole-cell biocatalytic approach using a recombinant *Pseudomonas putida* hosting an oxidoreductase from *Cupriavidus basilensis.* This process is able to produce 30 g/L of 2,5-furandicarboxylic acid from HMF having a yield of 0.97 mol/mol [216]. The productivity of the whole process is 0.21 g/(L·h) under aerobic fed-batch conditions. The second approach consists of the use of a chloroperoxidase and *C. antarctica* lipase B [217,218].

4. Enzymes in the Fine Chemicals Industry

The role of Biocatalysis in the pharmaceutical and fine-chemical industries is clearly expanding. There are estimated to be around 150 implemented biocatalytic processes in industry, and the majority of these are in the pharmaceutical sector [219]. As stereoisomerism is quite relevant in pharmaceutics and fine chemistry, and a single swap in hydrogen position may mean a great difference in the bioactive function of a compound, Biocatalysis fits well to meet these challenges. Besides the enantio-, regio- and chemoselectivity that enzymes exhibit, their ability to perform complex catalysis procedures in a simple step, which otherwise, under a chemical approach, might require laborious treatments, toxic reagents or high energy input, has provided a strong basis for their usage in the fine chemicals industry. Some examples of the contribution of Biocatalysis to this field are described below.

A precursor of aspartame, *N*-(benzyloxycarbonyl)-L-aspartyl-L,-phenylalanine methyl ester (Z-APM) has been synthesized from *N*-(benzyloxycarbonyl)-L-aspartic acid (Z-L-ASP) and L-phenylalanine methyl ester (L-PM) with thermolysin [220]. By reverting the peptide-bond hydrolytic nature of thermolysin, this precursor is synthesized on a multi-thousand ton per year scale by a DSM/Tosoh joint venture (Holland Sweetener Company, Geleen, The Netherlands).

Pyrethroids now constitute the majority of commercial household insecticides. A novel industrial application of a lyase enzyme is to produce a pyrethroid intermedieate, (*S*)-*m*-phenoxy-benzaldehyde cyanohydrin (sPBC). Hydroxynitrile lyase, is an enzyme that catalyzes the addition of HCN to aldehydes and ketones [221]. Thus, the production of sPBC from *m*-phenoxybenzaldehyde by hydroxynitril lyase has been carried out on a multi-ton scale by DSM.

L-carnitine is a quaternary ammonium compound biosynthesized from the amino acids lysine and methionine [222]. Within mammal cells, L-carnitine is required for the transport of fatty acids from the cytosol into the mitochondria during the lipids oxidation [223]. Due to this role in lipid oxidation, L-carnitine is often advertised to improve fat metabolism, reduce fat mass, and increase muscle mass, which is widely available as an over-the-counter nutritional. L-carnitine is produced by Lonza (Basel, Switzerland) on an industrial scale using an enzyme from strain *Agrobacterium* HK1349 in a whole cell biotransformation process [224]. The process actually comprises a dehydrogenation of γ-butyrobetaine to 4-(trimethylamino)-butenoic acid followed by selective addition of a water molecule by L-carnitine lyase.

L-*tert*-leucine is an important chiral building block and intermediate of many ligands in chemo-catalysis in the pharmaceutical industry [225]. Menzel *et al.* [226] presented a process to produce L-*tert*-leucine in a whole-cell Biocatalysis system using a recombinant *E. coli* host expressing the two enzymes required by the process: a leucine hydrogenase and a formate dehydrogenase. The former catalyzes the main reaction, reductive amination of an α-keto acid, and the latter serves as a NADH regenerator. A yield of 84% in 24 h and an enantiomeric purity of ee > 99% was obtained. The process has been scaled up to tons level. Degussa-Hüls has used a similar whole-cell Biocatalysis system to produce L-*tert*-leucine, which has taken to a commercial scale [227].

Lonza produces 5-methylpyrazine-2-carboxylic acid on a commercial scale from the *p*-xylene analogue 2,5-dimethylpyrazine. This compound is used as a blood-lowering drug of the sulfenyl urea class commercialized as Glycotrol. The synthesis is carried out in *P. putida* ATCC33015 as a whole-cell biocatalyst, expressing a series of enzymes (a monooxygenase and two dehydrogenases) [224].

Japan Energy (Saitama, Japan) produces alkyloxiranes and phenyloxiranes on an industrial scale, especially the chiral building block 2-(trifluoromethyl)oxirane from 3,3,3- trifluoropropene through an oxidative reaction, using a whole-cell biocatalyst from *Nocardia corallina* [228].

Vanillin, the main component of vanillin extract, a natural flavoring, whose demand has long exceeded the supply, should be synthesized chemically from guaiacol, in order meet demand. Nevertheless, a whole-cell biocatalytic synthesis of vanillin from glucose has now been elaborated. Glucose is transformed into vanillic acid by a recombinant *E. coli* biocatalyst under fed-batch fermentor conditions. Then reduction of vanillic acid to vanillin is carried out by aryl aldehyde dehydrogenase isolated from *Neurospora crassa.* The biocatalytic route avoids the use of carcinogenic reagents and non-renewable petroleum derivatives (guaiacol) [229].

Aresta *et al.* used a carboxylase in the synthesis of 4-hydroxy benzoic acid, which is an intermediate for the synthesis of preservatives. The carboxylase enzyme is extracted from the bacteria *Thauera aromatica*, and it acts over the phenyl-phosphate moeity carboxylating it with 100% selectivity and 90% yield at 300 °K, $P_{CO2} = 0.1$ MPa [230].

Alfuzosin, a quinazoline derivative, against Benign prostatic hyperplasia (BPH), is synthesized through intermediate compund tetrahydro-*N*-[3-(methylamino)-propyl]-2-furan-carboxamide. This intermediate was traditionally produced in a three-steps chemical route from 2-tetrahydrofuroic acid involving a series of chemicals and distillation. The enzyme-catalyzed reaction with lipase B from *Candida antacrtica*, afforded the intermediate in two steps in which the same enzyme catalyzed the two reactions: esterification and amidation. This biocatalytic procedure simplyfied and made more environmentally friendly the previous chemical synthesis [231].

Another important intermediate in the synthesis of antiviral nucleosides such as 3'-deoxy-3'-azidothymidine (AZT), is thymidine. Production of thymidine is currently based in the hydrolysis of DNA from natural sources. A new biocatalytic approach for its production has been developed. It consists of transfering the 2'-deoxy-ribose moiety of 2'-deoxyinosine to thymine. The purine nucleoside phosphorylase of *Bacillus stearothermophilus* was used to carry out this reaction. A xantine oxidase is added in order to convert the concomitant hypoxantine produced to uric acid, and thus draw the thermodynamic equilibrium towards the synthesis of thymidine (Figure 9). The yield of thymidine of this whole-cell biocatalytic route reached 68% under mild conditions [232].

Figure 9. Biocatalytic synthesis of thymidine using a Purine Nucleoside Phosphorilase (PNPase) and the biocatalytical removal of hypoxantine with a Xanthine Oxidase (XAO) to obtain uric acid. Compounds: 1: 2'deoxyinosine; 2: Thymine; 3: Thymidine; 4: Hypoxanthine; 5: Uric acid (see also [233]).

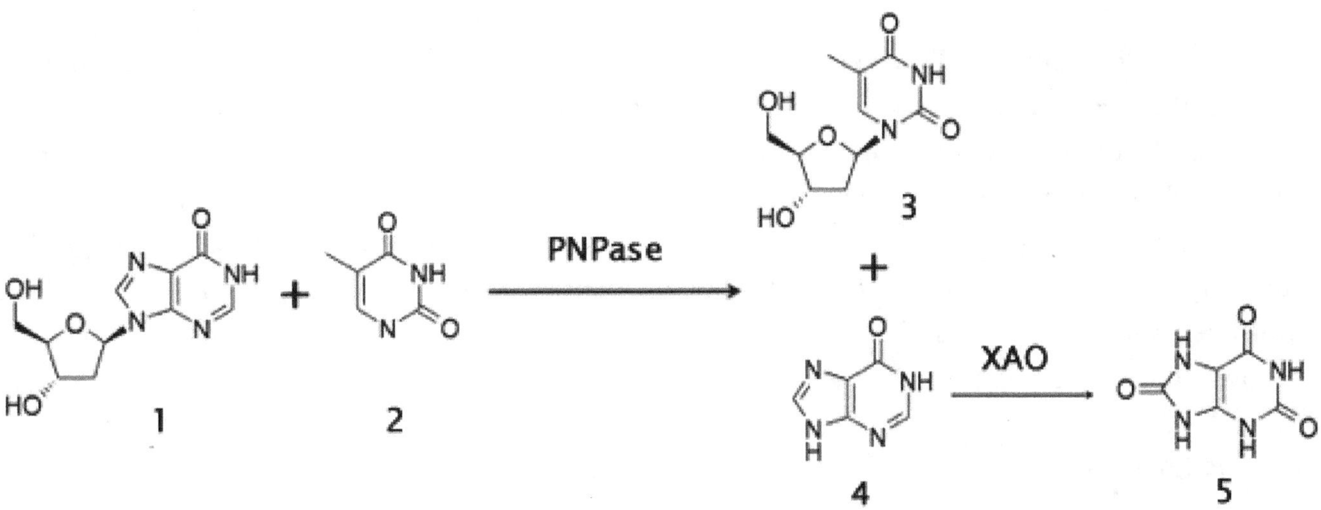

Many catalytic functions that enzymes perform in the pharmaceutical industry is in resolution of racemic mixtures, in which one enantiomer is acylated or hydrolyzed selectively affording an easier separation of both enantiomers. One recent example of this is the resolution of pro-drug (R,S)-2-bromophenylacetic acid octyl ester. Rivera *et al.* [234] accomplished its resolution with high enantioselectivity using lipases embedded in crude latex from *Carica papaya.* An enantioselectivity E-value E > 200 was achieved using decane as solvent, 50 mM of substrate and 50 mg/mL enzyme/reaction medium. Furthermore, the E-value doubled after purifying latex and removing proteases.

5. Conclusions

As outlined above, Biocatalysis is currently employed in a number of processes and products in diverse fields, and certainly new areas of application will be added. Advances in computational power have enabled the advent of bioinformatics, genomic sequencing and powerful analytic methods in physics, chemistry and molecular biology, which have allowed us to understand the dynamics, *in vivo* and *in vitro*, of biomacromolecules. The convergence of these tools has led to an improved access to more biocatalysts and to a deeper knowledge of them. In the case of enzymes, it has allowed us to explore their behavior, stability, specificity and even to begin to modify their very own nature through protein engineering. Larger availability of biocatalysts with superior qualities has been the main force behind the development of new industrial biocatalytic processes. The continued progress and interest in enzymes, serves as stimulus to make further efforts and ensure a steady success in meeting new synthetic challenges.

As per the environmental and economic benefits already reaped by the use of Biocatalysis in industry, it is possible to tell that the use of Biocatalysis is only expected to grow. This might be achieved with isolated enzymes but will probably be more successful using genetically engineered organisms where new pathways might be designed. The development of more robust, versatile, efficient enzymes via protein engineering, faster optimization of reaction conditions, microscale processing and better capabilities in design and machining new equipment will open new perspectives for the manufacture of many more products via Biocatalysis.

Future developments in Biocatalysis for industrial applications should be directed into the expansion of capabilities of current biocatalysts, to obtain new biocatalytic activities able to transform the vast amount of non-natural compounds found in industry and to achieve productivities comparable to current chemical processes. There is no doubt that modern biological tools, such as protein engineering and directed evolution will play an important role in meeting these challenges. As biodiversity remains largely unexplored, developing high-throughput screening methods to discover new efficient enzymes will also impact upon the development of biocatalyzed process for production of biobased chemicals.

Acknowledgments

We would like to thank to CONACYT for the scholarship provided to first author and for project CB 2008-01-104429.

References

1. Council of European Union. *En Route to the Knowledge-Based Bio-Economy*; German Presidency of the Council of European Union: Cologne, Germany, 2007.
2. Sandoval, G.; Plou, F.G. *Obtención enzimática de compuestos bioactivos a partir de recursos naturales iberoamericanos*; Consejo Superior de Instigaciones Cientificas (CSIC): Madrid, Spain, 2012; pp. 1–336.

3. Jahnz, U.; Schubert, M.; Baars, H.; Klaus, V. Process for producing the potential food ingredient DFA III from inulin-screening, genetic engineering, fermentation and immobilisation of inulase II. *Int. J. Pharm.* **2003**, *256*, 199–206.
4. Uchiyama, T.; Niwa, S.; Tanaka, K. Purification and properties of *Arthrobacter ureafaciens* inulase II. *Biochim. Biophys. Acta* **1973**, *315*, 412–420.
5. Kim, G.E.; Lee, S. Efficient production of DFA III (di-D-fructofuranose-dianhydride) from chicory root. In *Abstracts of the World Congress on Biotechnology*; DECHEMA: Berlin, Germany, 2000; Volume 272.
6. Yokota, A.; Hirayama, S.; Enomoto, K.; Miura, Y.; Takao, S.; Tomita, F. Production of inulin fructotransferase (depolymerizing) by *Arthrobacter* sp. H65–7 and preparation of DFA III from inulin by the enzyme. *J. Ferm. Bioeng.* **1991**, *72*, 258–261.
7. Kawamura, M.; Takahashi, S.; Uchiyama, T. Purification and some properties of inulin fructotransferase (depolymerizing) from Arthrobacter ilicis. *Agric. Biol. Chem.* **1988**, *52*, 3209–3210.
8. Gosling, A.; Stevens, G.W.; Barber, A.R.; Kentish, S.E.; Gras, S.L. Recent advances refining galactooligosaccharide production from lactose. *Food Chem.* **2010**, *121*, 307–318.
9. Rastall, R.A.; Gibson, G.R.; Gill, H.S.; Guarner, F.; Klaenhammer, T.R.; Pot, B.; Reid, G.; Rowland, I.R.; Sanders, M.E. Modulation of the microbial ecology of the human colon by probiotics, prebiotics and synbiotics to enhance human health: An overview of enabling science and potential applications. *FEMS Microbiol. Ecol.* **2005**, *52*, 145–152.
10. Sanz, J.I. Production of Glactooligo-Sacchariddes from Lactose by Immobilized β-Galactosidase and Posterioir Chromatographic Spearation. Ph.D. Thesis, Ohio State University, Columbus, OH, USA, 2009.
11. Gullón, B.; Gómez, B.; Martínez-Sabajanesb, M.; Yáñez, R.; Parajó, J.C.; Alonso, J.L. Pectic oligosaccharides: Manufacture and functional properties. *Trends Food Sci. Technol.* **2013**, *30*, 153–161.
12. Neri, D.F.M.; Balcão, V.M.; Costa, R.S.; Rocha, I.C.A.P.; Ferreira, E.M.F.C.; Torres, D.P.M.; Rodrigues, L.R.M.; Carvalho, L.B., Jr.; Teixeira, J.A. Galacto-oligosaccharides production during lactose hydrolysis by free *Aspergillus oryzae* β-galactosidase and immobilized on magnetic polysiloxane-polyvinyl alcohol. *Food Chem.* **2009**, *115*, 92–99.
13. Barreteau, H.; Delattre, C.; Michaud, P. Production of oligosaccharides as promising new food additive generation. *Food Technol. Biotechnol.* **2006**, *44*, 323–333.
14. Gaur, R.; Pant, H.; Jain, R.; Khare, S.K. Galacto-oligosaccharide synthesis by immobilized *Aspergillus oryzae* β-Galactosidase. *Food Chem.* **2006**, *97*, 426–430.
15. Whisner, C.M.; Weaver, C.M. Galacto-oligosaccharides: Prebiotic effects on calcium absorption and bone health. In *Nutritional Influences on Bone Health*; Burckhardt, P., Dawson-Hughes, B., Weaver, C.M., Eds.; Springer: London, UK, 2013; pp. 315–323.
16. Perugino, G.; Trincone, A.; Rossi, M.; Moracci, M. Oligosaccharide synthesis by glycosynthases. *Trends Biotech.* **2004**, *22*, 31−37.
17. Dias, L.G.; Veloso, A.C.A.; Correia, D.M.; Rocha, O.; Torres, D.; Rocha, I.; Rodrigues, L.R.;

Peres, A.M. UV spectrophotometry method for the monitoring of galacto-oligosaccharides production. *Food Chem.* **2009**, *113*, 246–252.

18. Rodríguez, B.; Poveda, A.; Jiménez, J.; Ballesteros, A.O.; Plou, F.J. Galacto-oligosaccharide synthesis from lactose solution or skim milk using the β-galactosidase from *Bacillus circulans*. *J. Agric. Food Chem.* **2012**, *60*, 6391–6398.
19. Torres, D.; Gonçalves, M.D.P.F.; Teixeira, J.A.; Rodrigues, L. Galacto-oligosaccharides: Production, properties, applications, and significance as prebiotics. *Compr. Rev. Food Sci. F.* **2010**, *9*, 438–454.
20. Tzortzis, G.; Vulevic, J. Galacto-oligosaccharide prebiotics. In *Prebiotics and Probiotics Science and Technology*; Charalampopoulos, D., Rastall, R.A., Eds.; Springer: New York, NY, USA, 2009; pp. 207–244.
21. Park, H.Y.; Kim, H.J.; Lee, J.K.; Kim, D.; Oh, D.K. Galactooligosaccharide production by a thermostable β-galactosidase from *Sulfolobus solfataricus*. *World J. Microb. Biot.* **2008**, *24*, 1553–1558.
22. Rodriguez, B.; de Abreu, M.A.; Fernandez, L.; de Beer, R.; Poveda, A.; Jimenez, J.; Haltrich, D.; Ballesteros, A.O.; Fernandez, M.; Plou, F.J. Production of galacto-oligosaccharides by the β-galactosidase from *Kluyveromyces lactis*: Comparative analysis of permeabilized cells *versus* soluble enzyme. *J. Agric. Food Chem.* **2011**, *59*, 10477–10484.
23. Hsu, C.A.; Lee, S.L.; Chou, C.C. Enzymatic production of galactooligosaccharides by β-galactosidase from *Bifidobacterium longum* BCRC 15708. *J. Agric. Food Chem.* **2007**, *55*, 2225–2230.
24. Hansson, T.; Adlercreutz, P. Optimization of galactooligosaccharide production from lactose using β-glycosidases from hyperthermophiles. *Food Biotechnol.* **2001**, *15*, 79–97.
25. UBIC Europe Marketing Consulting. *The World GOS Market*; UBIC Europe Press: Sierre, Switzerland, 2010.
26. Padalino, M.; Perez-Conesa, D.; López-Nicolás, R.; Frontela-Saseta, C.; Berruezo, G. Effect of fructooligosaccharides and galactooligosaccharides on the folate production of some folate-producing bacteria in media cultures or milk. *Int. Dairy J.* **2012**, *27*, 27–33.
27. Fanaro, S.; Boehm, G.; Garssen, J.; Knol, J.; Mosca, F.; Stahl, B.; Vigi, V. Galacto-oligosaccharides and long-chain fructo-oligosaccharides as prebiotics in infant formulas: A review. *Acta Paediatr. Suppl.* **2005**, *94*, 22–26.
28. Watzl, B.; Girrbach, S.; Roller, M. Inulin, oligofructose and immunomodulation. *Br. J. Nutr.* **2005**, *93*, S49–S55.
29. Bornet, F.R.; Brouns, F.; Tashiro, Y.; Duvillier, V. Nutritional aspects of short-chain fructooligosaccharides: Natural occurrence, chemistry, physiology and health implications. *Dig. Liver Dis.* **2002**, *34*, S111–S120.
30. Yun, J.W. Fructooligosaccharides—Occurrence, preparation, and application. *Enzyme Microb. Tech.* **1996**, *19*, 107–117.
31. Arrizon, J.; Urias-Silvas, J.E.; Sandoval, G.; Mancilla-Margalli, N.A.; Gschaedler, A.C.; Morel, S.; Monsan, P. Production and bioactivity of fructan-type oligosaccharides. In *Food Oligosaccharides:*

Production, Analysis and Bioactivity; Moreno, F.J., Sanz, M.L., Eds.; Wiley-Blackwell: Hoboken, NJ, USA, 2013; in press.

32. Beine, R.; Morarua, R.; Nimtz, M.; Na'amniehc, S.; Pawlowskic, A.; Buchholz, K.; Seibel, J. Synthesis of novel fructooligosaccharides by substrate and enzyme engineering. *J. Biotechnol.* **2008**, *138*, 33–41.
33. Castillo, E.; López-Munguía, A. Synthesis of levan in water-miscible organic solvents. *J. Biotechnol.* **2004**, *114*, 209–217.
34. Vega, R.J.; Zúniga, M.E. Potential application of commercial enzyme preparations for industrial production of short-chain fructooligosaccharides. *J. Mol. Catal. B Enzym.* **2012**, *76*, 44–51.
35. Singh, R.S.; Singh, R.P. Production of fructooligosaccharides from inulin by endoinulinases and their prebiotic potential. *Food Technol. Biotech*. **2010**, *48*, 435–450.
36. Tanriseven, A.; Aslan, Y. Immobilization of Pectinex Ultra SP-L to produce fructooligosaccharides. *Enzyme Microb. Technol.* **2005**, *36*, 550–554.
37. Guío, F.; Rodríguez, M.A.; Alméciga, C.J.; Sánchez, O.F. Recent trends in fructooligosaccharides production. *Recent Pat. Food Nutr. Agric.* **2009**, *1*, 221–230.
38. Jala, R.C.R.; Hu, P.; Yang, T.; Jiang, Y.; Zheng, Y; Xu, X. Lipases as biocatalysts for the synthesis of structured lipids. In *Lipases and Phospholipases. Methods in Molecular Biology*; Sandoval, G., Ed.; Springer-Humana Press: New York, USA, 2012; Volume 861, Chapter 23.
39. Xu, X.; Akoh, C.C. Enzymatic production of Betapol and other specialty fats. In *Lipid Biotechnology*; Marcel Dekker: New York, NY, USA, 2002; pp. 461–478.
40. Odle, J. New insights into the utilization of medium-chain triglycerides by the neonate: Observations from a piglet model. *J. Nutr.* **1997**, *127*, 1061–1067.
41. Bugaut, M. Occurrence, absorption and metabolism of short chain fatty acids in the digestive tract of mammals. *Comp. Biochem. Physiol. B Biochem. Mol. Biol.* **1987**, *86*, 439–472.
42. Osborn, H.T.; Akoh, C. Structured lipids: Novel fats with medical, nutraceutical, and food applications. *Compr. Rev. Food Sci. F.* **2002**, *1*, 110–120.
43. Xu, X.; Hoy, C.E.; Balchen, S.; Adler-Nissen, J. Specific-Structured Lipid: Nutritional Perspectives and Production Potentials. In Proceedings of International Symposium on the Approach to Functional Cereals and Oils. CCOA, Beijing, China, 9–14 November 1997.
44. Trivedi, R.; Singh, R.P. Modification of oils and fats to produce structured lipids. *J. Oleo Sci.* **2005**, *54*, 423–430.
45. Hellner, G.; Tőke, E.R.; Nagy, V.; Szakács, G.; Poppe, L. Integrated enzymatic production of specific structured lipid and phytosterol ester compositions. *Process Biochem.* **2010**, *45*, 1245–1250.
46. Lee, K.T.; Akoh, C.C. Characterization of enzymatically synthesized structured lipids containing eicosapentaenoic, docosahexaenoic, and caprylic acids. *J. Am. Oil Chem. Soc.* **1998**, *75*, 495–499.
47. Akoh, C.C. Structured lipids. In *Food Lipids Chemistry, Nutrition, and Biotechnology*; Akoh, C.C., Min, D.B., Eds.; Marcel Dekker: New York, NY, USA, 1998; pp. 699–727.
48. Iwasaki, Y.; Yamane, T. Enzymatic synthesis of structured lipids. *J. Mol. Catal. B Enzym.* **2000**, *10*, 129–140.

49. Khodadadi, M.; Aziz, S.; St-Louis, R.; Kermash, S. Lipase-catalyzed synthesis and characterization of flaxseed oil-based structured lipids. *J. Funct. Foods* **2013**, *5*, 424–433.
50. Casas-Godoy, L.; Marty, A.; Sandoval, G.; Ferreira-Dias, S. Optimization of medium chain length fatty acid incorporation into olive oil catalyzed by immobilized Lip2 from *Yarrowia lipolytica*. *Biochem. Eng. J.* **2013**, *77*, 20–27.
51. Villeneuve, P.; Barouh, N.; Baréa, B.; Piombo, G.; Figueroa-Espinoza, M.C.; Turon, M.C.; Pina, M.; Lago, R. Chemoenzymatic synthesis of structured triacylglycerols with conjugated linoleic acids (CLA) in central position. *Food Chem.* **2007**, *100*, 1443–1452.
52. Nagachinta, S.; Akoh, C.C. Enrichment of palm olein with long chain polyunsaturated fatty acids by enzymatic acidolysis. *LWT Food Sci. Technol.* **2012**, *46*, 29–35.
53. Tecelão, C.; Rivera, I.; Sandoval, G.; Ferreira-Dias, S. *Carica papaya* latex: A low-cost biocatalyst for human milk fat substitutes production. *Eur. J. Lipid Sci. Technol.* **2012**, *114*, 266–276.
54. Nagao, T.; Watanabe, Y.; Maruyama, M.; Momokawa, Y.; Kishimoto, N.; Shimada, Y. One-pot enzymatic synthesis of docosahexaenoic acid-rich triacylglycerols at the sn-1(3) position using by-product from selective hydrolysis of tuna oil. *New Biotechnol.* **2011**, *28*, 7–13.
55. Buchholtz, K.; Volker, K.; Borscheuer, U.T. *Biocatalysts and Enzyme Technology*; Wiley-VCH: Berlin, Germany, 2005; pp. 1–465.
56. Bornscheuer, U.T.; Kazlauskas, R.J. *Hydrolases in Organic Synthesis—Regio- and Stereoselective Biotransformations*; Wiley-VCH: Berlin, Germany, 1999; pp. 1–355.
57. Cao, L. *Carrier-Bound Immobilized Enzymes*; Wiley-VCH: Berlin, Germany, 2005; pp. 1–578.
58. Gog, A.; Roman, M.; Toşa, M.; Paizs, C.; Irimie, F.D. Biodiesel production using enzymatic transesterification—Current state and perspectives. *Renew. Energ.* **2012**, *39*, 10–16.
59. Véras, I.C.; Silva, F.A.; Ferrão-Gonzales, A.D.; Moreau, V.H. One-step enzymatic production of fatty acid ethyl ester from high-acidity waste feedstocks in solvent-free media. *Bioresour. Technol.* **2011**, *102*, 9653–9658.
60. Fan, X.; Niehus, X.; Sandoval, G. Lipases as biocatalyst for biodiesel production. In *Lipases and Phospholipases. Methods in Molecular Biology*; Sandoval, G., Ed.; Springer-Humana Press: New York, NY, USA, 2012; Volume 861, Chapter 27.
61. Fukuda, H.; Kondo, A.; Noda, H. Biodiesel fuel production by transesterification of oils. *J. Biosci. Bioeng.* **2001**, *92*, 405–416.
62. Maleki, E.; Aroua, M.K.; Sulaiman, N.M.N. Improved yield of solvent free enzymatic methanolysis of palm and jatropha oils blended with castor oil. *Appl. Energ.* **2013**, *104*, 905–909.
63. Hama, S.; Kondo, A. Enzymatic biodiesel production: An overview of potential feedstocks and process development. *Bioresour. Technol.* **2013**, *135*, 386–395.
64. Al-Zuhair, S.; Hasan, M.; Ramachandran, K.B. Kinetics of the enzymatic hydrolysis of palm oil by lipase. *Process Biochem.* 2003, *38*, 1155–1163.
65. Salis, A.; Pinna, M.; Monduzzi, M.; Solinas, V. Biodiesel production from triolein and short chain alcohols through biocatalysis. *J. Biotechnol.* **2005**, *119*, 291–299.
66. Shimada, Y.; Watanabe, Y.; Samukawa, T.; Sugihara, A.; Noda, H.; Fukuda, H.; Tominaga, Y. Conversion of vegetable oil to biodiesel using immobilized *Candida antarctica* lipase. *J. Am. Oil Chem. Soc.* **1999**, *76*, 789–793.

67. Li, Q.; Xu, J.; Du, W.; Li, Y.; Liu, D. Ethanol as the acyl acceptor for biodiesel production. *Renew. Sust. Energ. Rev.* **2013**, *25*, 742–748.

68. Modi, M.K.; Reddy, J.R.; Rao, B.V.; Prasad, R.B. Lipase mediated conversion of vegetable oils into biodiesel using ethyl acetate as acyl acceptor. *Bioresour. Technol.* **2007**, *98*, 1260–1264.

69. Rivera, I.; Villanueva, G.; Sandoval, G. Biodiesel production from animal grease wastes by enzymatic catalysis. *Grasas Aceites* **2009**, *60*, 468–474.

70. Royon, D.; Daz, M.; Ellenrieder, G.; Locatelli, S. Enzymatic production of biodiesel from cotton seed oil using t-butanol as a solvent. *Bioresour. Technol.* **2007**, *98*, 648–653.

71. Abigor, R.D.; Uadia, P.O.; Foglia, T.A.; Haas, M.J.; Jones, K.C.; Okpefa, E.; Obibuzor, J.U.; Bafor, M.E. Lipase-catalysed production of biodiesel fuel from some Nigerian lauric oils. *Biochem. Soc. Trans.* **2000**, *28*, 979–981.

72. Nelson, L.; Foglia, T.; Marmer, W.N. Lipase-catalyzed production of biodiesel. *J. Am. Oil Chem. Soc.* **1996**, *73*, 1191–1195.

73. Linko, Y.Y.; Lamsä, M.; Huhtala, A.; Rantanen, O. Lipase biocatalysis in the production of esters. *J. Am. Oil Chem. Soc.* **1995**, *72*, 1293–1299.

74. Du, W.; Xu, Y.; Liu, D.; Zeng, J. Comparative study on lipase-catalyzed transformation of soybean oil for biodiesel production with different acyl acceptors. *J. Mol. Catal. B Enzym.* **2004**, *30*, 125–129.

75. Watanabe, Y.; Shimada, Y.; Sugihara, A.; Noda, H.; Fukuda, H.; Tominaga, Y. Continuous production of biodiesel fuel from vegetable oil using immobilized *Candida antarctica* lipase. *J. Am. Oil Chem. Soc.* **2000**, *77*, 355–360.

76. Zhang, B.; Weng, Y.; Xu, H.; Mao, Z. Enzyme immobilization for biodiesel production. *Appl. Microbiol. Biotechnol.* **2012**, *93*, 61–70.

77. Soumanou, M.M.; Bornscheuer, U.T. Lipase-catalyzed alcoholysis of vegetable oils. *Eur. J. Lipid Sci. Technol.* **2003**, *105*, 656–660.

78. Zhu, X.; Zhou, T.; Wu, X.; Cai, Y.; Yao, D.; Xie, C.; Liu, D. Covalent immobilization of enzymes within micro-aqueous organic media. *J. Mol. Catal. B Enzym.* **2011**, *3–4*, 145–149.

79. Zaks, A.; Klibanov, A.M. Enzyme-catalyzed processes in organic solvents. *Proc. Natl. Acad. Sci. USA* **1985**, *82*, 3192–3196.

80. Iso, M.; Chen, B.; Eguchi, M.; Kudo, T.; Shrestha, S. Production of biodiesel fuel from triglycerides and alcohol using immobilized lipase. *J. Mol. Catal. B Enzym.* **2001**, *16*, 53–58.

81. Chen, J.W.; Wu, W.T. Regeneration of immobilized *Candida antarctica* lipase for transesterification. *J. Biosci. Bioeng.* **2003**, *95*, 466–469.

82. Nie, K.; Xie, F.; Wang, F.; Tan, T. Lipase catalyzed methanolysis to produce biodiesel: optimization of the biodiesel production. *J. Mol. Catal. B Enzym.* **2006**, *43*, 142–147.

83. Ciftci, O.N.; Temelli, F. Enzymatic conversion of corn oil into biodiesel in a batch supercritical carbon dioxide reactor and kinetic modeling. *J. Supercrit. Fluids.* **2013**, *75*, 172–180.

84. Lin, Y.C.; Yang, P.M.; Chen, S.C.; Lin, J.F. Improving biodiesel yields from waste cooking oil using ionic liquids as catalysts with a microwave heating system. *Fuel Process. Technol.* **2013**, *115*, 57–62.

85. Earle, M.J.; Plechkova, N.V.; Seddon, K.R. Green synthesis of biodiesel using ionic liquids. *Pure Appl. Chem.* **2009**, *81*, 2045–2057.

86. Ha, S.H.; Lan, M.N.; Lee, S.H.; Hwang, S.M.; Koo, Y.M. Lipase-catalyzed biodiesel production from soybean oil in ionic liquids. *Enzyme Microb. Technol.* **2007**, *41*, 480–483.
87. Koda, R.; Numata, T.; Hama, S.; Tamalampudi, S.; Nakashima, K.; Tanaka, T; Ogino, C.; Fukuda, H.; Kondo, A. Ethanolysis of rapeseed oil to produce biodiesel fuel catalyzed by *Fusarium heterosporum* lipase-expressing fungus immobilized whole-cell biocatalysts. *J. Mol. Catal. B Enzym.* **2010**, *68*, 101–104.
88. Adachi, D.; Koha, F.; Hamab, S.; Ogino, C.; Kondo, A. A robust whole-cell biocatalyst that introduces a thermo- and solvent-tolerant lipase into *Aspergillus oryzae* cells: Characterization and application to enzymatic biodiesel production. *Enzyme Microb. Technol.* **2013**, *52*, 331–335.
89. Fukuda, H.; Hama, S.; Tamalampudi, S.; Noda, H. Whole-cell biocatalysts for biodiesel fuel production. *Trends Biotechnol.* **2008**, *26*, 668–673.
90. Atkinson, B.; Black, G.M.; Lewis, P.J.S.; Pinches, A. Biological particles of given size, shape, and density for use in biological reactors. *Biotechnol. Bioeng.* **1979**, *21*, 193–200.
91. Meng, X.; Yang, J.; Xu, X.; Zhang, L.; Nie, Q.; Xian, M. Biodiesel production from oleaginous microorganisms. *Renew. Energ.* **2009**, *34*, 1–5.
92. Turner, N.J.; Truppo, M.D. Biocatalysis enters a new era. *Curr. Opin. Chem. Biol.* **2013**, *17*, 212–214.
93. Woodley, J. New opportunities for biocatalysis: Making pharmaceutical processes greener. *Trends Biotechnol.* **2008**, *26*, 321–327.
94. Kittl, R.; Withers, S.G. New approaches to enzymatic glycoside synthesis through directed evolution. *Carbohyd. Res.* **2010**, *345*, 1272–1279.
95. Hay, A.S. Polymerization by oxidative coupling: Discovery and commercialization of PPO®. *J. Polym. Sci. Pol. Chem.* **1998**, *36*, 505–517.
96. He, F.; Li, S.; Garreau, H.; Vert, M.; Zhuo, R. Enzyme-catalyzed polymerization and degradation of copolyesters of ε-caprolactone and γ-butyrolactone. *Polymer* **2005**, *46*, 12682–12688.
97. Salvachúa, D.; Prieto, A.; Mattinen, M.L.; Tamminen, T.; Liitiä, T.; Lille, M.; Willför, S.; Martínez, A.T.; Martínez, M.J.; Faulds, C.B. Versatile peroxidase as a valuable tool for generating new biomolecules by homogeneous and heterogeneous cross-linking. *Enzyme Microb. Technol.* **2013**, *52*, 303–311.
98. Oguchi, T.; Tawaki, S.; Uyama, H.; Kobayashi, S. Enzymatic synthesis of soluble polyphenol. *Bull. Chem. Soc. Jpn.* **2000**, *73*, 1389–1396.
99. Dordick, J.S.; Marletta, M.A.; Klibanov, A.M. Polymerization of phenols catalyzed by peroxidase in nonaqueous media. *Biotech. Bioeng.* **1987**, *30*, 31–36.
100. Oguchi, T.; Tawaki, S.; Uyama, H.; Kobayashi, S. Soluble polyphenol. *Macromol. Rapid Commun.* **1999**, *20*, 401–403.
101. Mita, N.; Oguchi, T.; Tawaki, S.; Uyama, H.; Kobayashi, S. Control of structure and molecular weight of polyphenols in enzymatic oxidative polymerization. *Polymer Prepr.* **2000**, *41*, 223–224.
102. Kurioka, H.; Komatsu, I.; Uyama, H.; Kobayashi, S. Enzymatic oxidative polymerization of alkylphenols. *Macromol. Rapid Commun.* **1994**, *15*, 507–510.
103. Uyama, H.; Kurioka, H.; Kobayashi, S. Preparation of polyphenol particles by dispersion polymerization using enzyme as catalyst. *Chem. Lett.* **1995**, *24*, 795–796.

104. Uyama, H. Enzymatic polymerization. In *Future Directions in Biocatalysis*; Matsuda, T., Ed.; Elsevier Science: Cambridge, MA, USA, 2007; pp. 1–2.

105. Reihmann, M.; Ritter, H. Synthesis of phenol polymers using peroxidases. *Adv. Polym. Sci.* **2006**, *194*, 1–49.

106. Kobayashi, S.; Uyama, H.; Ikeda, R. Artificial Urushi. *Chem. Eur. J.* **2001**, *7*, 4755–4760.

107. Kim, Y.H.; An, E.S.; Song, B.K.; Kim, D.S.; Chelikani, R. Polymerization of cardanol using soybean peroxidase and its potential application as anti-biofilm coating material. *Biotechnol. Lett.* **2003**, *25*, 1521–1524.

108. Tonami, H.; Uyama, H.; Kobayashi, S.; Kubota, M. Peroxidase-catalyzed oxidative polymerization of m-substituted phenol derivatives. *Macromol. Chem. Phys.* **1999**, *200*, 2365–2371.

109. Kadota, J.; Fukuoka, T.; Uyama, H.; Hasegawa, K.; Kobayashi, S. New Positive-type photoresists based on enzymatically synthesized polyphenols. *Macromol. Rapid Commun.* **2004**, *25*, 441–444.

110. Antoniotti, S.; Santhanam, L.; Ahuja, D.; Hogg, M.G.; Dordick, J.S. Structural diversity of peroxidase-catalyzed oxidation products of o-methoxyphenols. *Org. Lett.* **2004**, *6*, 1975–1978.

111. Mita, N.; Tawaki, S.I.; Uyama, H.; Kobayashi, S. Laccase-catalyzed oxidative polymerization of phenols. *Macromol. Biosci.* **2003**, *3*, 253–257.

112. Ikeda, R.; Sugihara, J.; Uyama, H.; Kobayashi, S. Poly(2,6-dihydroxy-l,4-oxyphenylene synthesis of a new poly(phenylene oxide) derivative. *Polym. Bull.* **1997**, *38*, 273–277.

113. Ikeda, R.; Sugihara, J.; Uyama, H.; Kobayashi, S. Enzymatic oxidative polymerization of 2,6-dimethylphenol. *Macromolecules* **1996**, *29*, 8702–8705.

114. Uyama, H.; Kurioka, H.; Sugihara, J.; Komatsu, I.; Kobayashi, S. Oxidative polymerization of p-alkylphenols catalyzed by horseradish peroxidase. *J. Polym. Sci. Pol. Chem.* **1997**, *35*, 1453–1459.

115. Wang, P.; Martin, B.D.; Parida, S.; Rethwisch, D.G.; Dordick, J.S. Multienzymic synthesis of poly(hydroquinone) for use as a redox polymer. *J. Am. Chem. Soc.* **1995**, *117*, 12885–12886.

116. Akkara, J.A.; Senecal, K.J.; Kaplan, D.L. Synthesis and characterization of polymers produced by horseradish peroxidase in dioxane. *J. Polym. Sci. Pol. Chem.* **1991**, *29*, 1561–1574.

117. Akita, M.; Tsutsumi, D.; Kobayashi, M.; Kise, H. Structural change and catalytic activity of horseradish peroxidase in oxidative polymerization of phenol. *Biosci. Biotech. Bioch.* **2001**, *65*, 1581–1588.

118. Angerer, P.S.; Studer, A.; Witholt, B.; Li, Z. Oxidative polymerization of a substituted phenol with ion-paired horseradish peroxidase in an organic solvent. *Macromolecules* **2005**, *38*, 6248–6250.

119. Eker, B.; Zagorevski, D.; Zhu, G.; Linhardt, R.J.; Dordick, J.S. Enzymatic polymerization of phenols in room-temperature ionic liquids. *J. Mol. Catal. B Enzym.* **2009**, *59*, 177–184.

120. Mita, N.; Tawaki, S.; Uyama, H.; Kobayashi, S. Enzymatic oxidative polymerization of phenol in an aqueous solution in the presence of a catalytic amount of cyclodextrin. *Macromol. Biosci.* **2002**, *3*, 127–130.

121. Kim, Y.J.; Uyama, H.; Kobayashi, S. Regioselective synthesis of poly(phenylene) as a complex with poly(ethylene glycol) by template polymerization of phenol in water. *Macromolecules* **2003**, *36*, 5058–5060.

122. Kommareddi, N.S.; Tata, M.; Karayigitoglu, C.; John, V.T.; McPherson, G.L.; Herman, M.F.; Oconnor, C.J.; Lee, Y.S.; Akkara, J.A.; Kaplan, D.L. Enzymatic polymerizations using surfactant microstructures and the preparation of polymer-ferrite composites. *Appl. Biochem. Biotechnol.* **1995**, *51–52*, 241–252.
123. Ghan, R.; Shutava, T.; Patel, A.; John, V.T.; Lvov, Y. Enzyme-catalyzed polymerization of phenols within polyelectrolyte microcapsules. *Macromolecules* **2004**, *37*, 4519–4524.
124. Marín, F.R.; Frutos, M.J.; Pérez-Alvarez, J.A.; Martinez-Sánchez, F.; del Río, J.A. Flavonoids as nutraceuticals: Structural related antioxidant properties and their role on ascorbic acid preservation. *Stud. Nat. Prod. Chem.* **2002**, *26*, 741–778.
125. Di Carlo, G.; Mascolo, N.; Izzo, A.A.; Capasso, F. Flavonoids: Old and new aspects of a class of natural therapeutic drugs. *Life Sci.* **1999**, *65*, 337–353.
126. Rice-Evans, C.A.; Miller, N.J.; Paganga, G. Structure-antioxidant activity relationships of flavonoids and phenolic acids. *Free Radic. Biol. Med.* **1996**, *20*, 933–956.
127. *Flavonoids in Biology and Medicine III—Current Issues in Flavonoids Research*; Das, N.P., Cheeseman, K.H., Eds.; Informa Healthcare: London, UK, 1991; Volume 14, pp. 77–78.
128. Mejias, L.; Reihmann, M.H.; Sepulveda-Boza, S.; Ritter, H. New polymers from natural phenols using horseradish or soybean peroxidase. *Macromol. Biosci.* **2002**, *2*, 24–32.
129. Kurisawa, M.; Chung, J.E.; Uyama, H.; Kobayashi, S. Laccase-catalyzed synthesis and antioxidant property of poly(catechin). *Macromol. Biosci.* **2003**, *3*, 758–764.
130. Kurisawa, M.; Chung, J.E.; Uyama, H.; Kobayashi, S. Oxidative coupling of epigallocatechin gallate amplifies antioxidant activity and inhibits xanthine oxidase activity. *Chem. Commun.* **2004**, doi:10.1039/B312311A.
131. Kurisawa, M.; Chung, J.E.; Kim, Y.J.; Uyama, H.; Kobayashi, S. Amplification of antioxidant activity and xanthine oxidase inhibition of catechin by enzymatic polymerization. *Biomacromolecules* **2003**, *4*, 469–471.
132. Gonçalves, I.; Matamá, T.; Cavaco-Paulo, A.; Silva, C. Laccase coating of catheters with poly(catechin) for biofilm reduction. *Biocatal. Biotransform.* **2013**, in press.
133. Kurisawa, M.; Chung, J.E.; Uyama, H.; Kobayashi, S. Enzymatic synthesis and antioxidant properties of poly(rutin). *Biomacromolecules* **2003**, *4*, 1394–1399.
134. Božič, M.; Gorgieva, S.; Kokol, V. Laccase-mediated functionalization of chitosan by caffeic and gallic acids for modulating antioxidant and antimicrobial properties. *Carbohyd. Polym.* **2012**, *87*, 2388–2398.
135. Brzonova, I.; Steiner, W.; Zankel, A.; Nyanhongo, G.S. Enzymatic synthesis of catechol and hydroxyl-carboxic acid functionalized chitosan microspheres for iron overload therapy. *Eur. J. Pharm. Biopharm.* **2011**, *79*, 294–303.
136. Fras-Zemljič, L.; Kokol, V.; Čakara, D. Antimicrobial and antioxidant properties of chitosan-based viscose fibres enzymatically functionalized with flavonoids. *Text. Res. J.* **2011**, *81*, 1532–1540.
137. Sousa, F.; Guebitz, G.M.; Kokol, V. Antimicrobial and antioxidant properties of chitosan enzymatically functionalized with flavonoids. *Process Biochem.* **2009**, *44*, 749–756.
138. Pina-Luis, G.; Rosquete-Pina, G.; Valdés, A.C.; Ochoa, A.; Rivero, I.; Díaz-García, M.E. Morin functionalized Merrifield's resin: A new material for enrichment and sensing heavy metals. *React. Funct. Polym.* **2012**, *72*, 61–68.

139. Donato, L.; Chiappetta, G.; Drioli, E. Naringin-imprinted polymer layer using photo polymerization method. *Sep. Sci. Technol.* **2011**, *46*, 1555–1562.
140. Spizzirri, U.G.; Parisi, O.I.; Iemma, F.; Cirillo, G.; Puoci, F.; Curcio, M.; Picci, N. Antioxidant-polysaccharide conjugates for food application by eco-friendly grafting procedure. *Carbohyd. Polym.* **2010**, *79*, 333–340.
141. Jaeger, K.E.; Eggert, T. Lipases for biotechnology. *Curr. Opin. Biotechnol.* **2002**, *13*, 390–397.
142. Sharma, R.; Chisti, Y.; Banerjee, U.C. Production, purification, characterization and applications of lipases. *Biotechnol. Adv.* **2001**, *19*, 627–662.
143. Yu, Y.; Wu, D.; Liu, C.; Zhao, Z.; Yang, Y.; Li, Q. Lipase/esterase-catalyzed synthesis of aliphatic polyesters via polycondensation: A review. *Process Biochem.* **2012**, *47*, 1027–1036.
144. Yang, Y.; Yu, Y.; Zhang, Y.; Liu, C.; Shi, W.; Li, Q. Lipase/esterase-catalyzed ring-opening polymerization: A green polyester synthesis technique. *Process Biochem.* **2011**, *46*, 1900–1908.
145. He, F.; Wang, Y.P.; Liu, G.; Jia, H.L.; Feng, J.; Zhuo, R.X. Synthesis, characterization and ring-opening polymerization of a novel six-membered cyclic carbonate bearing pendent allyl ether group. *Polymer* **2008**, *49*, 1185–1190.
146. Runge, M.; O'Hagan, D.; Haufe, G. Lipase-catalyzed polymerization of fluorinated lactones and fluorinated hydroxycarboxylic acids. *J. Polym. Sci. Pol. Chem.* **2000**, *38*, 2004–2012.
147. Zinck, P. One-step synthesis of polyesters specialties for biomedical applications. *Rev. Environ. Sci. Biotechnol.* **2009**, *8*, 231–234.
148. Kobayashi, S.; Makino, A. Enzymatic polymer synthesis: An opportunity for green polymer chemistry. *Chem. Rev.* **2009**, *109*, 5288–5353.
149. Uyama, H.; Takamoto, T.; Kobayashi, S. Enzymatic synthesis of polyesters in ionic liquids. *Polym. J.* **2002**, *34*, 94–96.
150. Chaudhary, A.K.; Beckman, E.J.; Russell, A.J.; Rational control of polymer molecular weight and dispersity during enzyme-catalyzed polyester synthesis in supercritical fluids. *J. Am. Chem. Soc.* **1995**, *117*, 3728–3733.
151. Okumura, S.; Iwai, M.; Tominaga, T. Synthesis of ester oligomer by *Aspergillus niger* lipase. *Agr. Biol. Chem. (Tokyo)* **1984**, *48*, 2805–2813.
152. Binns, F.; Roberts, S.M.; Taylor, A.; Williams, C.F. Enzymic polymerization of unactivated diol/diacid system. *J. Chem. Soc. Perkin Trans. 1* **1993**, *1*, 899–904.
153. Linko, Y.Y.; Seppala, J. Producing high molecular weight biodegradable polyesters. *Chem. Tech.* **1996**, *26*, 25–31.
154. Gómez-Patiño, M.B.; Cassani, J.; Jaramillo-Flores, M.E.; Zepeda-Vallejo, L.G.; Sandoval, G.; Jimenez-Estrada, M.; Arrieta-Baez, D. Oligomerization of 10,16-dihydroxyhexadecanoic acid and methyl 10,16-dihydroxyhexadecanoate catalyzed by lipases. *Molecules* **2013**, *18*, 9317–9333.
155. Kobayashi, S.; Uyama, H.; Suda, S.; Namekawa, S. Dehydration polymerization in aqueous medium catalyzed by lipase. *Chem. Lett.* **1997**, *26*, 105–107.
156. Poojari, Y.; Palsule, A.S.; Cai, M.; Clarson, S.J.; Gross, R.A. Synthesis of organosiloxane copolymers using enzymatic polyesterification. *Eur. Polym. J.* **2008**, *44*, 4139–4145.
157. Yang, Y.; Lu, W.; Cai, J.; Hou, Y.; Ouyang, S.; Xie, W.; Gross, R.A. Poly(oleicdiacid-co-glycerol): Comparison of polymer structure resulting from chemical and lipase catalysis. *Macromolecules* **2011**, *44*, 1977–1985.

158. Korupp, C.; Weberskirch, R.; Muller, J.J.; Liese, A.; Hilterhaus, L. Scaleup of lipase-catalyzed polyester synthesis. *Org. Process. Res. Dev.* **2010**, *14*, 1118–1124.
159. Fehling, E.; Bergander, K.; Klein, E.; Weber, N.; Vosmann, K. Thiol-functionalized copolymeric polyesters by lipase-catalyzed esterification and transesterification of 1,12-dodecanedioic acid and its diethyl ester, respectively, with 1-thioglycerol. *Biotechnol. Lett.* **2010**, *32*, 1463–1471.
160. Hu, J.; Gao, W.; Kulshrestha, A.S.; Gross, R.A. Sweet polyesters: Lipase-catalyzed condensation polymerization of alditols. *Macromolecules* **2006**, *39*, 6789–6792.
161. Uyama, H.; Kobayashi, S. Enzymatic synthesis of polyesters via polycondensation. *Adv. Polym. Sci.* **2006**, *194*, 133–158.
162. Brazwell, E.M.; Filos, D.Y.; Morrow, C.J. Biocatalytic synthesis of polymers. III. Formation of a high molecular weight polyester through limitation of hydrolysis by enzyme-bound water and through equilibrium control. *J. Polym. Sci. Pol. Chem.* **1995**, *33*, 89–95.
163. Kobayashi, S. Recent developments in lipase-catalyzed synthesis of polyesters. *Macromol. Rapid Commun.* **2009**, *30*, 237–266.
164. Azim, H.; Dekhterman, A.; Jiang, Z.; Gross, R.A. *Candida antarctica* lipase B-catalyzed synthesis of poly(butylene succinate): Shorter chain building blocks also work. *Biomacromolecules* **2006**, *7*, 3093–3097.
165. Mesiano, A.J.; Beckman, E.J.; Russell, A.J. Biocatalytic synthesis of fluorinated polyesters. *Biotechnol. Prog.* **2000**, *16*, 64–68.
166. Uyama, H.; Yaguchi, S.; Kobayashi, S. Lipase-catalyzed polycondensation of dicarboxylic acid-divinyl esters and glycols to aliphatic polyesters. *J. Polym. Sci. Pol. Chem.* **1999**, *37*, 2737–2745.
167. Yao, D.; Li, G.; Kuila, T.; Li, P.; Kim, N.H.; Kim, S.I.; Lee, J.H. Lipase-catalyzed synthesis and characterization of biodegradable polyester containing l-malic acid unit in solvent system. *J. Appl. Polym. Sci.* **2011**, *120*, 1114–1120.
168. Steunenberg, P.; Uiterweerd, M.; Sijm, M.; Scott, E.L.; Zuilhof, H.; Sanders, J.P.M.; Franssen, M.C.R. Enzyme-catalyzed polymerization of β-alanine esters, a sustainable route towards the formation of poly-β-alanine. *Curr. Org. Chem.* **2013**, *17*, 682–690.
169. Kumar, D.; Bhalla, T.C. Microbial proteases in peptide synthesis: Approaches and applications. *Appl. Microbiol. Biotechnol.* **2005**, *68*, 726–736.
170. Kuhn, D.; Durrschmidt, P.; Mansfeld, J.; Ulbrich-Hofmann, R. Boilysin and thermolysin in dipeptide synthesis: A comparative study. *Biotechnol. Appl. Biochem.* **2002**, *36*, 71–76.
171. Bille, V.; Ripak, C.; Assche, I.; Forni, L.; Degelaen, J.; Searso, J. Semi-enzymic synthesis of somatostatin. In Proceedings of 21st European Peptide Symposium, Platja d'Aro, Spain, 2–8 September 1990.
172. Gu, Q.M.; Maslanka, W.W.; Cheng, H.N. Enzyme-catalyzed polyamides and their derivatives. In *Polymer Biocatalysis and Biomaterials II*; *ACS Symposium Series 999*; Cheng, H.N., Gross, R.A., Eds.; Oxford University Press: Washington, DC, USA, 2008; pp 309–319.
173. Poulhès, F.; Mouysset, D.; Gil, G.; Bertrand, M.P.; Gastaldi, S. Speeding-up enzyme-catalyzed synthesis of polyamides using ω-amino-a-alkoxy-acetate as monomer. *Polymer* **2013**, *54*, 3467–3471.

174. Ikeda, R.; Tanaka, H.; Uyama, H.; Kobayashi, S. Laccase-catalyzed polymerization of acrylamide. *Macromol. Rapid Commun.* **1998**, *19*, 423–425.

175. Hollmann, F.; Gumulya, Y.; Tölle, C.; Liese, A.; Thum, O. Evaluation of the laccase from *Myceliophthora thermophila* as industrial biocatalyst for polymerization reactions. *Macromolecules* **2008**, *41*, 8520–8524.

176. Emery, O.; Lalot, T.; Brigodiot, M.; Maréchal, E. Free-radical polymerization of acrylamide by horseradish peroxidase-mediated initiation. *J. Polym. Sci. Pol. Chem.* **1997**, *35*, 3331–3333.

177. Mucientes, A.E.; Santiago, F.; Carrero, A.M.; Talavera, B. Superabsorbent hydrogels of poly(sodium acrylate) with crude and exfoliated vermiculites. *J. Polym. Eng.* **2013**, *33*, 61–69.

178. Jones, N.A.; Atkins, E.D.T.; Hill, M.J.; Cooper, S.J.; Franco, L. Chain-folded lamellar crystals of aliphatic polyamides. comparisons between nylons 44, 64, 84, 104, and 124. *Macromolecules* **1996**, *29*, 6011–6018.

179. Ragupathy, L.; Ziener, U.; Dyllick-Brenzinger, R.; von Vacano, B.; Landfester, K. Enzyme-catalyzed polymerizations at higher temperatures: Synthetic methods to produce polyamides and new poly(amide-co-ester)s. *J. Mol. Catal. B Enzym.* **2012**, *76*, 94–105.

180. Kalra, B.; Gross, R.A. Horseradish peroxidase mediated free radical polymerization of methyl methacrylate. *Biomacromolecules* **2000**, *1*, 501–505.

181. Singh, A.; Ma, D.C.; Kaplan, D.L. Enzyme-mediated free radical polymerization of styrene. *Biomacromolecules* **2000**, *1*, 592–596.

182. Sandoval, G.; Rivera, I.; Barrera-Rivera, K.A.; Martínez-Richa, A. Biopolymer synthesis catalyzed by tailored lipases. *Macromol. Symp.* **2010**, *289*, 135–139.

183. Varma, I.K.; Albertsson, A.C.; Rajkhowa, R.; Srivastava, R.K. Enzyme catalyzed synthesis of polyesters. *Prog. Polym. Sci.* **2005**, *30*, 949–981.

184. Kobayashi, S. Enzymatic polymerization: A new method of polymer synthesis. *J. Polym. Sci. Pol. Chem.* **1999**, *37*, 3041–3056.

185. Kobayashi, S.; Uyama, H.; Ohmae, M. Enzymatic polymerization for precision polymer synthesis. *Bull. Chem. Soc. Jpn.* **2001**, *74*, 613–635.

186. Uyama, H.; Kikuchi, H.; Takeya, K.; Kobayashi, S. Lipase-catalyzed ring-opening polymerization and copolymerization of 15-pentadecanolide. *Acta Polym.* **1996**, *47*, 357–360.

187. Uyama, H.; Takeya, K.; Kobayashi, S. Enzymatic ring-opening polymerization of lactones to polyesters by lipase catalyst: Unusually high reactivity of macrolides. *Bull. Chem. Soc. Jpn.* **1995**, *68*, 56–61.

188. Uyama, H.; Takeya, K.; Hoshi, N.; Kobayashi, S. Lipase-catalyzed ring-opening polymerization of 12-dodecanolide. *Macromolecules* **1995**, *28*, 7046–7050.

189. Bisht, K.S.; Henderson, L.A.; Gross, R.A. Enzyme-catalyzed ring-opening polymerization of ω-pentadecalactone. *Macromolecules* **1997**, *30*, 2705–2711.

190. Hunsen, M.; Azim, A.; Mang, H.; Wallner, S.R.; Ronkvist, A.; Xie, W.C.; Gross, R.A. A cutinase with polyester synthesis activity. *Macromolecules* **2007**, *40*, 148–150.

191. Nishida, H.; Yamashita, M.; Nagashima, M.; Endo, T.; Tokiwa, Y. Synthesis of metal-free poly(1,4-dioxan-2-one) by enzyme-catalyzed ring-opening polymerization. *J. Polym. Sci. Pol. Chem.* **2000**, *38*, 1560–1567.

192. Jiang, Z.; Azim, H.; Gross. R.A. Lipase-catalyzed copolymerization of ω-pentadecalactone with *p*-dioxanone and characterization of copolymer thermal and crystalline properties. *Biomacromolecules* **2007**, *8*, 2262–2269.
193. Fickers, P.; Marty, A.; Nicaud, J.M. The lipases from *Yarrowia lipolytica*: Genetics, production, regulation, biochemical characterization and biotechnological applications. *Biotechnol. Adv.* **2011**, *29*, 632–644.
194. Uyama, H.; Kikuchi, H.; Takeya, K.; Hoshi, N.; Kobayashi, S. Immobilized lipase showing high catalytic activity toward enzymatic ring-opening polymerization of macrolides. *Chem. Lett.* **1996**, *25*, 107–108.
195. Kobayashi, S.; Uyama, H. Precision enzymatic polymerization to polyesters with lipase catalysts. *Macromol. Symp.* **1999**, *144*, 237–246.
196. Kobayashi, S.; Uyama, H.; Namekawa, S. *In vitro* biosynthesis of polyesters with isolated enzymes in aqueous systems and organic solvents. *Polym. Degrad. Stabil.* **1998**, *59*, 195–201.
197. Kumar, A.; Kalra, B.; Dekhterman, A.; Gross, R.A. Efficient ring-opening polymerization and copolymerization of ε-caprolactone and ω-pentadecalactone catalyzed by *Candida antartica* lipase B. *Macromolecules* **2000**, *33*, 6303–6309.
198. Kobayashi, S.; Uyama, H.; Namekawa, S.; Hayakawa, H. Enzymatic ring-opening polymerization and copolymerization of 8-octanolide by lipase catalyst. *Macromolecules* **1998**, *31*, 5655–5659.
199. Namekawa, S.; Uyama, H.; Kobayashi, S. Lipase-catalyzed ring-opening polymerization of 16-hexadecanolide. *P. Jpn. Acad. B* **1998**, *74*, 65–68.
200. Ebata, H.; Toshima, K.; Matsumura, S. Lipase-catalyzed transformation of poly(ε-caprolactone) into cyclic dicaprolactone. *Biomacromolecules* **2000**, *1*, 511–514.
201. Sugihara, S.; Toshima, K.; Matsumura, S. New strategy for enzymatic synthesis of high-molecular-weight poly(butylene succinate) via cyclic oligomers. *Macromol. Rapid Commun.* **2006**, *27*, 203–207.
202. Kikuchi, H.; Uyama, H.; Kobayashi, S. Lipase-catalyzed ring-opening polymerization of substituted lactones. *Polym. J.* **2002**, *34*, 835–893.
203. Küllmer, K.; Kikuchi, H.; Uyama, H.; Kobayashi, S. Lipase-catalyzed ring-opening polymerization of α-methyl-δ-valerolactone and α-methyl-ε-caprolacton. *Macromol. Rapid. Commun.* **1998**, *19*, 127–130.
204. Cordova, A.; Iversen, T.; Martinelle, M. Lipase-catalysed formation of macrocycles by ring-opening polymerisation of ϵ-caprolactone. *Polymer* **1998**, *39*, 6519–6524.
205. Namekawa, S.; Uyama, H.; Kobayashi, S. Lipase-catalyzed ring-opening and copolymerization of β-propiolactione. *Polym. J.* **1996**, *28*, 730–731.
206. Suzuki, Y.; Taguchi, S.; Hisano, T.; Toshima, K.; Matsumura, S.; Doi, Y. Correlation between structure of the lactones and substrate specificity in enzyme-catalyzed polymerization for the synthesis of polyesters. *Biomacromolecules* **2003**, *4*, 537–543.
207. Nagasawa, T.; Shimizu, H.; Yamada, H. The superiority of the third-generation catalyst, *Rhodococcus rhodochrous* J1 nitrile hydratase, for industrial production of acrylamide. *Appl. Microbiol. Biotechnol.* **1993**, *40*, 189–195.
208. Thomas, S.M.; DiCosimo, R.; Nagarajan, V. Biocatalysis: Applications and potentials for the chemical industry. *Trends Biotechnol.* **2002**, *20*, 238–242.

209. Chassin, C. A biotechnological process for the production of nicotinamide. *Chim. Oggi.* **1996**, *14*, 9–12.
210. Nagasawa, T.; Matthew, C.D.; Mauger, J.; Yamada, H. Nitrile hydratase-catalyzed production of nicotinamide from 3-cyanopyridine in *Rhodococcus rhodochrous* J1. *Appl. Environ. Microbiol.* **1988**, *54*, 1766–1769.
211. Robins, K.T.; Nagasawa, T. Process for Preparing Amides. PCT Int. Appl. US 7,666,635 B2, 23 February 2010.
212. Heveling, J.; Armbruster, E.; Utiger, L.; Rhoner, M.; Dettwiler, H.R.; Chuck, R.J. Process for Preparing Nicotinamide. US Patent 5,719,045, 17 February 1998.
213. Straathof, A.J.J. Transformation of biomass into commodity chemicals using enzymes or cells. *Chem. Rev.* **2013**, doi:10.1021/cr400309c.
214. Vuyyuru, K.R.; Strasser, P. Oxidation of biomass derived 5-hydroxymethylfurfural using heterogeneous and electrochemical catalysis. *Catal. Today* **2012**, *195*, 144–154.
215. Casanova, O.; Iborra, S.; Corma, A. Biomass into chemicals: Aerobic oxidation of 5-hydroxymethyl-2-furfural into 2,5-furandicarboxylic acid with gold nanoparticle catalysts. *ChemSusChem* **2009**, *2*, 1138–1144.
216. Koopman, F.; Wierckx, N.; de Winde, J.H.; Ruijssenaars, H.J. Efficient whole-cell biotransformation of 5-(hydroxymethyl)furfural into FDCA, 2,5-furandicarboxylic acid *Bioresour. Technol.* **2010**, *101*, 6291–6296.
217. Krystof, M.; Pérez-Sánchez, M.; Domínguez de María, P. Lipase-mediated selective oxidation of furfural and 5-hydroxymethylfurfural. *ChemSusChem* **2013**, *6*, 826–830.
218. Van Deurzen, M.P.J.; van Rantwijk, F.; Sheldon, R.A. Chloroperoxidase-catalyzed oxidation of 5-hydroxymethylfurfural. *J. Carbohydr. Chem.* **1997**, *16*, 299–309.
219. Panke, S.; Wubbolts, M. Advances in biocatalytic synthesis of pharmaceutical intermediates. *Curr. Opin. Chem. Biol.* **2005**, *9*, 188–194.
220. Ooshima, H.; Mori, H.; Harano, Y. Synthesis of aspartame precursor by solid thermolysin in organic solvent. *Biotechnol. Lett.* **1985**, *7*, 789–792.
221. Griengl, H.; Klempier, N.; Pochlauer, P.; Schmidt, M.; Shi, N.Y.; Zabelinskaja-Mackova, A.A. Enzyme catalysed formation of (*S*)-cyanohydrins derived from aldehydes and ketones in a biphasic solvent system. *Tetrahedron* **1998**, *54*, 14477–14486.
222. Bieber, L.L. Carnitine. *Annu. Rev. Biochem.* **1988**, *57*, 261–283.
223. Fritz, I.B. Action of carnitine on long chain fatty acid oxidation by liver. *Am. J. Physiol.* **1959**, *197*, 297–304.
224. Meyer, H.P.; Kiener, A.; Imwinkelried, R.; Shaw, N. Biotransformations for fine chemicals production. *Chimia* **1997**, *51*, 287–289.
225. Bommarius, A.S.; Schwarm, M.; Stingl, K.; Kottenhahn, M.; Huthmacher, K.; Drauz, K. Synthesis and use of enantiomerically pure *tert*-leucine. *Tetrahedron Asymmetry* **1995**, *6*, 2851–2888.
226. Menzel, A.; Werner, H.; Altenbuchner, J.; Gröger, H. From enzymes to "designer bugs" in reductive amination: A new process for the synthesis of L-*tert*-leucine using a whole cell-catalyst. *Eng. Life Sci.* **2004**, *4*, 573–576.

227. Krix, G.; Bommarius, A.S.; Drauz, K.; Kottenhahn, M.; Schwarm, M.; Kula, M.R. Enzymatic reduction of α-keto acids leading to L-amino acids, D- or L-hydroxy acids. *J. Biotechnol.* **1997**, *53*, 29–39.
228. Furuhashi, K. Biological routes to optically active epoxides. In *Chirality in Industry*; Collins, A.N., Sheldrake, G.N., Crosby, J.C., Eds.; John Wiley & Sons: NJ, USA, 1992; pp. 167–186.
229. Li, K.; Frost, J. Synthesis of vanillin from glucose. *J. Am. Chem. Soc.* **1998**, *120*, 10545–10546.
230. Aresta, M.; Quaranta, E.; Liberio, R.; Dileo, C.; Tommasi, I. Enzymatic synthesis of 4-OH-benzoic acid from phenol and CO_2: The first example of a biotechnological application of a carboxylase enzyme. *Tetrahedron* **1998**, *54*, 8841–8846.
231. Baldessari, A. Lipases as catalysts in synthesis of fine chemicals. In *Lipases and Phospholipases. Methods in Molecular Biology*; Sandoval, G., Ed.; Springer-Humana Press: New York, NY, USA, 2012; Volume 861, pp. 445–448.
232. Pal, S.; Nair, V. Enzymatic synthesis of thymidine using bacterial whole cells and isolated purine nucleoside phosphorylase. *Biocatal. Biotransform.* **1997**, *15*, 147–158.
233. Liese, A.; Villela-Filho, M. Production of fine chemicals using biocatalysis. *Curr. Opin. Biotechnol.* 1999, *10*, 595–603.
234. Rivera, I.; Mateos, J.C.; Marty, A.; Sandoval, G.; Duquesne, S. Lipase from Carica papaya latex presents high enantioselectivity toward the resolution of prodrug (*R,S*)-2-bromophenylacetic acid octyl ester. *Tetrahedron Lett.* **2013**, *54*, 5523–5526.

9

Decarboxylation of Pyruvate to Acetaldehyde for Ethanol Production by Hyperthermophiles

Mohammad S. Eram † **and Kesen Ma** *

Department of Biology, University of Waterloo, 200 University Avenue West, Waterloo, Ontario N2L 3G1, Canada; E-Mail: smeram@uwaterloo.ca

† Current address: Structural Genomics Consortium, University of Toronto, Toronto, Ontario M5G 1L7, Canada.

* Author to whom correspondence should be addressed; E-Mail: kma@uwaterloo.ca

---

Abstract: Pyruvate decarboxylase (PDC encoded by *pdc*) is a thiamine pyrophosphate (TPP)-containing enzyme responsible for the conversion of pyruvate to acetaldehyde in many mesophilic organisms. However, no *pdc*/PDC homolog has yet been found in fully sequenced genomes and proteomes of hyper/thermophiles. The only PDC activity reported in hyperthermophiles was a bifunctional, TPP- and CoA-dependent pyruvate ferredoxin oxidoreductase (POR)/PDC enzyme from the hyperthermophilic archaeon *Pyrococcus furiosus*. Another enzyme known to be involved in catalysis of acetaldehyde production from pyruvate is CoA-acetylating acetaldehyde dehydrogenase (AcDH encoded by *mhpF* and *adhE*). Pyruvate is oxidized into acetyl-CoA by either POR or pyruvate formate lyase (PFL), and AcDH catalyzes the reduction of acetyl-CoA to acetaldehyde in mesophilic organisms. AcDH is present in some mesophilic (such as clostridia) and thermophilic bacteria (e.g., *Geobacillus* and *Thermoanaerobacter*). However, no AcDH gene or protein homologs could be found in the released genomes and proteomes of hyperthermophiles. Moreover, no such activity was detectable from the cell-free extracts of different hyperthermophiles under different assay conditions. In conclusion, no commonly-known PDCs was found in hyperthermophiles. Instead of the commonly-known PDC, it appears that at least one multifunctional enzyme is responsible for catalyzing the non-oxidative decarboxylation of pyruvate to acetaldehyde in hyperthermophiles.

Keywords: pyruvate decarboxylase; hyperthermophiles; alcohol fermentation; alcohol dehydrogenase; pyruvate ferredoxin oxidoreductase; pyruvate; acetaldehyde; ethanol

1. Introduction

Thermophilic microorganisms can be categorized into several groups: moderate thermophiles, or simply thermophiles, are those that grow optimally between 50–64 °C, extreme thermophiles are those with optimal growth temperatures between 65–79 °C. Finally, the organisms that can grow optimally above 80 °C are called hyperthermophiles. Hyperthermophiles can survive at room temperature for long periods of time, but cannot propagate at temperatures lower than 50 °C [1–5].

Hyperthermophilic microorganisms are widely studied for their remarkable scientific values and industrial potential. It is generally accepted that hyperthermophilic enzymes have very similar functions and catalytic mechanisms to their mesophilic ones. However, most of the hyperthermophilic enzymes characterized so far have optimum temperatures close to the host organism's growth requirements; thus, due to their intrinsic properties, the enzymes are stable and active under conditions that are detrimental to their mesophilic counterparts. Interestingly, enzymes from such extremophiles usually show increased stability not to one, but to several environmental factors. There are a number of advantages for using the hyper/thermophilic enzymes (especially for industrial applications) over their mesophilic partners, including the reduced risk of contamination during industrial processes, the possibility of self-distillation of the products at high temperatures, decreased viscosity and increased solubility/bioavailability of both the enzyme and the substrate(s) leading to minimization of the diffusion limitations, and elimination of the costly transportation under cold temperature-controlled environment [6–10].

The aforementioned properties along with high demand from the biotech industries for the development of "tailor-made" bio-catalysts have created significant attention on the biochemistry and physiology of these organisms. Enzymes such as proteases, polymerases, hydrolases, isomerases, lipases, and oxidases are studied for their potential biotechnological exploitation, with the ultimate goal of using them or their products (mainly enzymatic) for biotechnological applications.

It is also appealing to determine the molecular, biochemical, physiological, and evolutionary mechanisms that enable these organisms to adapt to such hostile environments. Furthermore, hyperthermophilic proteins serve as models to study enzyme evolution, structure-function relationships and catalytic mechanisms. The findings of these studies can benefit the design of highly stable and active enzymes to be used for many applications [6,11,12].

2. Microbial Production of Ethanol

Demand for biofuel as substitutes for oil-based fuels is increasing due to concerns related to national security, economic stability, environmental impacts, and global warming. The national research council of the United States has predicted that, by 2020, half of all organic chemicals and materials will be produced by bioconversion. Bio-ethanol can also be used as a precursor for many other commodity chemicals, such as acetaldehyde, acetic acid and their derivatives [13–15].

The most commonly used ethanologenic organisms being intensively studied or already in use for industrial-scale production are *Zymomonas mobilis*, *Saccharomyces cervisiae*, *Escherichia coli*, and *Klebsiella oxytoca*. Substantial attention and effort have been dedicated to redirecting the metabolic pathways of these and others towards higher yield of ethanol production, by means of metabolic engineering [16–18].

However, lack of suitable microorganisms that can efficiently convert the raw biomaterials to bio-ethanol has been one of the main obstacles to widespread use of bio-fuels. In addition to the ability to ferment a wide variety of sugars, some other features must be considered when choosing organisms for industrial-scale bio-ethanol production. These important features include but are not limited to the ability to have high ethanol yield, tolerance to fermentation products/by-products, simple growth requirements, and the ability to grow under conditions that prevent contaminating organisms from growing [14,16].

Production of bio-ethanol using thermophilic and hyperthermophilic organisms is the focus of many research groups. Extremophiles in general and hyperthermophiles in particular are outstanding organisms that produce highly stable enzymes due to their natural habitats, and many of them are able to tolerate changes in environment; making them good candidates for bio-ethanol production [19].

Several distinct advantages are associated with using thermophiles over mesophiles, including high temperatures and the mostly anaerobic nature of thermophilic organisms, which result in elimination of oxygenation and cooling of the fermenter. Another aspect is improved solubility of many reaction components at elevated temperatures [20]. In addition, the high temperature of the process leads to lowering the viscosity of reaction mixtures, causing improved production yields. Various thermophiles can ferment hexose and/or pentose sugars, as well as more complex substrates such as cellulose and xylan in some cases. Many of these organisms and their enzymes are relatively resistant to sudden pH or temperature changes and high concentrations of solvents [21–23]. High temperatures can result in lower gas solubility and significantly decrease the risk of process failure and product loss due to contamination that is the common problem in the yeast-based fermentation system. At the same time, high temperatures lower the cost of ethanol recovery due to the high volatility of ethanol at high fermentation temperatures [14,19,24,25]. However, there are some disadvantages associated with using hyperthermophiles, the most important one being their intrinsic low substrate tolerance and product/by-product inhibition [26,27]. Moreover, some of these organisms are mixed-fermenters that result in production of sometimes too many types of products during growth [14,16].

Application of metabolic engineering approaches has had a great impact on elimination of the problems associated with using thermophiles, and led to development of strains with bio-ethanol yields that are almost equal to those of the yeast-based systems. Members of the genus *Clostridium*, especially thermophilic members such as *Clostridium thermocellum*, have been studied intensively due to their competence in production substantial amounts of ethanol, butanol and hydrogen [24,28,29]. Members of the genus *Thermoanaerobacter*, including *T. ethanolicus*, *T. tengcongensis*, and *T. pentosaceus* are extremely thermophilic bacteria that are well studied for their high ethanol production potential especially from pentoses [30–34]. The genus *Geobacillus* has been studied widely for bio-ethanol production potential [19,29,35,36]. Production of ethanol, although at lower concentrations, has also been reported for the extremely thermophilic *Caldicellulosiruptor* species that includes *C. owensensis* [37], *C. kristjanssonii* [38], and *C. saccharolyticus* [39].

Compared to the thermophilic ethanol producers, very little is known about the ethanol production levels and pathways in the extremely thermophilic and hyperthermophilic microorganisms. It was shown that the peptide- and carbohydrate-fermenting hyperthermophilic archaeon *Pyrococcus furiosus* can produce H_2, CO_2, acetate, alanine, and small amounts of ethanol [40]. The strictly anaerobic archaeon *Thermococcus* sp. strain ES1 produced some ethanol and butanol when cultures were grown at low concentrations of elemental sulfur [41]. The production of ethanol as an end product of fermentation was also shown in the hyperthermophilic anaerobic archaeon *Thermococcus guaymasensis* [42] and more recently in the autotrophic hyperthermophile, *Thermococcus onnurineus* [43]. Within the bacterial hyperthermophiles, traces of ethanol have been reported in cultures of different Thermotogales including *T. hypogea* [44], *T. lettingae* [45], *T. neapolitana* [46], *Kosmotoga olearia* [47], and *Thermosipho affectus* [48].

3. Key Enzymes Involved in Ethanol Production

One of the key enzymes in both ethanol production pathways is alcohol dehydrogenase. Alcohol dehydrogenases are members of the oxidoreductase family and are present in all three domains of life [49,50]. They belong to the dehydrogenase/reductase superfamily of enzymes and catalyze the reversible inter-conversion of alcohols to corresponding aldehydes or ketons. ADHs can be classified based on their cofactor requirements: (I) the flavin adenine di-nucleotide (FAD)-dependent ADHs; (II) the pyrollo-quinoline quinone (PQQ), heme or cofactor F_{420} dependent ADHs; (III) NAD(P)-dependent ADHs [49,51]. Alternatively, they can be divided into three major groups based on their molecular size and metal contents: the first group is known as zinc-dependent long chain alcohol dehydrogenase; which have sizes of 300–900 amino acids, the second group is the short chain alcohol dehydrogenase: which contain no metal ions and have approximate lengths of 250 amino acids; and the third group is the long-chain iron dependent ADHs; with a length of 385–900 residues [49–52].

Many different ADHs have been characterized from various thermophilic and hyperthermophilic bacteria and archaea, with a majority of them being NAD(P)-dependent. Some of the more recently characterized hyper/thermophilic ADHs are those from *P. furiosus* [53,54], *Thermococcus hydrothermalis* [55], *Thermococcus kodakarensis* [56–58], *Thermococcus sibiricus* [59,60], *Thermococcus guaymasensis* [42], *Sulfolobus acidocaldarius* [61], *Thermococcus* strain ES1 [62], *Aeropyrum pernix* [63], *Thermotoga hypogea* [64], and *Pyrobaculum aerophilum* [65].

Although there is a relatively long list of ADHs isolated and characterized from thermophilic and hyperthermophilic archaea and bacteria, with the physiological roles of several proposed to be in the reduction of aldehydes to alcohols, other enzymes involved in the ethanol production pathways are not well characterized, especially the enzyme(s) that catalyze the production of acetaldehyde from pyruvate.

4. Pathways for the Production of Acetaldehyde from Pyruvate

Pyruvate is an intermediate in the central metabolism of carbohydrates [66,67], and it can be converted to acetaldehyde that will eventually be reduced to ethanol using one of the following two pathways:

(1) A two-step pathway that is used by yeast and a few bacteria like *Zymomonas mobilis* [68] and *Sarcina ventriculi* [69]. In this pathway pyruvate is non-oxidatively decarboxylated to acetaldehyde

and carbon dioxide, which is catalyzed by pyruvate decarboxylase (PDC). Acetaldehyde is then converted to ethanol that is catalyzed by ADH (Figure 1);

(2) A three-step pathway that is more widespread in bacteria. Pyruvate is oxidatively decarboxylated to acetyl-coenzyme A (acetyl-CoA) by the metalloenzyme pyruvate ferredoxin oxidoreductase (POR) and/or pyruvate formate lyase (PFL). Acetyl-CoA is then converted to acetaldehyde by a CoA-dependent-acetylating acetaldehyde dehydrogenase (AcDH). Finally, acetaldehyde is reduced to ethanol by ADH.

The key metabolite for the two known pathways is acetaldehyde. The thiamine pyrophosphate (TPP)-dependent enzyme pyruvate decarboxylase is the only enzyme proficient at direct conversion of pyruvate to acetaldehyde. Interestingly, a majority (but not all) of the enzymes which are involved in the acetaldehyde production pathways are members of the superfamily of TPP-dependent enzymes, which includes PDC, POR, and PFL [70,71].

Figure 1. Two pathways of ethanol production from pyruvate. POR; Pyruvate ferredoxin oxidoreductase; PFL; Pyruvate formate lyase, AcDH; Acetaldehyde dehydrogenase, ADH; Alcohol dehydrogenase, PDC; pyruvate decarboxylase; CoASH; coenzyme A, Fd_{ox}; oxidized ferredoxin, Fd_{red}; reduced ferredoxin.

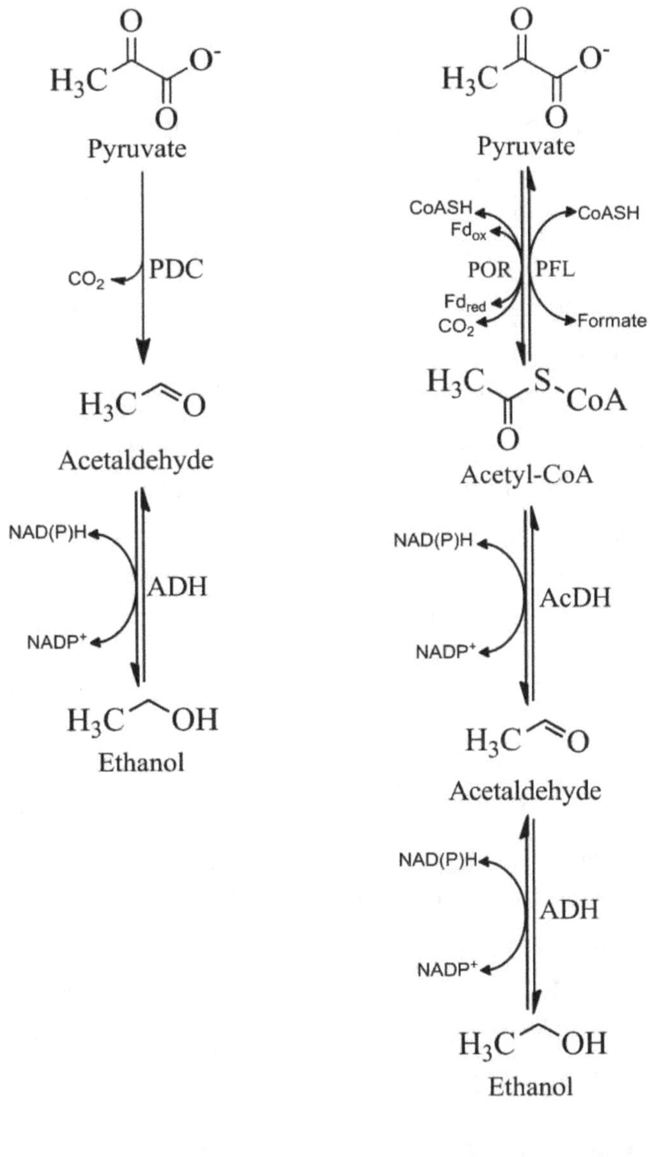

TPP, also known as thiamine diphosphate (ThDP), is composed of an aromatic methylaminopyrimidine ring, linked to a methyl thiazolium ring via. a methylene group with a pyrophosphate group attached to a hydroxylethyl side chain. TPP is derived from the water-soluble vitamin B1 and is the most common cofactor for enzymes that catalyze the cleavage and formation of carbon-carbon bonds next to a carbonyl group; hence TPP-dependent enzymes are involved in a wide range of metabolic pathways. Unlike many other cofactors (e.g., nicotinamide adenine dinucleotide, NADH) which are basically co-reactants, TPP remains at the enzymes' catalytic center and is directly involved in the catalysis of the reaction [72]. The reactions catalyzed by TPP-dependent enzymes can be divided into at least three groups: the oxidative reactions, non-oxidative reactions, and carboligation reactions [73].

5. Pyruvate Decarboxylase (PDC)

In 1911, Neuberg and Karczag for the first time described the decarboxylation of pyruvate to acetaldehyde in *S. cerevisiae*. In 1922, the same research group detected the potential of yeast in the formation of C-C bonds. Neuberg named the new enzyme "carboligase", and assumed it to exist apart from "α-carboxylase" (PDC) in yeast [74]. However, the preliminary characterization of the enzymes' cofactor was delayed until 1937, when Lohmann and Schuster analyzed structure of the enzymes' cofactor to be "cocarboxylase" or "aneurinpyrophosphate" or thiamin diphosphate [73].

The enzyme catalyzes non-oxidative decarboxylation of α-keto acids to produce a corresponding aldehyde and carbon dioxide. The most extensively examined enzymes of this group are the ones from *Saccharomyces cerevisiae* and its bacterial counterpart *Z. mobilis*. In addition to decarboxylation of pyruvate, PDC also catalyzes the enantio-selective formation of 2-hydroxy ketons via. carboligase side reactions.

PDC, or its gene (*pdc*), is found to be widely distributed in fungi and higher plants but it is relatively rare in prokaryotes and unknown in animals. In fungi, PDC is found in *Saccharomyces cerevisiae*, *Saccharomyces carlsbergensis* (also known as *S. pastorianus*) and *Saccharomyces uvarum*, *Neurospora crassa*, members of the *Kluyveromyces* species, members of the *Aspergillus* species, *Hanseniaspora uvarum*, *Schizosaccharomyces pombe*, and in *Candida* (*Torulopsis*) *glabrata*. PDC is present in a variety of plants, including maize (*Zea maize*), parsnip, orange, pea (*Pisum sativum*), jack bean, sweet potato, wheat, cotton wood, soybean and rice (*Oryza sativa*). In prokaryotes, PDC is found and studied in *Z. mobilis*, *Sarcina ventriculi*, *Clostridium botulinum*, *Acetobacter* species, *Zymobacter palmae*, and in *Erwinia amylovora* [75–79]. So far there has been no report on finding PDC/*pdc* homolog in thermophilic or hyperthermophilic bacteria or in any of the members of the third major evolutionary lineage of life, archaea as a whole [76–78,80].

PDCs from different organisms show at least a 30% identity at the amino acid level and most of them are composed of subunits of 562–610 amino acid residues. The holoenzyme is usually composed of four identical or non-identical subunits of approximately 60 kDa (ensuing in a total mass of about 240 kDa) in which every two subunits binds tightly (but not covalently) to a set of cofactors including TPP and Mg^{2+} ion. PDCs with four subunits are often arranged as a dimer of dimers, with multiple close contacts within the dimers and several contacts between the dimers. The contact area between two related dimers forms the "V" conformation that is a common property of all TPP-dependent enzymes studied so far, and it also has an essential role in cofactor binding for this group of enzymes [81,82].

The catalytic mechanism of PDC for the most part follows the principles of catalytic mechanisms of other TPP-dependent enzymes: in brief, carbonyl addition of pyruvate to the reactive C2 atom of the cofactor thiazolium ring [73] yields the intermediate 2-(2-lactyl)-TDP (LTDP). The subsequent release of carbon dioxide produces resonating carbanion/enamine forms of 2-(1-hydroxyethyl)-TDP (HETDP, also known as hydroxyethylidene-TPP). The resonating form is considered to be a central and highly reactive intermediate state in TPP-dependent enzymes acting on pyruvate. However, unlike most other TPP-dependent enzymes in which the intermediate is oxidized, the carbanion/enamine in PDC is protonated at the C2α position, yielding C2α-hydroxylethylthiamine diphosphate (HETDP) before the final release of acetaldehyde completes the reaction [72,83,84].

Crystal structures of several pyruvate decarboxylases are solved particularly from yeasts and *Z. mobilis* [81,85–87]. The active sites of these enzymes are also studied comprehensively using site-directed mutagenesis [88–90].

6. Pyruvate Ferredoxin Oxidoreductase (POR)

The enzyme pyruvate ferredoxin oxidoreductase (also known as pyruvate synthase as the reaction is reversible) is one of the best studied members of the 2-oxoacid oxidoreductase family [91–93]. The enzyme catalyzes coenzyme A and TPP-dependent oxidative decarboxylation of pyruvate to acetyl-CoA, releasing a molecule of CO_2 and transferring the reducing equivalents to the electron acceptor ferredoxin or flavodoxin. Alternatively, in other pyruvate oxidizing enzymes, the reducing equivalents are transferred to NAD^+ (in the case of pyruvate dehydrogenase using lipoate as oxidizing agent for the production of acetyl-CoA), to molecular-oxygen-producing hydrogen peroxide (in the case of pyruvate oxidase), or to the carbonyl groups producing formate (in case of pyruvate formate lyase) [94–96]. In acetaldehyde- and ethanol-producing organisms, acetyl-CoA is usually converted to acetaldehyde via. the CoA-dependent (acetylating) acetaldehyde dehydrogenase.

POR uses iron-sulfur cluster chemistry to catalyze the pyruvate decarboxylation and release of acetyl-CoA. POR is an ancient molecule, and it seems to have existed even before the divergence of the domains of the bacteria and archaea [97]. The enzyme is present in all three domains of life. All archaea catalyze the conversion of pyruvate to acetyl-CoA using POR, and all of the archaeal genomes sequenced so far contain hetero-tetrameric PORs, which have been proposed to be the closest to the POR common ancestor [97,98].

POR is prevalent mainly in anaerobic bacteria and infrequently found in anaerobic protozoa, for example, in *Giardia duodenalis* [99] and *Enthamoeba histolytica* [100,101]. The enzyme has been isolated and studied from many different anaerobic or microaerophilic microorganisms including anaerobic bacteria like the genera *Clostridium* [102], *Moorella thermoacetica* [103] and anaerobic sulphate-reducing bacteria *Desulfovibrio africanus* [104–107]. In hyperthermophiles, PORs are characterized from the hyperthermophilic bacterium *Thermotoga maritima* [108] and hyperthermophilic archaea *Pyrococcus furiosus* [109] and *Archaeoglobus fulgidus* [110], as well as the methanogenic archaea *Methanosarcina barkeri* [111,112] and *Methanobacterium thermoautotrophicum* [113].

The quaternary oligomeric structure of the POR is variable depending on the source microorganism and can be homo-dimeric (e.g., most bacterial PORs), hetero-dimeric (e.g., POR of *Halobacterium salinarium*), hetero-tetrameric (archaeal PORs), and heteropentameric (anabolic PORs), although all of the PORs studied so far, regardless of their source and structure, seem to be phylogenetically related and derived from a common archaeal-type heterotetrameric ancestor [97,98].

The crystal structures of several POR have been determined. PORs from *Desulfovibrio africanus* (with and without bound substrate) and *Desulfovibrio vulgaris* [114–116] are among the most extensively studied PORs. POR is a metalloenzyme and all PORs studied so far contained between one and three [4Fe-4S] clusters arranged in a spatial order from the TPP located at the active center of the enzyme toward its surface, suggesting that they are part of an electron transfer pathway [117,118].

POR can also catalyze the reaction to form pyruvate from acetyl-CoA and carbon dioxide, which is the basis of the carbon dioxide fixation in many autotrophic microorganisms [119]. This type represents the so-called "anabolic" PORs that are studied from the thermophilic facultative aerobic bacterium *Hydrogenobacter thermophilus* [120–122], as well as the hydrogenotrophic methanoarchaeon *Methanococcus maripaludis* [123,124]. Interestingly, in the case of the heteropentameric POR of *M. maripaludis*, four subunits are very closely related to the archaeal heterotetrameric (ancestral) PORs, the fifth subunit has no known homologue within PORs.

The general steps of the POR catalytic reactions follow the same principles as those of other TPP-dependent enzymes. However, the enzyme is unique in one aspect: unlike most other TPP-dependent enzymes, POR takes advantage of free radical chemistry to catalyze the decarboxylation reaction [94,125].

Pyruvate dehydrogenase complexes (PDH), which also catalyze the oxidative decarboxylation of pyruvate to acetyl-CoA using NAD^+ as an electron acceptor, are normally present in aerobic organisms [126,127], which have been found in some thermopiles [128–130]. But no PDH has been identified in hyperthermophiles [131]. Since PDH does not play any significant role in the production of acetaldehyde and ethanol, no further description of this enzyme complex will be given in this review.

7. POR/PDC Bi-Functional Enzyme

In 1997, it was reported that the POR was also capable of converting pyruvate to acetaldehyde in the hyperthermophilic anaerobic archaeon *Pyrococcus furiosus* [80]. Unlike the commonly-known PDCs, which employ chemical rather than radical intermediates and therefore are oxygen insensitive, the reported PDC activity was highly oxygen sensitive. Both the POR and PDC activities of the hyperthermophilic enzyme were TPP- and coenzyme A-dependent. By using the coenzyme A analogue (desulfocoenzyme A), it was shown that coenzyme A has only a structural, and not a catalytic role in the catalyzed PDC reaction. Consequently, a "switch" mechanism was proposed for the enzyme's bi-functionality, suggesting the conversion of active aldehyde to either acetyl-CoA or acetaldehyde, depending on the binding of CoA. According to the proposed model, binding of coenzyme A causes conformational changes in the intermediate structure, causing its protonation and generation of hydroxyethyl-TPP (HETDP). This reaction leads to release of acetaldehyde, allowing for the regeneration of TPP and possible release of CoA [80]. Ferredoxin is not required for its full PDC activity nor has any inhibitory effect when tested under *in vitro* conditions. However, it is likely that the *in vivo* PDC activity might be dependent on the availability of oxidized ferredoxin, which means that a lower ratio of oxidized to reduced ferredoxin may favor the PDC activity and *vice versa*. Therefore, it can be predicted that more acetaldehyde would be produced for being reduced to ethanol if the ratio of the oxidized to reduced ferredoxin would be kept very low under the anaerobic growth condition. Such a low ratio may also require a relatively low activity of the ferredoxin-oxidizing hydrogenases present in *P. furiosus* [132,133].

To date there has been no further study on the bi-functionality of the POR enzyme or the physiological relevance of such bifunctionality in any other organisms. It is not clear whether this bi-functionality is only a trait of *Pyrococcales*' POR or a common property of all hyper/thermophilic PORs.

8. Acetaldehyde Dehydrogenase (CoA-Acetylating)

Acetaldehyde dehydrogenase (CoA-acetylating, EC 1.2.1.10) is a member of a very divergent superfamily of enzymes known as the "aldehyde dehydrogenases". The prototype enzyme (*adhE*) was first discovered in *Escherichia coli* and is required for its anaerobic growth [134]. It was then discovered in the strictly anaerobic bacterium *Clostridium kluyveri* [135]. The enzyme is responsible for the conversion of acetyl-coenzyme A (acetyl-CoA) to acetaldehyde that is eventually converted to ethanol. Two forms of the enzyme are available: one is the monofunctional enzyme with only AcDH activity (*mhpF*) and the other is the bifunctional enzyme with both AcDH and ADH activities (*adhE*). The latter group is composed of an ADH active C-terminal and an AcDH active N-terminal, a structure believed to be the result of gene fusion between the genes encoding for each single enzyme [136,137].

Reports are available on isolation and characterization of the bifunctional NADP-dependent alcohol/acetaldehyde dehydrogenase (CoA-acetylating) from mesophilic microorganisms including *Giardia lamblia* [138] and *Enthamoeba histolytica* [101,139]. They are also present in some thermophiles, including *T. ethanolicus* [140,141], *T. mathranii* [32] and members of the genus *Geobacillus* [19]. However, no mono- or bi-functional AcDH activity was characterized from hyperthermophiles. Survey of the fully sequenced genomes of hyperthermophilic archaea and bacteria has shown no *adhE* or *mhpF* homologue either (Eram and Ma, unpublished data).

9. Conclusions

Many hyperthermophilic microorganisms produce ethanol as an end metabolic product. Although alcohol dehydrogenase and pyruvate ferredoxin oxidoreductase are found to be present, enzymes catalyzing the production of acetaldehyde from pyruvate are not well characterized. The commonly-known pyruvate decarboxylase and coenzyme A-dependent aldehyde dehydrogenase have not been identified. The only report of a bi-functional pyruvate decarboxylase is the POR/PDC from *P. furiosus*, which is thermostable but oxygen-sensitive. Therefore, it is likely that they use a two-step pathway to convert pyruvate to ethanol. The regulation of each of the POR/PDC activities is not clear but it may be related to the redox states inside the cells. To date, it appears that at least one multifunctional enzyme is responsible for catalyzing the non-oxidative decarboxylation of pyruvate to acetaldehyde in hyperthermophiles, and further study is needed to understand the catalysis of acetaldehyde production from pyruvate at high temperatures.

Acknowledgments

This work was supported by research grants from the Natural Sciences and Engineering Research Council (Canada), Genome Canada/Genome Prairie and Canada Foundation for Innovation to KM.

References

1. Wiegel, J. Temperature spans for growth: Hypothesis and discussion. *FEMS Microbiol. Lett.* **1990**, *75*, 155–169.
2. Charlier, D.; Droogmans, L. Microbial Life at high temperature, the challenges, the strategies. *Cell. Mol. Life Sci.* **2005**, *62*, 2974–2984.
3. Stetter, K. History of discovery of the first hyperthermophiles. *Extremophiles* **2006**, *10*, 357–362.
4. Lebedinsky, A.; Chernyh, N.; Bonch-Osmolovskaya, E. Phylogenetic systematics of microorganisms inhabiting thermal environments. *Biochemistry (Mosc.)* **2007**, *72*, 1299–1312.
5. Wagner, I.D.; Wiegel, J. Diversity of Thermophilic Anaerobes. *Ann. NY Acad. Sci.* **2008**, *1125*, 1–43.
6. Morozkina, E.; Slutskaya, E.; Fedorova, T.; Tugay, T.; Golubeva, L.; Koroleva, O. Extremophilic microorganisms: Biochemical adaptation and biotechnological application. *Appl. Biochem. Microbiol.* **2010**, *46*, 1–14.
7. Sommer, P.; Georgieva, T.; Ahring, B.K. Potential for using thermophilic anaerobic bacteria for bioethanol production from hemicellulose. *Biochem. Soc. Trans.* **2004**, *32*, 283–289.
8. Vieille, C.; Zeikus, G.J. Hyperthermophilic enzymes: Sources, uses, and molecular mechanisms for thermostability. *Microbiol. Mol. Biol. Rev.* **2001**, *65*, 1–43.
9. Haki, G.D.; Rakshit, S.K. Developments in industrially important thermostable enzymes: A review. *Bioresour. Technol.* **2003**, *89*, 17–34.
10. Egorova, K.; Antranikian, G. Industrial relevance of thermophilic Archaea. *Curr. Opin. Microbiol.* **2005**, *8*, 649–655.
11. Van den Burg, B. Extremophiles as a source for novel enzymes. *Curr. Opin. Microbiol.* **2003**, *6*, 213–218.
12. Atomi, H.; Sato, T.; Kanai, T. Application of hyperthermophiles and their enzymes. *Curr. Opin. Biotechnol.* **2011**, *22*, 618–626.
13. Lynd, L.R.; Wyman, C.E.; Gerngross, T.U. Biocommodity engineering. *Biotechnol. Prog.* **1999**, *15*, 777–793.
14. Zaldivar, J.; Nielsen, J.; Olsson, L. Fuel ethanol production from lignocellulose: A challenge for metabolic engineering and process integration. *Appl. Microbiol. Biotechnol.* **2001**, *56*, 17–34.
15. Mabee, W.E.; Saddler, J.N. Bioethanol from lignocellulosics: Status and perspectives in Canada. *Bioresour. Technol.* **2010**, *101*, 4806–4813.
16. Dien, B.S.; Cotta, M.A.; Jeffries, T.W. Bacteria engineered for fuel ethanol production: Current status. *Appl. Microbiol. Biotechnol.* **2003**, *63*, 258–266.

17. Buschke, N.; Schäfer, R.; Becker, J.; Wittmann, C. Metabolic engineering of industrial platform microorganisms for biorefinery applications-optimization of substrate spectrum and process robustness by rational and evolutive strategies. *Bioresour. Technol.* **2012**, *135*, 544–554.
18. Jang, Y.-S.; Park, J.M.; Choi, S.; Choi, Y.J.; Seung, D.Y.; Cho, J.H.; Lee, S.Y. Engineering of microorganisms for the production of biofuels and perspectives based on systems metabolic engineering approaches. *Biotechnol. Adv.* **2012**, *30*, 989–1000.
19. Taylor, M.P.; Eley, K.L.; Martin, S.; Tuffin, M.I.; Burton, S.G.; Cowan, D.A. Thermophilic ethanologenesis: Future prospects for second-generation bioethanol production. *Trends Biotechnol.* **2009**, *27*, 398–405.
20. Bustard, M.T.; Burgess, J.G.; Meeyoo, V.; Wright, P.C. Novel opportunities for marine hyperthermophiles in emerging biotechnology and engineering industries. *J. Chem. Technol. Biotechnol.* **2000**, *75*, 1095–1109.
21. Huber, H.; Stetter, K.O. Hyperthermophiles and their possible potential in biotechnology. *J. Biotechnol.* **1998**, *64*, 39–52.
22. Schiraldi, C.; de Rosa, M. The production of biocatalysts and biomolecules from extremophiles. *Trends Biotechnol.* **2002**, *20*, 515–521.
23. Hough, D.W.; Danson, M.J. Extremozymes. *Curr. Opin. Chem. Biol.* **1999**, *3*, 39–46.
24. Lamed, R.; Zeikus, J.G. Ethanol production by thermophilic bacteria: Relationship between fermentation product yields of and catabolic enzyme activities in *Clostridium thermocellum* and *Thermoanaerobium brockii. J. Bacteriol.* **1980**, *144*, 569–578.
25. Klapatch, T.R.; Hogsett, D.A.L.; Baskaran, S.; Pal, S.; Lynd, L.R. Organism development and characterization for ethanol production using thermophilic bacteria. *Appl. Biochem. Biotechnol.* **1994**, *45–46*, 209–223.
26. Chang, T.; Yao, S. Thermophilic, lignocellulolytic bacteria for ethanol production: Current state and perspectives. *Appl. Microbiol. Biotechnol.* **2011**, *92*, 13–27.
27. Zeikus, J.G.; Ben-Bassat, A.; Ng, T.K.; Lamed, R.J. Thermophilic ethanol fermentations. In *Trends in the Biology of Fermentations for Fuels and Chemicals*; Hollaender, A., Rabson, R., Rogers, P., Pietro, A.S., Valentine, R., Wolfe, R., Eds.; Springer: New York, NY, USA, 1981; pp. 441–461.
28. Demain, A.L.; Newcomb, M.; Wu, J.H.D. Cellulase, clostridia, and ethanol. *Microbiol. Mol. Biol. Rev.* **2005**, *69*, 124–154.
29. Barnard, D.; Casanueva, A.; Tuffin, M.; Cowan, D. Extremophiles in biofuel synthesis. *Environ. Technol.* **2010**, *31*, 871–888.
30. Tomás, A.F.; Karagöz, P.; Karakashev, D.; Angelidaki, I. Extreme thermophilic ethanol production from rapeseed straw: Using the newly isolated *Thermoanaerobacter pentosaceus* and combining it with *Saccharomyces cerevisiae* in a two-step process. *Biotechnol. Bioeng.* **2013**, *110*, 1574–1582.
31. Shaw, A.J.; Podkaminer, K.K.; Desai, S.G.; Bardsley, J.S.; Rogers, S.R.; Thorne, P.G.; Hogsett, D.A.; Lynd, L.R. Metabolic engineering of a thermophilic bacterium to produce ethanol at high yield. *Proc. Natl. Acad. Sci. USA* **2008**, *105*, 13769–13774.

32. Yao, S.; Mikkelsen, M.J. Identification and overexpression of a bifunctional aldehyde/alcohol dehydrogenase responsible for ethanol production in *Thermoanaerobacter mathranii*. *J. Mol. Microbiol. Biotechnol.* **2010**, *19*, 123–133.
33. Svetlitchnyi, V.; Kensch, O.; Falkenhan, D.; Korseska, S.; Lippert, N.; Prinz, M.; Sassi, J.; Schickor, A.; Curvers, S. Single-step ethanol production from lignocellulose using novel extremely thermophilic bacteria. *Biotechnol. Biofuels* **2013**, *6*, 1–15.
34. Yao, S.; Mikkelsen, M. Metabolic engineering to improve ethanol production in *Thermoanaerobacter mathranii*. *Appl. Microbiol. Biotechnol.* **2010**, *88*, 199–208.
35. Thompson, A.; Studholme, D.; Green, E.; Leak, D. Heterologous expression of pyruvate decarboxylase in *Geobacillus thermoglucosidasius*. *Biotechnol. Lett.* **2008**, *30*, 1359–1365.
36. Cripps, R.E.; Eley, K.; Leak, D.J.; Rudd, B.; Taylor, M.; Todd, M.; Boakes, S.; Martin, S.; Atkinson, T. Metabolic engineering of *Geobacillus thermoglucosidasius* for high yield ethanol production. *Metab. Eng.* **2009**, *11*, 398–408.
37. Huang, C.-Y.; Patel, B.K.; Mah, R.A.; Baresi, L. *Caldicellulosiruptor owensensis* sp. nov., an anaerobic, extremely thermophilic, xylanolytic bacterium. *Int. J. Syst. Bacteriol.* **1998**, *48*, 91–97.
38. Bredholt, S.; Sonne-Hansen, J.; Nielsen, P.; Mathrani, I.M.; Ahring, B.K. *Caldicellulosiruptor kristjanssonii* sp. nov., a cellulolytic, extremely thermophilic, anaerobic bacterium. *Int. J. Syst. Bacteriol.* **1999**, *49*, 991–996.
39. Van Niel, E.W.J.; Claassen, P.A.M.; Stams, A.J.M. Substrate and product inhibition of hydrogen production by the extreme thermophile, *Caldicellulosiruptor saccharolyticus*. *Biotechnol. Bioeng.* **2003**, *81*, 255–262.
40. Kengen, S.; de Bok, F.; van Loo, N.; Dijkema, C.; Stams, A.; de Vos, W. Evidence for the operation of a novel Embden-Meyerhof pathway that involves ADP-dependent kinases during sugar fermentation by *Pyrococcus furiosus*. *J. Biol. Chem.* **1994**, *269*, 17537–17541.
41. Ma, K.; Loessner, H.; Heider, J.; Johnson, M.; Adams, M. Effects of elemental sulfur on the metabolism of the deep-sea hyperthermophilic archaeon *Thermococcus* strain ES-1: Characterization of a sulfur-regulated, non-heme iron alcohol dehydrogenase. *J. Bacteriol.* **1995**, *177*, 4748–4756.
42. Ying, X.; Ma, K. Characterization of a zinc-containing alcohol dehydrogenase with stereoselectivity from the hyperthermophilic archaeon *Thermococcus guaymasensis*. *J. Bacteriol.* **2011**, *193*, 3009–3019.
43. Moon, Y.-J.; Kwon, J.; Yun, S.-H.; Lim, H.L.; Kim, M.-S.; Kang, S.G.; Lee, J.-H.; Choi, J.-S.; Kim, S.L.; Chung, Y.-H. Proteome analyses of hydrogen-producing hyperthermophilic archaeon *Thermococcus onnurineus* NA1 in different one-carbon substrate culture conditions. *Mol. Cell. Proteomics* **2012**, doi:10.1074/mcp.M111.015420.
44. Fardeau, M.L.; Ollivier, B.; Patel, B.K.C.; Magot, M.; Thomas, P.; Rimbault, A.; Rocchiccioli, F.; Garcia, J.L. *Thermotoga hypogea* sp. nov., a xylanolytic, thermophilic bacterium from an oil-producing well. *Int. J. Syst. Bacteriol.* **1997**, *47*, 1013–1019.
45. Balk, M.; Weijma, J.; Stams, A.J.M. *Thermotoga lettingae* sp. nov., a novel thermophilic, methanol-degrading bacterium isolated from a thermophilic anaerobic reactor. *Int. J. Syst. Evol. Microbiol.* **2002**, *52*, 1361–1368.

46. De Vrije, T.; Bakker, R.; Budde, M.; Lai, M.; Mars, A.; Claassen, P. Efficient hydrogen production from the lignocellulosic energy crop *Miscanthus* by the extreme thermophilic bacteria *Caldicellulosiruptor saccharolyticus* and *Thermotoga neapolitana. Biotechnol. Biofuels* **2009**, *2*, e12.
47. DiPippo, J.L.; Nesbo, C.L.; Dahle, H.; Doolittle, W.F.; Birkland, N.-K.; Noll, K.M. *Kosmotoga olearia* gen. nov., sp. nov., a thermophilic, anaerobic heterotroph isolated from an oil production fluid. *Int. J. Syst. Evol. Microbiol.* **2009**, *59*, 2991–3000.
48. Podosokorskaya, O.A.; Kublanov, I.V.; Reysenbach, A.L.; Kolganova, T.V.; Bonch-Osmolovskaya, E.A. *Thermosipho affectus* sp. nov., a thermophilic, anaerobic, cellulolytic bacterium isolated from a Mid-Atlantic Ridge hydrothermal vent. *Int. J. Syst. Evol. Microbiol.* **2011**, *61*, 1160–1164.
49. Reid, M.F.; Fewson, C.A. Molecular characterization of microbial alcohol dehydrogenases. *Crit. Rev. Microbiol.* **1994**, *20*, 13–56.
50. Littlechild, J.A.; Guy, J.E.; Isupov, M.N. Hyperthermophilic dehydrogenase enzymes. *Biochem. Soc. Trans.* **2004**, *32*, 255–258.
51. Radianingtyas, H.; Wright, P.C. Alcohol dehydrogenases from thermophilic and hyperthermophilic archaea and bacteria. *FEMS Microbiol. Rev.* **2003**, *27*, 593–616.
52. Korkhin, Y.; Kalb, A.J.; Peretz, M.; Bogin, O.; Burstein, Y.; Frolow, F. NADP-dependent bacterial alcohol dehydrogenases: Crystal structure, cofactor-binding and cofactor specificity of the ADHs of *Clostridium beijerinckii* and *Thermoanaerobacter brockii. J. Mol. Biol.* **1998**, *278*, 967–981.
53. Machielsen, R.; Uria, A.R.; Kengen, S.W.M.; van der Oost, J. Production and characterization of a thermostable alcohol dehydrogenase that belongs to the aldo-keto reductase superfamily. *Appl. Environ. Microbiol.* **2006**, *72*, 233–238.
54. Van der Oost, J.; Voorhorst, W.G.B.; Kengen, S.W.M.; Geerling, A.C.M.; Wittenhorst, V.; Gueguen, Y.; de Vos, W.M. Genetic and biochemical characterization of a short-chain alcohol dehydrogenase from the hyperthermophilic archaeon *Pyrococcus furiosus. Eur. J. Biochem.* **2001**, *268*, 3062–3068.
55. Antoine, E.; Rolland, J.-L.; Raffin, J.-P.; Dietrich, J. Cloning and over-expression in *Escherichia coli* of the gene encoding NADPH group III alcohol dehydrogenase from *Thermococcus hydrothermalis. Eur. J. Biochem.* **1999**, *264*, 880–889.
56. Bashir, Q.; Rashid, N.; Jamil, F.; Imanaka, T.; Akhtar, M. Highly thermostable L-threonine dehydrogenase from the hyperthermophilic archaeon *Thermococcus kodakaraensis. J. Biochem. (Tokyo)* **2009**, *146*, 95–102.
57. Bowyer, A.; Mikolajek, H.; Stuart, J.W.; Wood, S.P.; Jamil, F.; Rashid, N.; Akhtar, M.; Cooper, J.B. Structure and function of the l-threonine dehydrogenase (TkTDH) from the hyperthermophilic archaeon *Thermococcus kodakaraensis. J. Struct. Biol.* **2009**, *168*, 294–304.
58. Wu, X.; Zhang, C.; Orita, I.; Imanaka, T.; Fukui, T.; Xing, X.-H. Thermostable alcohol dehydrogenase from *Thermococcus kodakarensis* KOD1 for enantioselective bioconversion of aromatic secondary alcohols. *Appl. Environ. Microbiol.* **2013**, *79*, 2209–2217.

59. Stekhanova, T.N.; Mardanov, A.V.; Bezsudnova, E.Y.; Gumerov, V.M.; Ravin, N.V.; Skryabin, K.G.; Popov, V.O. Characterization of a thermostable short-chain alcohol dehydrogenase from the hyperthermophilic archaeon *Thermococcus sibiricus. Appl. Environ. Microbiol.* **2010**, *76*, 4096–4098.
60. Lyashenko, A.V.; Bezsudnova, E.Y.; Gumerov, V.M.; Lashkov, A.A.; Mardanov, A.V.; Mikhailov, A.M.; Polyakov, K.M.; Popov, V.O.; Ravin, N.V.; Skryabin, K.G.; *et al.* Expression, purification and crystallization of a thermostable short-chain alcohol dehydrogenase from the archaeon *Thermococcus sibiricus. Acta Crystallogr. F Struct. Biol. Cryst. Commun.* **2010**, *66*, 655–657.
61. Pennacchio, A.; Giordano, A.; Pucci, B.; Rossi, M.; Raia, C. Biochemical characterization of a recombinant short-chain NAD(H)-dependent dehydrogenase/reductase from *Sulfolobus acidocaldarius. Extremophiles* **2010**, *14*, 193–204.
62. Ying, X.; Grunden, A.; Nie, L.; Adams, M.; Ma, K. Molecular characterization of the recombinant iron-containing alcohol dehydrogenase from the hyperthermophilic Archaeon, *Thermococcus* strain ES1. *Extremophiles* **2009**, *13*, 299–311.
63. Guy, J.E.; Isupov, M.N.; Littlechild, J.A. The structure of an alcohol dehydrogenase from the hyperthermophilic archaeon *Aeropyrum pernix. J. Mol. Biol.* **2003**, *331*, 1041–1051.
64. Ying, X.; Wang, Y.; Badiei, H.; Karanassios, V.; Ma, K. Purification and characterization of an iron-containing alcohol dehydrogenase in extremely thermophilic bacterium *Thermotoga hypogea. Arch. Microbiol.* **2007**, *187*, 499–510.
65. Vitale, A.; Thorne, N.; Lovell, S.; Battaile, K.P.; Hu, X.; Shen, M.; D'Auria, S.; Auld, D.S. Physicochemical characterization of a thermostable alcohol dehydrogenase from *Pyrobaculum aerophilum. PLoS One* **2013**, *8*, e63828.
66. Verhees, C.H.; Kengen, S.W.M.; Tuininga, J.E.; Schut, G.J.; Adams, M.W.W.; de Vos, W.M.; van der Oost, J. The unique features of glycolytic pathways in Archaea. *Biochem. J.* **2003**, *375*, 231–246.
67. Siebers, B.; Schönheit, P. Unusual pathways and enzymes of central carbohydrate metabolism in Archaea. *Curr. Opin. Microbiol.* **2005**, *8*, 695–705.
68. Buchholz, S.E.; Dooley, M.M.; Eveleigh, D.E. *Zymomonas*—An alcoholic enigma. *Trends Biotechnol.* **1987**, *5*, 199–204.
69. Canale-Parola, E. Biology of the sugar-fermenting *Sarcinae. Microbiol. Mol. Biol. Rev.* **1970**, *34*, 82–97.
70. Duggleby, R.G. Domain relationships in thiamine diphosphate-dependent enzymes. *Acc. Chem. Res.* **2006**, *39*, 550–557.
71. Costelloe, S.; Ward, J.; Dalby, P. Evolutionary analysis of the TPP-dependent enzyme family. *J. Mol. Evol.* **2008**, *66*, 36–49.
72. Kluger, R.; Tittmann, K. Thiamin diphosphate catalysis: Enzymic and nonenzymic covalent intermediates. *Chem. Rev.* **2008**, *108*, 1797–1833.
73. Schellenberger, A. Sixty years of thiamin diphosphate biochemistry. *Biochim. Biophys. Acta* **1998**, *1385*, 177–186.
74. Iding, H.; Siegert, P.; Mesch, K.; Pohl, M. Application of alpha-keto acid decarboxylases in biotransformations. *Biochim. Biophys. Acta* **1998**, *1385*, 307–322.

75. Bringer-Meyer, S.; Schimz, K.L.; Sahm, H. Pyruvate decarboxylase from *Zymomonas mobilis*. Isolation and partial characterization. *Arch. Microbiol.* **1986**, *146*, 105–110.
76. Raj, K.C.; Ingram, L.O.; Maupin-Furlow, J.A. Pyruvate decarboxylase: A key enzyme for the oxidative metabolism of lactic acid by *Acetobacter pasteurianus. Arch. Microbiol.* **2001**, *176*, 443–451.
77. Raj, K.C.; Talarico, L.A.; Ingram, L.O.; Maupin-Furlow, J.A. Cloning and characterization of the *Zymobacter palmae* pyruvate decarboxylase gene (*pdc*) and comparison to bacterial homologues. *Appl. Environ. Microbiol.* **2002**, *68*, 2869–2876.
78. Talarico, L.A.; Ingram, L.O.; Maupin-Furlow, J.A. Production of the Gram-positive *Sarcina ventriculi* pyruvate decarboxylase in *Escherichia coli. Microbiology* **2001**, *147*, 2425–2435.
79. Wang, Q.; He, P.; Lu, D.; Shen, A.; Jiang, N. Purification, characterization, cloning and expression of pyruvate decarboxylase from *Torulopsis glabrata* IFO005. *J. Biochem. (Tokyo)* **2004**, *136*, 447–455.
80. Ma, K.; Hutchins, A.; Sung, S.-J.S.; Adams, M.W.W. Pyruvate ferredoxin oxidoreductase from the hyperthermophilic archaeon, *Pyrococcus furiosus*, functions as a CoA-dependent pyruvate decarboxylase. *Proc. Natl. Acad. Sci. USA* **1997**, *94*, 9608–9613.
81. Dobritzsch, D.; Konig, S.; Schneider, G.; Lu, G. High resolution crystal structure of pyruvate decarboxylase from *Zymomonas mobilis*. Implications for substrate activation in pyruvate decarboxylases. *J. Biol. Chem.* **1998**, *273*, 20196–20204.
82. Jordan, F. Current mechanistic understanding of thiamin diphosphate-dependent enzymatic reactions. *Nat. Prod. Rep.* **2003**, *20*, 184–201.
83. Kluger, R. Thiamin diphosphate: A mechanistic update on enzymic and nonenzymic catalysis of decarboxylation. *Chem. Rev.* **1987**, *87*, 863–876.
84. Candy, J.M.; Duggleby, R.G. Structure and properties of pyruvate decarboxylase and site-directed mutagenesis of the *Zymomonas mobilis* enzyme. *Biochim. Biophys. Acta* **1998**, *1385*, 323–338.
85. Kutter, S.; Weiss, M.S.; Wille, G.; Golbik, R.; Spinka, M.; König, S. Covalently bound substrate at the regulatory site of yeast pyruvate decarboxylases triggers allosteric enzyme activation. *J. Biol. Chem.* **2009**, *284*, 12136–12144.
86. Siegert, P.; McLeish, M.J.; Baumann, M.; Iding, H.; Kneen, M.M.; Kenyon, G.L.; Pohl, M. Exchanging the substrate specificities of pyruvate decarboxylase from *Zymomonas mobilis* and benzoylformate decarboxylase from *Pseudomonas putida. Protein Eng. Des. Sel.* **2005**, *18*, 345–357.
87. Arjunan, P.; Umland, T.; Dyda, F.; Swaminathan, S.; Furey, W.; Sax, M.; Farrenkopf, B.; Gao, Y.; Zhang, D.; Jordan, F. Crystal structure of the thiamin diphosphate-dependent enzyme pyruvate decarboxylase from the yeast *Saccharomyces cerevisiae* at 2.3 Å resolution. *J. Mol. Biol.* **1996**, *256*, 590–600.
88. Pohl, M. Protein design on pyruvate decarboxylase (PDC) by site-directed mutagenesis. *Adv. Biochem. Eng. Biotechnol.* **1997**, *58*, 15–43.
89. Liu, M.; Sergienko, E.A.; Guo, F.; Wang, J.; Tittmann, K.; Hubner, G.; Furey, W.; Jordan, F. Catalytic acid-base groups in yeast pyruvate decarboxylase. 1. Site-directed mutagenesis and steady-state kinetic studies on the enzyme with the D28A, H114F, H115F, and E477Q substitutions. *Biochemistry (Mosc.)* **2001**, *40*, 7355–7368.

90. Schenk, G.; Layfield, R.; Candy, J.M.; Duggleby, R.G.; Nixon, P.F. Molecular evolutionary analysis of the thiamine-diphosphate-dependent enzyme, transketolase. *J. Mol. Evol.* **1997**, *44*, 552–572.
91. Raeburn, S.; Rabinowitz, J.C. Pyruvate: Ferredoxin oxidoreductase: II. Characteristics of the forward and reverse reactions and properties of the enzyme. *Arch. Biochem. Biophys.* **1971**, *146*, 21–33.
92. Uyeda, K.; Rabinowitz, J.C. Pyruvate-ferredoxin oxidoreductase. IV. Studies on the reaction mechanism. *J. Biol. Chem.* **1971**, *246*, 3120–3125.
93. Uyeda, K.; Rabinowitz, J.C. Pyruvate-ferredoxin oxidoreductase. III. Purification and properties of the enzyme. *J. Biol. Chem.* **1971**, *246*, 3111–3119.
94. Ragsdale, S.W. Pyruvate ferredoxin oxidoreductase and its radical intermediate. *Chem. Rev.* **2003**, *103*, 2333–2346.
95. Ragsdale, S.W.; Pierce, E. Acetogenesis and the Wood-Ljungdahl pathway of CO_2 fixation. *Biochim. Biophys. Acta* **2008**, *1784*, 1873–1898.
96. Tittmann, K. Reaction mechanisms of thiamin diphosphate enzymes: Redox reactions. *FEBS J.* **2009**, *276*, 2454–2468.
97. Kletzin, A.; Adams, M. Molecular and phylogenetic characterization of pyruvate and 2- ketoisovalerate ferredoxin oxidoreductases from *Pyrococcus furiosus* and pyruvate ferredoxin oxidoreductase from *Thermotoga maritima. J. Bacteriol.* **1996**, *178*, 248–257.
98. Zhang, Q.; Iwasaki, T.; Wakagi, T.; Oshima, T. 2-Oxoacid: Ferredoxin oxidoreductase from the thermoacidophilic archaeon, *Sulfolobus* sp. strain 7. *J. Biochem. (Tokyo)* **1996**, *120*, 587–599.
99. Townson, S.M.; Upcroft, J.A.; Upcroft, P. Characterisation and purification of pyruvate: Ferredoxin oxidoreductase from *Giardia duodenalis. Mol. Biochem. Parasitol.* **1996**, *79*, 183–193.
100. Horner, D.S.; Hirt, R.P.; Embley, T.M. A single eubacterial origin of eukaryotic pyruvate: Ferredoxin oxidoreductase genes: Implications for the evolution of anaerobic eukaryotes. *Mol. Biol. Evol.* **1999**, *16*, 1280–1291.
101. Pineda, E.; Encalada, R.; Rodríguez-Zavala, J.S.; Olivos-García, A.; Moreno-Sánchez, R.; Saavedra, E. Pyruvate: Ferredoxin oxidoreductase and bifunctional aldehyde-alcohol dehydrogenase are essential for energy metabolism under oxidative stress in *Entamoeba histolytica. FEBS J.* **2010**, *277*, 3382–3395.
102. Wahl, R.C.; Orme-Johnson, W.H. Clostridial pyruvate oxidoreductase and the pyruvate-oxidizing enzyme specific to nitrogen fixation in *Klebsiella pneumoniae* are similar enzymes. *J. Biol. Chem.* **1987**, *262*, 10489–10496.
103. Meinecke, B.; Bertram, J.; Gottschalk, G. Purification and characterization of the pyruvate-ferredoxin oxidoreductase from *Clostridium acetobutylicum. Arch. Microbiol.* **1989**, *152*, 244–250.
104. Pieulle, L.; Magro, V.; Hatchikian, E. Isolation and analysis of the gene encoding the pyruvate-ferredoxin oxidoreductase of *Desulfovibrio africanus*, production of the recombinant enzyme in *Escherichia coli*, and effect of carboxy-terminal deletions on its stability. *J. Bacteriol.* **1997**, *179*, 5684–5692.
105. Pieulle, L.; Guigliarelli, B.; Asso, M.; Dole, F.; Bernadac, A.; Hatchikian, E.C. Isolation and characterization of the pyruvate-ferredoxin oxidoreductase from the sulfate-reducing bacterium *Desulfovibrio africanus. Biochim. Biophys. Acta* **1995**, *1250*, 49–59.

106. Pieulle, L.; Charon, M.-H.; Bianco, P.; Bonicel, J.; Petillot, Y.; Hatchikian, E.C. Structural and kinetic studies of the pyruvate-ferredoxin oxidoreductase/ferredoxin complex from *Desulfovibrio africanus*. *Eur. J. Biochem.* **1999**, *264*, 500–508.

107. Pieulle, L.; Chabriere, E.; Hatchikian, C.; Fontecilla-Camps, J.C.; Charon, M.-H. Crystallization and preliminary crystallographic analysis of the pyruvate-ferredoxin oxidoreductase from *Desulfovibrio africanus*. *Acta Crystallogr. D Biol. Crystallogr.* **1999**, *55*, 329–331.

108. Blamey, J.M.; Adams, M.W. Characterization of an ancestral type of pyruvate ferredoxin oxidoreductase from the hyperthermophilic bacterium, *Thermotoga maritima*. *Biochemistry (Mosc.)* **1994**, *33*, 1000–1007.

109. Blamey, J.M.; Adams, M.W.W. Purification and characterization of pyruvate ferredoxin oxidoreductase from the hyperthermophilic archaeon *Pyrococcus furiosus*. *Biochim. Biophys. Acta* **1993**, *1161*, 19–27.

110. Kunow, J.; Linder, D.; Thauer, R.K. Pyruvate: Ferredoxin oxidoreductase from the sulfate-reducing *Archaeoglobus fulgidus*: Molecular composition, catalytic properties, and sequence alignments. *Arch. Microbiol.* **1995**, *163*, 21–28.

111. Bock, A.-K.; Prieger-Kraft, A.; Schönheit, P. Pyruvate a novel substrate for growth and methane formation in *Methanosarcina barkeri*. *Arch. Microbiol.* **1994**, *161*, 33–46.

112. Bock, A.K.; Kunow, J.; Glasemacher, J.; Schönheit, P. Catalytic properties, molecular composition and sequence alignments of pyruvate: Ferredoxin oxidoreductase from the methanogenic archaeon *Methanosarcina Barkeri* (Strain Fusaro). *Eur. J. Biochem.* **1996**, *237*, 35–44.

113. Tersteegen, A.; Dietmar, L.R.K.; Thauer, R.H. Structures and functions of four anabolic 2-oxoacid oxidoreductases in *Methanobacterium thermoautotrophicum*. *Eur. J. Biochem.* **1997**, *244*, 862–868.

114. Chabrière, E.; Vernede, X.; Guigliarelli, B.; Charon, M.-H.; Hatchikian, E.C.; Fontecilla-Camps, J.C. Crystal structure of the free radical intermediate of pyruvate:ferredoxin oxidoreductase. *Science* **2001**, *294*, 2559–2563.

115. Chabrière, E.; Charon, M.-H.; Volbeda, A.; Pieulle, L.; Hatchikian, E.C.; Fontecilla-Camps, J.-C. Crystal structures of the key anaerobic enzyme pyruvate: Ferredoxin oxidoreductase, free and in complex with pyruvate. *Nat. Struct. Mol. Biol.* **1999**, *6*, 182–190.

116. Garczarek, F.; Dong, M.; Typke, D.; Witkowska, H.E.; Hazen, T.C.; Nogales, E.; Biggin, M.D.; Glaeser, R.M. Octomeric pyruvate-ferredoxin oxidoreductase from *Desulfovibrio vulgaris*. *J. Struct. Biol.* **2007**, *159*, 9–18.

117. Bock, A.-K.; Schönheit, P.; Teixeira, M. The iron-sulfur centers of the pyruvate:ferredoxin oxidoreductase from *Methanosarcina barkeri* (Fusaro). *FEBS Lett.* **1997**, *414*, 209–212.

118. Charon, M.-H.; Volbeda, A.; Chabriere, E.; Pieulle, L.; Fontecilla-Camps, J.C. Structure and electron transfer mechanism of pyruvate: Ferredoxin oxidoreductase. *Curr. Opin. Struct. Biol.* **1999**, *9*, 663–669.

119. Shiba, H.; Kawasumi, T.; Igarashi, Y.; Kodama, T.; Minoda, Y. The CO_2 assimilation via. the reductive tricarboxylic acid cycle in an obligately autotrophic, aerobic hydrogen-oxidizing bacterium, *Hydrogenobacter thermophilus*. *Arch. Microbiol.* **1985**, *141*, 198–203.

120. Ikeda, T.; Ochiai, T.; Morita, S.; Nishiyama, A.; Yamada, E.; Arai, H.; Ishii, M.; Igarashi, Y. Anabolic five subunit-type pyruvate: Ferredoxin oxidoreductase from *Hydrogenobacter thermophilus* TK-6. *Biochem. Biophys. Res. Commun.* **2006**, *340*, 76–82.
121. Ikeda, T.; Yamamoto, M.; Arai, H.; Ohmori, D.; Ishii, M.; Igarashi, Y. Enzymatic and electron paramagnetic resonance studies of anabolic pyruvate synthesis by pyruvate: Ferredoxin oxidoreductase from *Hydrogenobacter thermophilus. FEBS J.* **2010**, *277*, 501–510.
122. Yamamoto, M.; Ikeda, T.; Arai, H.; Ishii, M.; Igarashi, Y. Carboxylation reaction catalyzed by 2-oxoglutarate: Ferredoxin oxidoreductases from *Hydrogenobacter thermophilus. Extremophiles* **2010**, *14*, 79–85.
123. Lin, W.C.; Yang, Y.-L.; Whitman, W.B. The anabolic pyruvate oxidoreductase from *Methanococcus maripaludis. Arch. Microbiol.* **2003**, *179*, 444–456.
124. Lin, W.; Whitman, W. The importance of *por*E and *por*F in the anabolic pyruvate oxidoreductase of *Methanococcus maripaludis. Arch. Microbiol.* **2004**, *181*, 68–73.
125. Imlay, A.J. Iron-sulphur clusters and the problem with oxygen. *Mol. Microbiol.* **2006**, *59*, 1073–1082.
126. Linn, T.C.; Pettit, F.H.; Reed, L.J. α-Keto acid dehydrogenase complexes, X. Regulation of the activity of the pyruvate dehydrogenase complex from beef kidney mitochondria by phosphorylation and dephosphorylation. *Proc. Natl. Acad. Sci. USA* **1969**, *62*, 234–241.
127. Patel, M.S.; Roche, T.E. Molecular biology and biochemistry of pyruvate dehydrogenase complexes. *FASEB J.* **1990**, *4*, 3224–3233.
128. Witzmann, S.; Bisswanger, H. The pyruvate dehydrogenase complex from thermophilic organisms: Thermal stability and re-association from the enzyme components. *Biochim. Biophys. Acta* **1998**, *1385*, 341–352.
129. Potter, S.; Fothergill-Gilmore, L.A. Purification and properties of pyruvate kinase from *Thermoplasma acidophilum. FEMS Microbiol. Lett.* **1992**, *94*, 235–239.
130. Heath, C.; Jeffries, A.C.; Hough, D.W.; Danson, M.J. Discovery of the catalytic function of a putative 2-oxoacid dehydrogenase multienzyme complex in the thermophilic archaeon *Thermoplasma acidophilum. FEBS Lett.* **2004**, *577*, 523–527.
131. Selig, M.; Schönheit, P. Oxidation of organic compounds to CO_2 with sulfur or thiosulfate as electron acceptor in the anaerobic hyperthermophilic archaea *Thermoproteus tenax* and *Pyrobaculum islandicum* proceeds via. the citric acid cycle. *Arch. Microbiol.* **1994**, *162*, 286–294.
132. Ma, K.; Weiss, R.; Adams, M.W.W. Characterization of hydrogenase II from the hyperthermophilic archaeon *Pyrococcus furiosus* and assessment of its role in sulfur reduction. *J. Bacteriol.* **2000**, *182*, 1864–1871.
133. Jenney, F.E.; Adams, M.W.W. Hydrogenases of the model hyperthermophiles. *Ann. NY Acad. Sci.* **2008**, *1125*, 252–266.
134. Rudolph, F.B.; Purich, D.L.; Fromm, H.J. Coenzyme A-linked aldehyde dehydrogenase from *Escherichia coli. J. Biol. Chem.* **1968**, *243*, 5539–5545.
135. Lurz, R.; Mayer, F.; Gottschalk, G. Electron microscopic study on the quaternary structure of the isolated particulate alcohol-acetaldehyde dehydrogenase complex and on its identity with the polygonal bodies of *Clostridium kluyveri. Arch. Microbiol.* **1979**, *120*, 255–262.

136. Nair, R.V.; Bennett, G.N.; Papoutsakis, E.T. Molecular characterization of an aldehyde/alcohol dehydrogenase gene from *Clostridium acetobutylicum* ATCC 824. *J. Bacteriol.* **1994**, *176*, 871–885.
137. Tóth, J.; Ismaiel, A.A.; Chen, J.-S. The *ald* gene, encoding a coenzyme A-acylating aldehyde dehydrogenase, distinguishes *Clostridium beijerinckii* and two other solvent-producing clostridia from *Clostridium acetobutylicum*. *Appl. Environ. Microbiol.* **1999**, *65*, 4973–4980.
138. Sánchez, L.B. Aldehyde dehydrogenase (CoA-acetylating) and the mechanism of ethanol formation in the amitochondriate protist, *Giardia lamblia*. *Arch. Biochem. Biophys.* **1998**, *354*, 57–64.
139. Bruchhaus, I.; Tannich, E. Purification and molecular characterization of the NAD(+)-dependent acetaldehyde/alcohol dehydrogenase from *Entamoeba histolytica*. *Biochem. J.* **1994**, *303*, 743–748.
140. Burdette, D.; Zeikus, J.G. Purification of acetaldehyde dehydrogenase and alcohol dehydrogenases from *Thermoanaerobacter ethanolicus* 39E and characterization of the secondary-alcohol dehydrogenase (2 degrees Adh) as a bifunctional alcohol dehydrogenase--acetyl-CoA reductive thioesterase. *Biochem. J.* **1994**, *302*, 163–170.
141. Brown, S.D.; Guss, A.M.; Karpinets, T.V.; Parks, J.M.; Smolin, N.; Yang, S.; Land, M.L.; Klingeman, D.M.; Bhandiwad, A.; Rodriguez, M. Mutant alcohol dehydrogenase leads to improved ethanol tolerance in *Clostridium thermocellum*. *Proc. Natl. Acad. Sci. USA* **2011**, *108*, 13752–13757.

10

Biocatalytic Synthesis of Chiral Alcohols and Amino Acids for Development of Pharmaceuticals

Ramesh N. Patel

SLRP Associates Consultation in Biotechnology, 572 Cabot Hill Road, Bridgewater, NJ 08807, USA; E-Mail: rameshpatelphd@yahoo.com

---

Abstract: Chirality is a key factor in the safety and efficacy of many drug products and thus the production of single enantiomers of drug intermediates and drugs has become increasingly important in the pharmaceutical industry. There has been an increasing awareness of the enormous potential of microorganisms and enzymes derived there from for the transformation of synthetic chemicals with high chemo-, regio- and enatioselectivities. In this article, biocatalytic processes are described for the synthesis of chiral alcohols and unntural aminoacids for pharmaceuticals.

Keywords: biocatalysis; enzymatic processes; chiral intermediates; drugs development

---

1. Introduction

For preparation of drugs and their intermediates, the synthesis of single enantiomers has become increasingly important in the pharmaceutical industry [1]. Single enantiomers can be produced by either by chemical or biocatalytic routes. The advantages of biocatalysis over chemical synthesis are that enzyme-catalyzed reactions are often highly enantioselective and regioselective. They can be carried out under mild conditions at ambient temperature and atmospheric pressure, thus avoiding the use of more extreme reaction conditions which could cause problems with isomerization, racemization, epimerization, and rearrangement of compound. Microbial cells and wide variety and class of enzymes derived there from can be used for chiral synthesis. Enzymes can be immobilized and reused for many cycles. In addition, enzymes can be over expressed to make biocatalytic processes economically efficient, and enzymes with modified activity can be tailor-made. Directed evolution of biocatalysts can lead to increased enzyme activity, selectivity and stability [2–15]. A number of review articles [16–31] have been published on the use of enzymes in organic synthesis. This chapter provides some examples of the use of enzymes for the synthesis chiral alcohols, unnatural amino acids, and amines for synthesis of phamaceuticals.

2. Enzymatic Preparation of Chiral Alcohols

2.1. Hydroxy Buspirone (Antianxiety Drug): Enzymatic Preparation of 6-Hydroxybuspirone

Buspirone (Buspar®, **1**, Figure 1) is a drug used for treatment of anxiety and depression that is thought to produce its effects by binding to the serotonin 5HT1A receptor [32–34]. Mainly as a result of hydroxylation reactions, it is extensively converted to various metabolites and blood concentrations return to low levels a few hours after dosing [35]. A major metabolite, 6-hydroxybuspirone **2**, produced by the action of liver cytochrome P450 CYP3A4, is present at much higher concentrations in human blood than buspirone itself. For development of 6-hydroxybuspirone as a potential antianxiety drug, preparation and testing of the two enantiomers as well as the racemate was of interest. An enantioselective microbial reduction process was developed for reduction of 6-oxobuspirone **3**, to either (*R*)- and (*S*)-6-hydroxybuspirone **2**. About 150 microbial cultures were screened for the enantioselective reduction of **3**. *Rhizopus stolonifer* SC 13898, *Neurospora crassa* SC 13816, *Mucor racemosus* SC 16198, and *Pseudomonas putida* SC 13817 gave >50% reaction yields and >95% e.e.s of (*S*)-6-hydroxybuspirone. The yeast strains *Hansenula polymorpha* SC 13845 and *Candida maltosa* SC 16112 gave (*R*)-6-hydroxybuspirone **2** in >60% reaction yield and >97% e.e. [36]. The NADP-dependent (*R*)-reductase (RHBR) from *Hansenula polymorpha* SC 13845 was purified to homogeneity, its N-terminal and internal sequences were determined and cloned and expressed in *Escherichia coli*. To regenerate the cofactor NADPH required for reduction we have also cloned and expressed the glucose-6-phosphate dehydrogenase gene from *Saccharomyces cerevisiae* in *Escherichia coli*. Recombinant cultures expressing (*R*)-reductase (RHBR) catalyzed the reduction of 6-ketobuspirone to (*R*)-6-hydroxybuspirone in 99% yield and 99.9% e.e. at 50 g/L substrate input [37].

Figure 1. Hydroxy buspirone (antianxiety drug): Enzymatic preparation of 6-hydroxybuspirone.

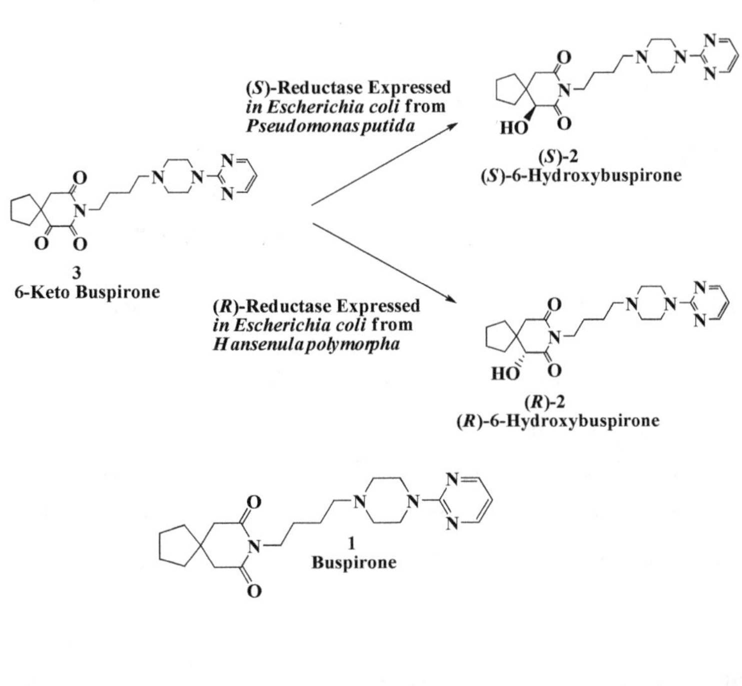

The NAD-dependent (*S*)-reductase (SHBR) from *Pseudomonas putida* SC 16269 was also purified to homogeneity, its N-terminal and internal sequences were determined and cloned and expressed in *Escherichia coli.* To regenerate the cofactor NADH required for reduction we have also cloned and expressed the NAD^+ dependent formate dehydrogenase gene from *Pichia pastoris* in *Escherichia coli.* Recombinant *Escherichia coli* expressing (*S*)-reductase was used to catalyze the reduction of 6-ketobuspirone to (*S*)-6-hydroxybuspirone, in >98% yield and >99.8% e.e. at 50 g/L substrate input [37].

*2.2. Cholesterol Lowering Agents: Enzymatic Preparation of (3S,5R)-Dihydroxy-6-(Benzyloxy) Hexanoic Acid, Ethyl Ester **4***

Compound **4** (Figure 2) is a key chiral intermediate required for the chemical synthesis of compound **5**, Arotvastatin **6**, and Rosuvastatin all are anticholesterol drugs which acts by inhibition of HMG CoA reductase [38–42].

Figure 2. Cholesterol lowering agents: Enzymatic preparation of (3*S*,5*R*)-dihydroxy-6-(benzyloxy) hexanoic acid, ethyl ester.

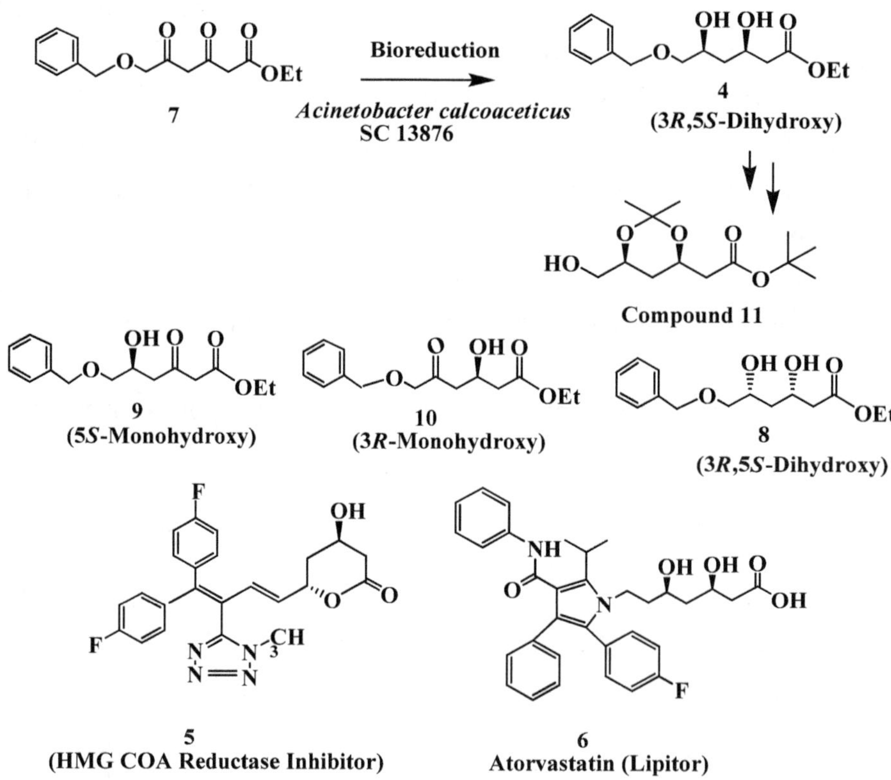

The enantioselective reduction of a diketone 3,5-dioxo-6-(benzyloxy) hexanoic acid, ethyl ester **7** to (3*R*,5*S*)-dihydroxy-6-(benzyloxy) hexanoic acid, ethyl ester **4** (Figure 2) was demonstrated by cell suspensions of *Acinetobacter calcoaceticus* SC 13876 [39,43]. On reduction of **7** by cell suspensions, the *syn*-**4** and *anti*-**8** dihydroxy esters were formed in the ratio of about 87:13, 83:17, 76:24 after 24 h at 2, 5 and 10 g/L of substrate input, respectively. There was no significant peak due to a monohydroxy ester. Chiral HPLC determined that the desired (3*R*,5*S*)-**4** was the major product with 99.4% e.e. Almost complete (>95%) conversion of the ethyl diketoester **7** to dihydroxy ester **4** in 24 h was seen up to a substrate concentration of 10 g/L and cell concentration of 200 g/L [39,43].

A mixture of ethyl 3-keto-5-hydroxy **9** (major) and 5-keto-3-hydroxy **10** (minor) was obtained from partial microbial reduction of ketoester **7**. These two mixtures were subjected to microbial reduction by *Acinetobacter sp* SC13874 cells for 6 h (incomplete reduction). The reduction provided the dihydroxy esters with the isomeric composition. The results indicated that the second reduction of the monohydroxy compound by SC13874 cells was quite enantiospecific. Reduction of the 3-keto-5-hydroxy **9** provided predominantly the (3*R)*-hydroxy, while reduction of the 3-hydroxy-5-keto ester **10** provided predominantly the (5*S)*-hydroxy compound [43].

Cell extracts of A. *calcoaceticus* SC 13876 in the presence of NAD^+, glucose, and glucose dehydrogenase reduced **7** to the corresponding monohydroxy compounds [3-hydroxy-5-oxo-6-(benzyloxy) hexanoic acid ethyl ester **9** and 5-hydroxy-3-oxo-6-(benzyloxy) hexanoic acid ethyl ester **10**]. Both **9** and **10** were further reduced to the (3*R*,5*S*)-dihydroxy compound **4** in 92% yield and 99% e.e. by cell extracts. (3*R*,5*S*)-**4** was converted to **11**, a key chiral intermediate for the synthesis of compound **5** and Atorvastatin **6**. Three different ketoreductases were purified to homogeneity from cell extracts of *A. calcoaceticus* SC 13876 and their biochemical properties were compared. Reductase I only catalyzes the reduction of ethyl diketoester **7** to its monohydroxy products whereas reductase II catalyzes the formation of dihydroxy products from monohydroxy substrates. A third reductase (III) was identified which catalyzes the reduction of diketoester **7** to *syn*-(3*R*,5*S*)-dihydroxy ester **4** [44], which now has been cloned and expressed in *E. coli* [44] and the reduction of diketoester 7 to *syn*-(3*R*,5*S*)-dihydroxy ester **4** was demonstrated by recombinant enzyme at 50 g/L substrate input with 10 g/L cell suspensions.

2.3. Atorvastatin: Enzymatic Preparation of (R)-4-Cyano-3-Hydroxybutyrate

An enzymatic process for the preparation of ethyl (*R*)-4-cyano-3-hydroxybutyric acid **12** (Figure 3), a key intermediate for the synthesis of Atorvastatin **6** was developed by Codexis [45]. In this process, first the enzymatic synthesis of ethyl (*S*)-4-chloro-3-hydroxybutyric acid derivatives **13** was carried out by ketoreductase-catalyzed conversion of 4-chloro-3-ketobutyric acid derivatives **14** [46]. The genes encoding halohydrin dehydrogenase from *Agrobacterium tumefaciens*, ketoreductase from *Candida magnoliae*, glucose dehydrogenase from *Bacillus subtilis* and formate dehydrogenase from *Candida boidinii* were separately cloned into *Escherichia coli* BL21. Each enzyme was then produced by fermentation, isolated and characterized. Then ethyl (*R*)-4-cyano-3-hydroxybutyrate **12** (Figure 3) was prepared from ethyl 4-chloroacetoacetate **14** by the following procedure: Ethyl 4-chloroacetoacetate 14 was incubated at pH 7.0 with ketoreductase, glucose dehydrogenase, and $NADP^+$ for 40 h to produce ethyl (*S*)-4-chloro-3-hydroxybutyrate **13** which was extracted with ethyl acetate, dried, filtered and concentrated to yield ~97% pure ester. The dried ethyl (*S*)-4-chloro-3-hydroxybutyrate **13** was dissolved in phosphate buffer and mixed with halohydrin dehalogenase and sodium cyanide at pH 8.0. After 57 h, essentially pure ethyl (*R*)-4-cyano-3-hydroxybutyrate **12**, an intermediate used in HMG-CoA reductase inhibitors syntheses, was recovered [45].

Figure 3. Atorvastatin: Enzymatic preparation of (*R*)-4-cyano-3-hydroxybutyrate.

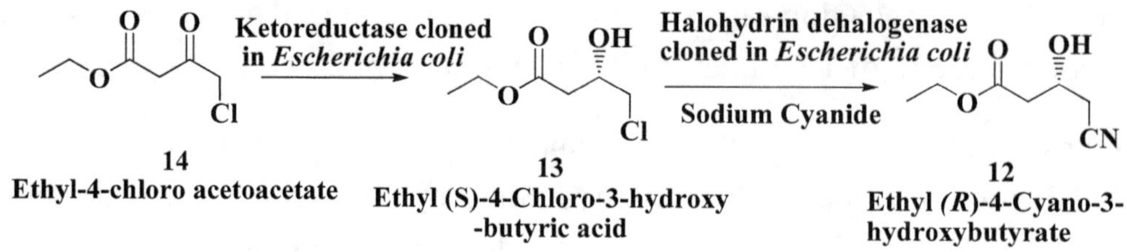

2.4. Preparation of (S)-4-Chloro-3-Hydroxybutanoic Acid Methyl Ester

(*S*)-4-chloro-3-hydroxybutanoic acid methyl ester **15** (Figure 4) is a key chiral intermediate in the total chemical synthesis of **16**, an inhibitor of HMG CoA reductase [46,47]. The reduction of 4-chloro-3-oxobutanoic acid methyl ester **17** to (*S*)-4-chloro-3-hydroxybutanoic acid methyl ester **15** (Figure 4) by cell suspensions of *Geotrichum candidum* SC 5469. In the biotransformation process, a reaction yield of 95% and e.e. of 96% were obtained for (*S*)-**15** by glucose-, acetate- or glycerol-grown cells (10% w/v) of *G. candidum* SC 5469 at 10 g/L substrate input. The e.e. of (*S*)-**15** was increased to 98% by heat-treatment of cell-suspensions (55 °C for 30 min) prior to conducting the bioreduction of **17** [48].

Figure 4. Chloesterol lowering agents: Preparation of (*S*)-4-chloro-3-hydroxybutanoic acid methyl ester.

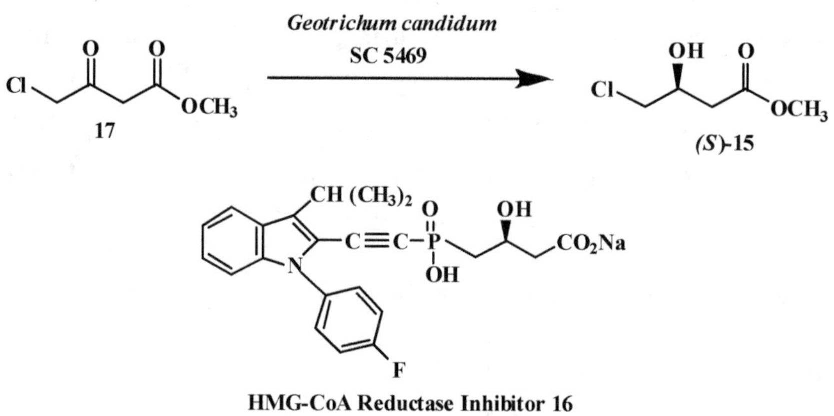

In an alternate approach, the asymmetric reduction of ethyl 4-chloroacetoacetate to (*S*)-4-chloro-3-hydroxybutonoate was demonstrated by a secondary alcohol dehydrogenase (PfODH) from *Pichia finlandica.* The gene encoding PfODH was cloned from *P. finlandica* and over expressed in *Escherichia coli*. Formate dehydrogenase was used to regenerate the cofactor NADH required for this reaction. Using recombinant *E. coli* coexpressing both PfODH and formate dehydrogenase from *Mycobacetrium* sp. produced to (*S*)-4-chloro-3-hydroxybutonoate in 98.5% yield and 99% e.e. at 32 g/L substrate input [49].

2.5. Rhinovirus Protease Inhibitor: Enzymatic Process for the Preparation of (R)-3-(4-Fluorophenyl)-2-Hydroxy Propionic Acid

(*R*)-3-(4-fluorophenyl)-2-hydroxy propionic acid **18** (Figure 5) is a building block for the synthesis of AG7088, a rhinovirus protease inhibitor **19** [50,51]. The preparation of **18** using a biocatalytic reduction of **20** in a membrane reactor [52]. A continuous enzymatic process for an efficient synthesis of (*R*)-3-(4-fluorophenyl)-2-hydroxy propionic acid at multikilogram scale with a high space-time yield (560 g/L/day) using a membrane reactor. The product was generated in excellent enantiomeric excess (e.e. > 99.9%) and good overall yield (68%–72%).

Figure 5. Rhinovirus protease inhibitor: Enzymatic process for the preparation of (*R*)-3-(4-fluorophenyl)-2-hydroxy propionic acid.

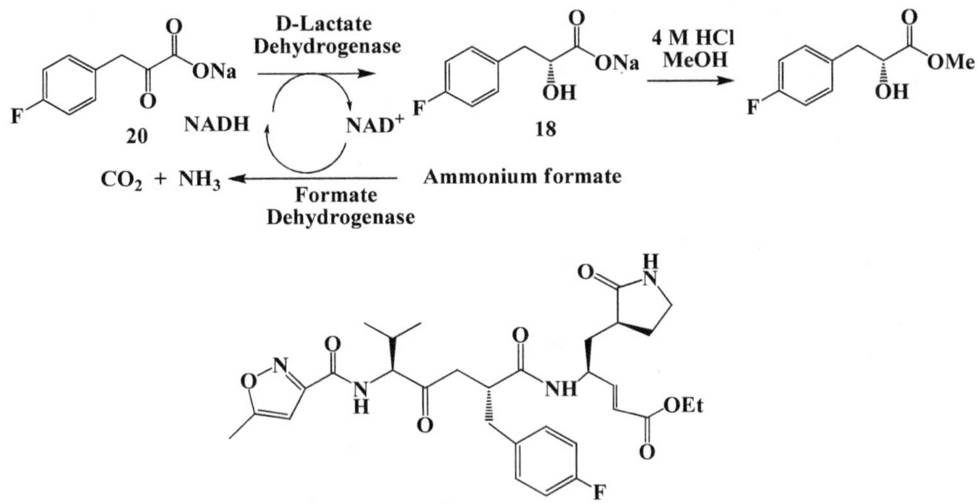

19
Rhinovirus Protease Inhibitor
Common Cold Treatment

Using this method, an overall quantity of 23 kg of **18** was prepared. The key step was an aqueous enzymatic reduction using D-lactate dehydrogenase (D-LDH) and formate dehydrogenase (FDH). Mechanistically, the keto acid salt **20** is stereoselectively reduced to the corresponding *R*-hydroxy acid **18** in the presence of D-lactate dehydrogenase by NADH. The cofactor itself is oxidized to NAD^+ in the process. Subsequently, in the presence of formate dehydrogenase, NAD^+ is reduced back to NADH by ammonium formate, which was oxidized to CO_2 and NH_3. In this fashion the expensive cofactor NAD^+ is regenerated by FDH, and only a catalytic amount of NAD^+ was required [52].

2.6. Enzymatic Preparation of Chiral Intermediates for Atazanavir

Atazanavir **21** (Figure 6) is an acyclic aza-peptidomimetic, a potent HIV protease inhibitor [53,54] approved by the Food and Drug Administration for treatment of Auto Immune Diseases (AIDS). An enzymatic process was developed for the preparation of (1*S*,2*R*)-[3-chloro-2-hydroxy-1-(phenylmethyl) propyl]carbamic acid, 1,1-dimethylethyl ester **22** , a key chiral intermediate required for the total synthesis of the HIV protease inhibitor atazanavir. The diastereoselective reduction of (1*S*)-[3-chloro-2-oxo-1-(phenylmethyl)propyl] carbamic acid, 1,1-dimethylethyl ester **23** was carried out using *Rhodococcus*, *Brevibacterium*, and *Hansenula* strains to provide **22**. Three strains of *Rhodococcus* gave >90% yield with a diastereomeric purity of >98% and an e.e. of 99.4% [55]. An efficient single-stage fermentation-biotransformation process was developed for the reduction of ketone **23** with cells of *Rhodococcus erythropolis* SC 13845 to yield **22** in 95% with a diasteromeric purity of 98.2% and an e.e. of 99.4% at substrate input of 10 g/L. The reduction process was further improved by generating mutants and selection of desired mutant for conversion of **23** to (1*S*,2*R*)-**22** at substrate input of 60 g/L [56]. (1*S*,2*R*)-22 was converted to epoxide **24** and used in the synthesis of atazanavir. Chemical reduction of chloroketone **23** using $NaBH_4$ produces the undesired chlorohydrin diastereomer [57].

Figure 6. Atazanavir (antiviral agent): Enzymatic reparation of (1*S*,2*R*)-[3-chloro-2-hydroxy-1-(phenylmethyl) propyl]-carbamic acid,1,1-dimethyl-ethyl ester.

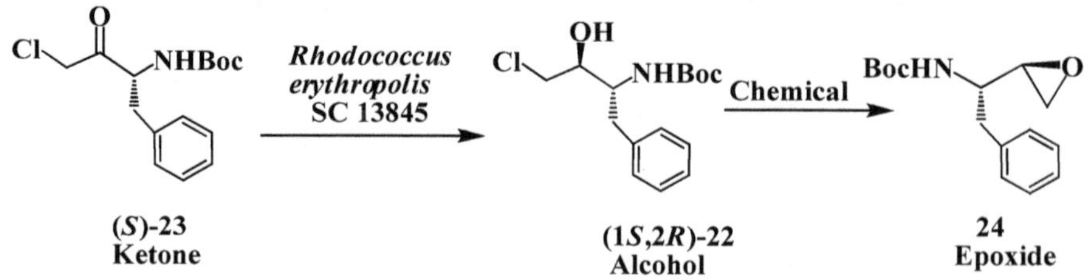

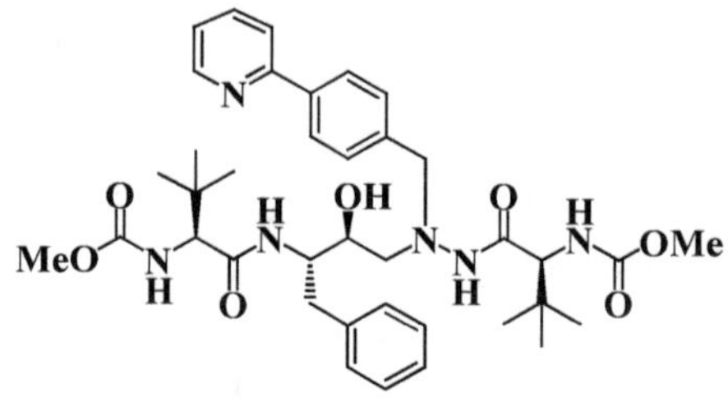

21
Atazanavir

2.7. Enzymatic Reduction Process for Synthesis of Montelukast Intermediate

The discovery of the biological activity of the slow reacting substance of anaphylaxis (SRS-A) and its relation to the leukotrienes (LTC4, LTD4, and LTE4) and asthma, the search for leukotriene antagonists has been intensive. As part of an ongoing program for the development of specific LTD4 antagonists for the treatment of asthma and other associated diseases at Merck have identified Montelukast **25** (Figure 7) as LTD4 antagonist [58–60].

Figure 7. Enzymatic reduction process for synthesis (*S*)-alcohol **27** for Montelukast intermediate.

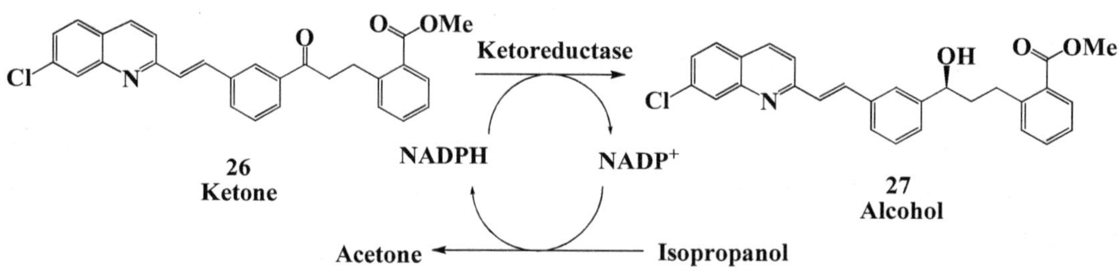

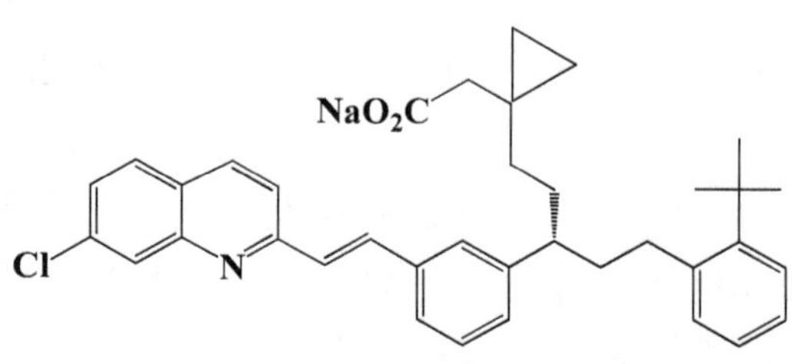

25
Montelukast (Singulair)

Merck has described the synthetic route for the production of montelukast, using a stereoselective reduction of a ketone **26** to the (*S*)-alcohol **27** as the key step. The alcohol subsequently undergoes a Sn2 displacement with a thiol to give the *R*-configured final product [59,60]. The reduction of the ketone **26** to produce the chiral alcohol **27** requires stoichiometric amounts of the chiral reducing agent (−)-flchlorodiisopino campheylborane [(−)-DIP-chloride]. (−)-DIP-chloride is selective and avoids the side reactions but it is corrosive and moisture-sensitive, causing burns if it is allowed to contact the skin. The reaction must be carried out at −20 to −25 °C to achieve the best stereoselectivity. The quench and extractive work-up generate large volumes of waste solvent, due to the product's low solubility. The potential advantages of biocatalytic transformation of ketone to alcohol were recognized early on by researchers at Merck. However, only two microorganisms were identified as having activity on the bulky and hydrophobic substrate [61]. Due to several reasons, an enzyme-catalyzed process for reduction of the ketone **26** was developed by Codexis. A ketoreductase was developed by directed evolution by high throughput screens using a slurry of the ketone substrate and high isopropanol concentration. Beneficial mutations among the various improved mutants were recombined in each round, and new mutations were made guided by ProSAR. The productivity of the final enzyme was improved 2,000-fold and stability was also substantially increased [62].

The final process was carried out as a slurry-to-slurry reaction at 45 °C, with the sparingly soluble ketone **26** being converted to an almost equally insoluble alcohol **27** at a concentration of 100 g/L substrate in aqueous isopropanol and toluene. A reaction yield of 99.3% and enantiomeric excess of 99.9% was obtained for alcohol 27 [62].

2.8. Anticancer Drug: Enzymatic Preparation of C-13 Paclitaxel Side-Chain Synthon

Among the antimitotic agents, paclitaxel (taxol®) **28** (Figure 8), a complex, polycyclic diterpene, exhibits a unique mode of action on microtubule proteins responsible for the formation of the spindle during cell division. Various types of cancers have been treated with paclitaxel and it was approved for use by the FDA for treatment of ovarian cancer and metastatic breast cancer [63–65]. A key precursor for the paclitaxel semi-synthetic process is the chiral C-13 paclitaxel side-chain **29**. An enzymatic enantioselective microbial reduction of 2-keto-3-(N-benzoylamino)-3-phenyl propionic acid ethyl ester **30** to yield (2*R*,3*S*)-*N*-benzoyl-3-phenyl isoserine ethyl ester **29** was demonstrated using two strains of *Hansenula* [66]. Preparative-scale bioreduction of ketone **30** was demonstrated using cell suspensions of *Hansenula polymorpha* SC 13865 and *Hansenula fabianii* SC 13894 in independent experiments. In both batches, a reaction yield of >80% and e.e.s of >94% were obtained for (2*R*,3*S*)-**29**. In a single-stage process, cells of *H. fabianii* were grown in a 15-L fermentor for 48 h, then the bioreduction process was initiated by addition of 30 g of substrate and 250 g of glucose and continued for 72 h. A reaction yield of 88% with an e.e. of 95% was obtained for (2*R*,3*S*)-**29**.

2.9. Antipsychotic Drug: Enzymatic Reduction of 1-(4-Fluorophenyl)4-[4-(5-Fluoro-2-Pyrimidinyl)1-Piperazinyl]-1-Butanone

The sigma receptor system in the brain and endocrine tissue has been target for development of new class of antipsychotic drugs [67,68]. Compound (*R*)-**31** (Figure 9) is a sigma ligand and has a high affinity for sigma binding site and antipsychotic efficacy. The enantioselective microbial reduction process was developed for the conversion of ketone **32** to both enantiomers of alcohol **31** [69]. Various microorganisms screened for the enatioselective reduction of 1-(4-fluorophenyl)4-[4-(5-fluoro-2-pyrimidinyl)1-piperazinyl]-1butanone **32**. From this screen, *Mortierella ramanniana* ATCC 38191 was identified to predominantly reduced compound **32** to (*R*)-**31**, while *Pullularia pullulans* ATCC 16623 was identified to predominantly reduced compound **32** to (*S*)-**31**. A single stage fermentation/biotransformation process was developed. Cells of *M. ramanniana* were grown in a 20-L fermentor and after 40 h growth period, the biotransformation process was initiated by addition of 40 g ketone **32** and 400 g glucose. The biotransformation process was completed in 24 h with a reaction yield of 100% and an e.e. of 98.9% for (*R*)-**31**. At the end of the biotransformation process, cells were removed by filtration and product was recovered from the filtrate in overall 80% yield [69].

Figure 8. Anticancer drug: Enzymatic preparation of C-13 paclitaxel side-chain synthon.

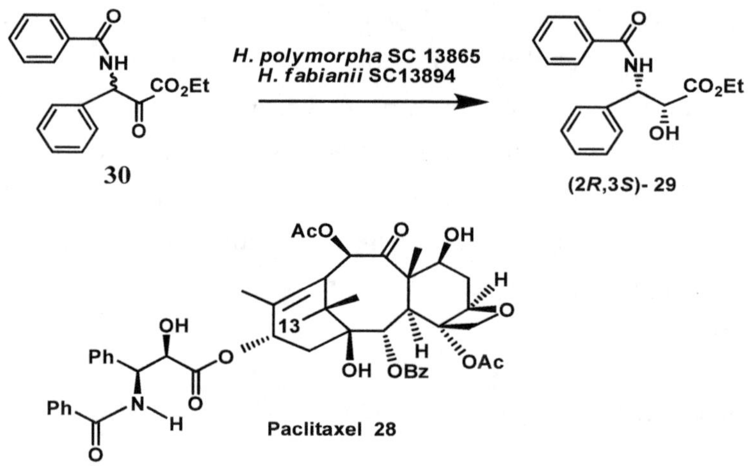

Figure 9. Antipsychotic drug: Enzymatic reduction of 1-(4-fluorophenyl)4-[4-(5-fluoro-2-pyrimidinyl)1-piperazinyl]-1-butanone.

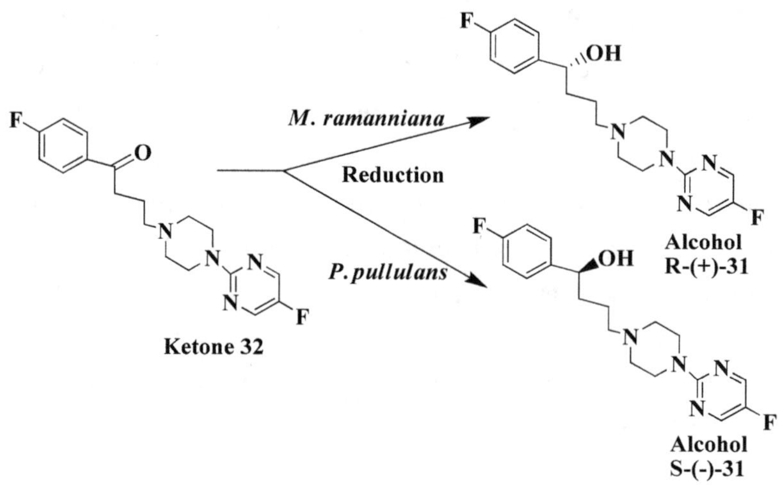

2.10. Retinoic Acid Receptpor Agonist: Enzymatic Preparation of 2-(R)-Hydroxy-2-(1',2',3',4'-Tetrahydro-1',1',4',4'-Tetramethyl-6'-Naphthalenyl)Acetate

Retinoic acid and its natural and synthetic analogs (retinoids) exert a wide variety of biological effects by binding to or activating a specific receptor or sets of receptors [70]. They have been shown to effect cellular growth and differentiation and are promising drugs for the treatment of cancers [71]. A few retinoids are already in clinical use for the treatment of dermatological diseases such as acne and psoriasis. (*R*)-3-Fluoro-4-[[hydroxy-(5,6,7,8-tetrahydro-5,5,8,8-tetramethyl-2-naphthalenyl)-acetyl]amino]benzoic acid **33** (Figure 10) is a retinoic acid receptor gamma-specific agonist potentially useful as a dermatological and anticancer drug [72].

Figure 10. Retinoic acid receptor agonist: Enzymatic preparation of 2-(*R*)-hydroxy-2-(1',2',3',4'-tetrahydro-1',1',4',4'-tetramethyl-6'-naphthalenyl)acetate.

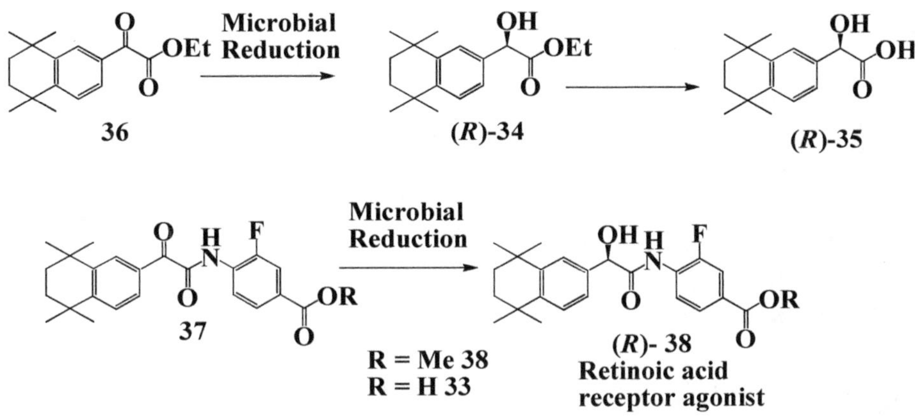

Ethyl 2-(*R*)-hydroxy-2-(1',2',3',4'-tetrahydro-1',1',4',4'-tetramethyl-6'-naphthalenyl)acetate **34** and the corresponding acid **35** were prepared as intermediates in the synthesis of the retinoic acid receptor gamma-specific agonist [73]. Enantioselective microbial reduction of ethyl 2-oxo-2-(1',2',3',4'-tetrahydro-1',1',4',4'-tetramethyl-6-naphthalenyl) acetate **36** to alcohol **34** was carried out using *Aureobasidium pullulans* SC 13849 in 98% yield and with an e.e. of 96%. At the end of the reaction, hydroxyester **34** was adsorbed onto XAD-16 resin and, after filtration, recovered in 94% yield from the resin with acetonitrile extraction. The recovered (*R*)-hydroxyester **34** was treated with Chirazyme L-2 or pig liver esterase to convert it to the corresponding (*R*)-hydroxyacid **35** in quantitative yield. The enantioselective microbial reduction of ketoamide **37** to the corresponding (*R*)-hydroxyamide **38** by *A. pullulans* SC 13849 has also been demonstrated [73].

2.11. Anti-Alzheimer's Drugs: Enzymatic Reduction of 5-Oxohexanoate and 5-Oxohexanenitrile

Ethyl-(*S*)-5-hydroxyhexanoate **39** and (*S*)-5-hydroxyhexanenitrile **40** (Figure 11) are key chiral intermediates in the synthesis of anti-Alzheimer's drugs [74]. Both chiral compounds have been prepared by enantioselective reduction of ethyl-5-oxohexanoate **41** and 5-oxohexanenitrile 42 by *Pichia methanolica* SC 16116 [75]. Reaction yields of 80%–90% and >95% e.e.s were obtained for each compound. In an alternate approach, the enzymatic resolution of racemic 5-hydroxyhexane nitrile **43** by enzymatic succinylation was demonstrated using immobilized lipase PS-30 to obtain (*S*)-5-hydroxyhexanenitrile **40** in 35% yield (maximum yield is 50%). (*S*)-5-Acetoxy-hexanenitrile **44** was prepared by enantioselective enzymatic hydrolysis of racemic 5-acetoxyhexanenitrile **45** by *Candida antarctica* lipase. A reaction yield of 42% and an e.e. of >99% were obtained [75].

Figure 11. Anti-Alzheimer's drugs: Enzymatic reduction of 5-oxohexanoate and 5-oxohexanenitrile.

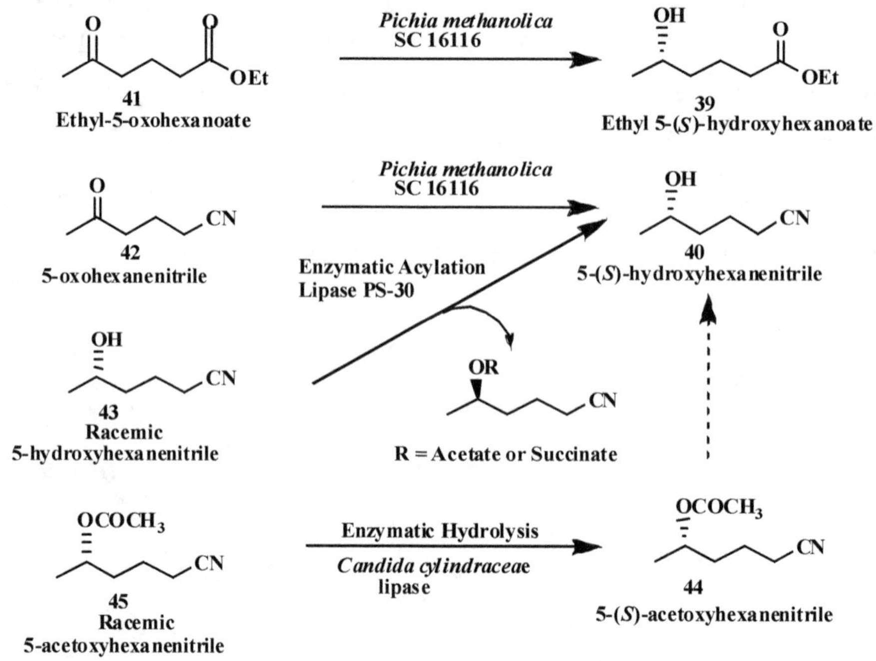

2.12. Enantioselective Microbial Reduction of Substituted Acetophenone

The chiral intermediates (*S*)-1-(2'-bromo-4'-fluorophenyl)ethanol 46 and (*S*)-methyl 4-(2'-acetyl-5'-fluorophenyl)-butanol **47** are potential intermediates for the synthesis of several potential anti-Alzheimer's drugs [76]. The chiral intermediate (*S*)-1-(2'-bromo-4'-fluoro phenyl)ethanol **46** (Figure 12A) was prepared by the enantioselective microbial reduction of 2-bromo-4-fluoro acetophenone **48** [77]. Organisms from genus *Candida, Hansenula, Pichia, Rhodotorula, Saccharomyces, Sphingomonas* and Baker's yeast reduced **48** to **46** in >90% yield and 99% enantiomeric excess (e.e.).

Figure 12. (**A**) Anti-Alzheimer's drugs: Enantioselective microbial reduction of substituted acetophenone; (**B**) Enantioselective microbial reduction of methyl-4-(2'-acetyl-5'-fluorophenyl) butanoates.

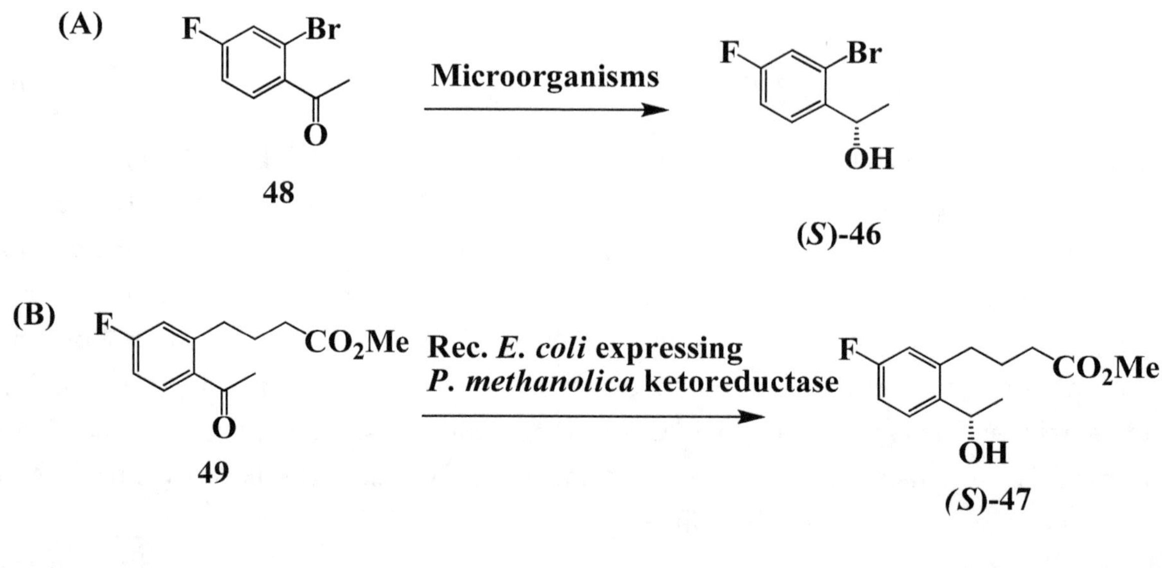

In an alternate approach, the enantioselective microbial reduction of methyl-4-(2'-acetyl-5'-fluorophenyl) butanoates **49** (Figure 12B) was demonstrated using strains of *Candida* and *Pichia*. Reaction yields of 40%–53% and e.e.s of 90%–99% were obtained for the corresponding (*S*)-hydroxy esters **47**. The reductase which catalyzed the enantioselective reduction of ketoesters was purified to homogeneity from cell extracts of *Pichia methanolica* SC 13825. It was cloned and expressed in *Escherichia coli* and recombinant cultures were used for the enantioselective reduction of the keto-methyl ester **49** to the corresponding (*S*)-hydroxy methyl ester **47**. On preparative scale, a reaction yield of 98% with an enantiomeric excess of 99% for **47** was obtained [77].

2.13. Anticancer Drug: Enzymatic Preparation of (S)-2-Chloro-1-(3-Chlorophenyl)Ethanol

The synthesis of the leading candidate compound **50** [78] in an anticancer program (IGF-1 receptor inhibitors) [79,80] required (*S*)-2-chloro-1-(3-chlorophenyl)ethanol **51** (Figure 13) as an intermediate. Other possible candidate compounds used are analogs of (*S*)-alcohol **51**. From microbial screen of the reduction of ketone **52** to (*S*)-alcohol **51**, two cultures namely *Hansenula polymorpha* SC13824 (73.8% enantiomeric excess) and *Rhodococcus globerulus* SC SC16305 (71.8% enantiomeric excess) were identified that had the highest enantioselectivity. A ketoreductase from *Hansenula polymorpha*, after purification to homogeneity, gave (*S*)-alcohol **51** with 100% ee [81]. The ketoreductase was cloned and expressed in *E. coli* together with a glucose-6-phosphate dehydrogenase from *Saccharomyces cerevisiae* to allow regeneration of the NADPH required for the reduction process. An extract of *E. coli* containing the two recombinant enzymes was used to reduce 2-chloro-1-(3-chloro-4fluorophenyl)ethanone **52**. Intact *E. coli* cells provided with glucose were used to prepare (*S*)-2-chloro-1-(3-chloro-4-fluorophenyl)ethanol **51** in 89% yield with 100% e.e. [81].

Figure 13. Anticancer drug: Enzymatic preparation of (*S*)-2-chloro-1-(3-chlorophenyl)ethanol.

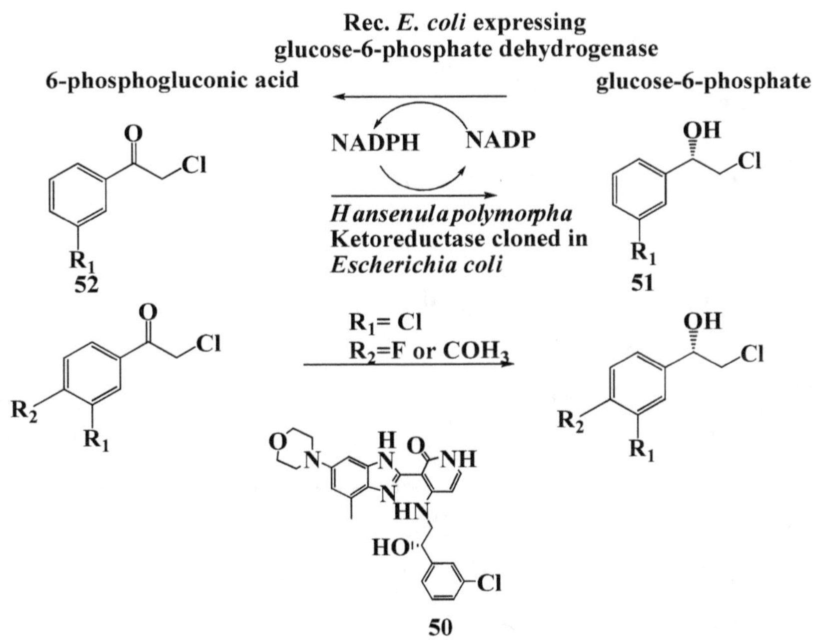

2.14. Thrombin Inhibitor: Enzymatic Preparation of (R)-2-Hydroxy-3,3-Dimethylbutanoic Acid

Thrombin is a trypsin-like protease enzyme that plays a critical role in intrinsic and extrinsic blood coagulation. As a result of the enzymatic activation of numerous coagulation factors, thrombin is activated to cleave fibrinogen, producing fibrin, which is directly responsible for blood clotting. An imbalance between these factors and their endogenous activators and inhibitors can give rise to a number of disease states such as myocardial infarction, unstable angina, stroke, ischemia, restenosis following angioplasty, pulmonary embolism, deep vein thrombosis, and arterial thrombosis [82,83]. Consequently, the aggressive search for a potent, selective, and bioavailable thrombin inhibitor is widespread [84]. An intensive effort by Merck has led to the identification of thrombin inhibitor **53** [85]. The synthesis of **53** required a key chiral intermediate (*R*)-hydroxy ester **54**. An enzymatic process was developed for the asymmetric reduction of ketoester **55** to (*R*)-**54** using commercially available ketoreductase KRED1001 (Figure 14). The cofactor NADPH required for this reaction was regenerated using glucose dehydrogenase. The hydroxy ester (*R*)-**54** was isolated as an oil and then saponified to the corresponding enantiomerically pure hydroxy acid (*R*)-**56** without epimerization [86]. The enantiomerically pure (*R*)-**56** was obtained in 82% isolated yield (>99.5% e.e.).

Figure 14. Thrombin inhibitor: Enzymatic preparation of (*R*)-2-Hydroxy-3,3-dimethylbutanoic acid.

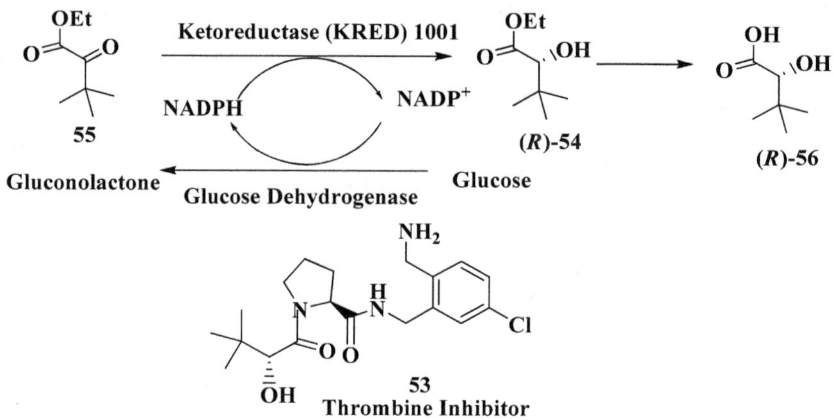

2.15. Endothelin Receptor Antagonist: Enantioselective Microbial Reduction of Keto Ester and Chloroketone

Endothelin is present in elevated levels in the blood of patients with hypertension, acute myocardial infarction and pulmonary hypertension. Two endothelin receptor sub-types have been identified which bind endothelin, thus causing vasoconstriction [87,88]. Endothelin receptor antagonists such as compound **57** (Figure 15) have potential therapeutic value. Synthesis of compound **57** required two key chiral intermediates (*S*)-alcohols **58** and **59**. Enantioselective microbial reduction of a ketoester **60** and a chlorinated ketone **61** to their corresponding (*S*)-alcohols **58** and **59** was demonstrated using *Pichia delftensis* MY 1569 and *Rhodotorula piliminae* ATCC 32762 to afford desired products in >98% e.e. and >99% e.e, respectively [89]. Reductions were scaled up to 23 L to produce the desired (*S*)-alcohols in 88% and 97% yields, respectively.

Figure 15. Endothelin receptor antagonist: Enantioselective microbial reduction of keto ester and chloroketone.

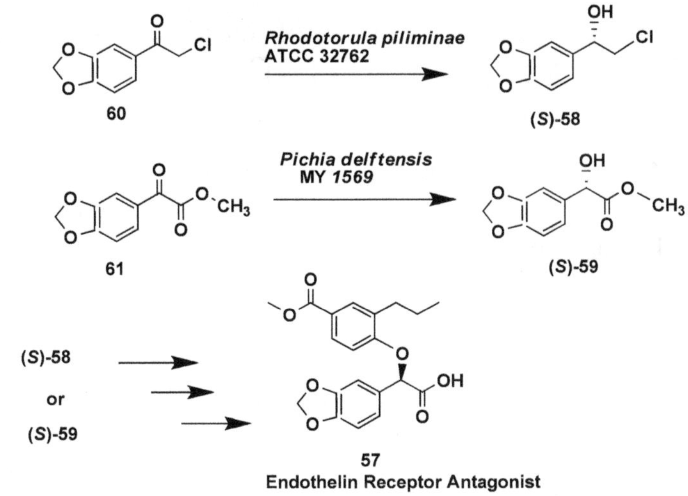

2.16. Calcium Channel Blocker: Preparation of [(3R-cis)-1,3,4,5-Tetrahydro-3-Hydroxy-4-(4-Methoxyphenyl)-6-(Trifluromethyl)-2H-1-Benzazepin-2-One]

Diltiazem **62** (Figure 16) a benzothiazepinone calcium channel blocking agent that inhibits influx of extracellular calcium through L-type voltage-operated calcium channels, has been widely used clinically in the treatment of hypertension and angina [90]. Since diltiazem has a relatively short duration of action [91], an 8-chloro derivative recently has been introduced into the clinic as a more potent analogue [92]. Lack of extended duration of action and little information on structure-activity relationships in this class of compounds led Floyd *et al.* [93] to prepare isosteric 1-benzazepin-2-ones; this led to identification of (*cis*)-3-(acetoxy)-1-[2-(dimethylamino)ethyl]-1,3,4,5-tetrahydro-4-(4-methoxyphenyl)-6-trifluoromethyl)-2H-1-benzazepin-2-one **63** as a longer lasting and more potent antihypertensive agent. A key intermediate in the synthesis of this compound was (3*R*-*cis*)-1,3,4,5-tetrahydro-3-hydroxy-4-(4-methoxyphenyl)-6-(trifluromethyl)-2*H*-1-benzazepin-2-one **64**. An enantioselective process was developed for the reduction of 4,5-dihydro-4-(4-methoxyphenyl)-6-(trifluoromethyl)-1*H*-1-benzazepin-2,3-dione **65** to **64** using *Nocardia salmonicolor* SC 6310, in 96% reaction yield with 99.8% e.e. [94].

Figure 16. Calcium channel blocker: Preparation of [(3*R*-*cis*)-1,3,4,5-tetrahydro-3-hydroxy-4-(4-methoxyphenyl)-6-(trifluromethyl)-2*H*-1-benzazepin-2-one].

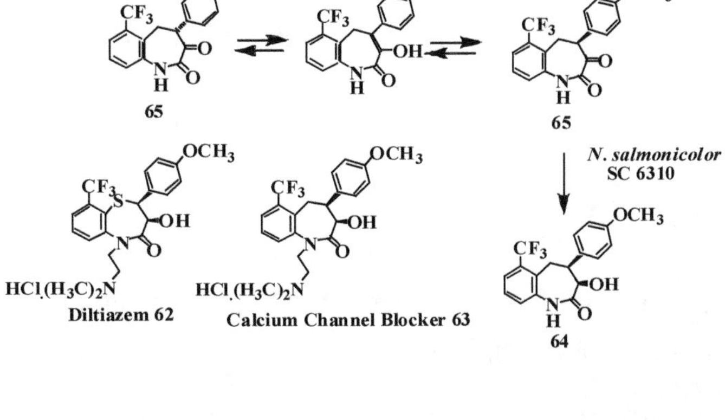

2.17. β3-Receptor Agonist: Reduction of 4-Benzyloxy-3-Methanesulfonylamino-2'-Bromo-Acetophenone

β3-Adrenergic receptors are found on the cell surfaces of both white and brown adipocytes and are responsible for lipolysis, thermogenesis, and relaxation of intestinal smooth muscle [95]. Consequently, several research groups are engaged in developing selective β3 agonists for the treatment of gastrointestinal disorders, type II diabetes, and obesity [96,97]. Biocatalytic syntheses of chiral intermediates required for the total synthesis of β3 receptor agonists **66** (Figure 17) has been demonstrated [98].

Figure 17. β3-Receptor agonist: Reduction of 4-benzyloxy-3-methanesulfonylamino-2-bromo-acetophenone.

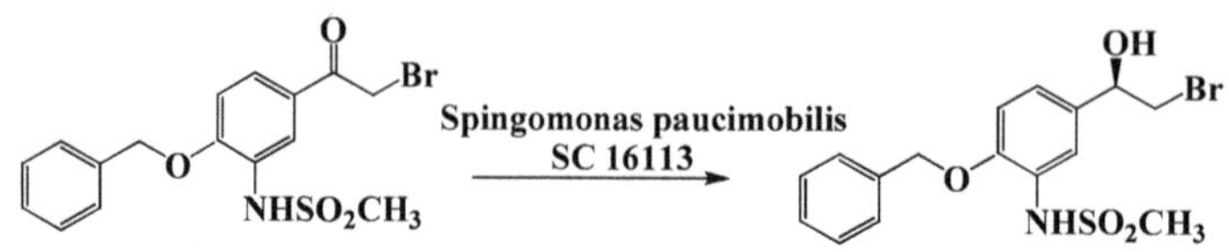

Ketone 67 (*R*)-Alcohol 68

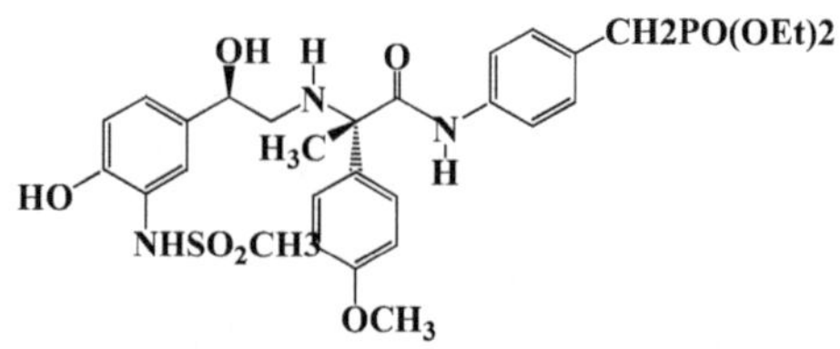

66

The microbial reduction of 4-benzyloxy-3-methanesulfonylamino-2'-bromo-acetophenone **67** to the corresponding (*R*)-alcohol **68** has been demonstrated [98] using *Sphingomonas. paucimobilis* SC 16113. The growth of *S. paucimobilis* SC 16113 was carried out in a 750-L fermentor and harvested cells (60 kg) were used to conduct the biotransformation in 10-L and 200-L preparative batches using 20% (wt/vol, wet cells). In some batches, the fermentation broth was concentrated 3-fold by microfilteration and subsequently washed with buffer by diafilteration and used directly in the bioreduction process. In all the batches, reaction yields of >85% and e.e.s. of >98% were obtained. The isolation of alcohol **68** from the 200-L batch gave 320 g (80% yield) of product with an e.e. of 99.5%.

In an alternate process, frozen cells of *S. paucimobilis* SC 16113 were used with XAD-16 hydrophobic resin (50 g/L) adsorbed substrate at 10 g/L concentration. In this process, an average reaction yield of 85% and an e.e. of >99% were obtained for alcohol **68**. At the end of the biotransformation, the reaction mixture was filtered on a 100 mesh (150 μm) stainless steel screen, and the resin retained by the screen was washed with water. The product was then desorbed from the resin with acetonitrile and crystallized in 75% overall yield with a 99.8% e.e.[98].

2.18. Penem and Carbapenem: Enzymatic Preparation of (R)-1,3-Butanediol and (R)-4-Chloro-3-Hydroxybutonoate

(*R*)-1,3-Butanediol **69** (Figure 18) is a key starting material of azetidinone derivatives **70**, which are key chiral intermediates for the synthesis of penem **71** and carbapenem antibiotics [99]. From a microbial screen, the *Candida parapsilosis* strain IFO 1396 was identified which produced

(*R*)-1,3-butanediol from the racemate. The (*S*)-1,3-butanediol oxidizing enzyme (CpSADH) which produced (*R*)-1,3-butanediol from the racemate was cloned in *Escherichia coli*. The recombinant culture catalyzed the enantioselective oxidation of secondary alcohols and also catalyzed the asymmetric reduction of aromatic and aliphatic ketones to their corresponding (*S*)-secondary alcohols. Using the recombinant enzyme, (*R*)-1,3-butanediol was produced in 97% yield and 95% e.e. using 150 g/L input of the racemate. Recombinant enzyme (CpSADH) was also used for reduction of ethyl 4-chloroacetoacetate **72** to produce ethyl-(*R*)-4-chloro-3-hydroxybutonoate **73** in 95% yield and 99% e.e. using 36 g/L substrate input. Isopropanol was used to regenerate the NADH required for this reduction. Ethyl-(*R*)-4-chloro-3-hydroxybutonoate is useful for the synthesis of L-carnitine **74** and (*R*)-4-hydroxyl pyrrolidone **75** [100,101]).

Figure 18. Penem and carbapenem: Enzymatic preparation of (*R*)-1,3-butanediol and (*R*)-4-chloro-3-hydroxybutonoate.

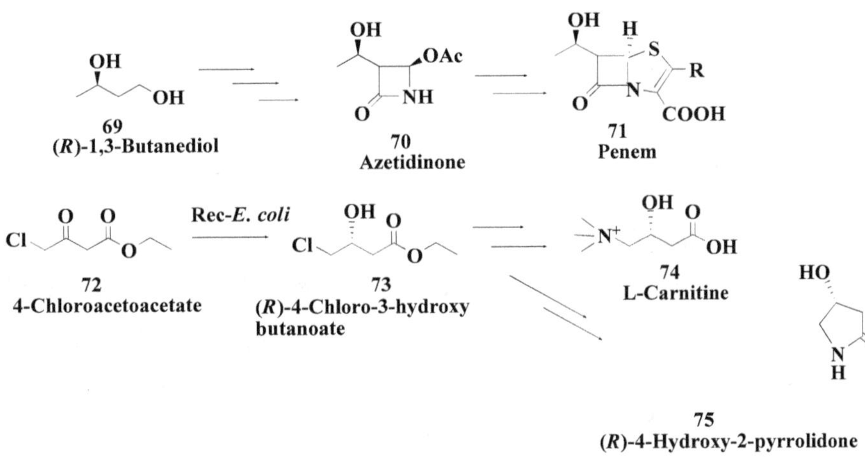

2.19. Integrin Receptor Agonist: Enzymatic Preparation of (R)-Allylic Alcohol

(*R*)-allylic alcohol **76** (Figure 19) was required as an intermediate for the synthesis of a desired monanoic derivate useful as an integrin receptor antagonist for the inhibition of bone desorption and treatment of osteoporosis [102]. A pilot scale whole cell process was developed for the enantioselective 1,2-reduction of prochiral alpha,beta-unsaturated ketone **77** to (*R*) allylic alcohol, (*R*)-**76** using *Candida chilensis* [103]. Initial development showed high enantiomeric excess (>95%) but low product yield (10%). Further process development, using a combination of statistically designed screening and optimization experiments, improved the desired alcohol yield to 90%. The fermentation growth stage, particularly medium composition and growth pH, had a significant impact on the bioconversion while process characterization identified diverse challenges including the presence of multiple enzymes, substrate/product toxicity, and biphasic cellular morphology. Manipulating the fermentation media allowed control of the whole cell morphology to a predominantly unicellular broth, away from the viscous pseudohyphae, which were detrimental to the bioconversion. The activity of a competing enzyme, which produced the undesired saturated ketone **78** and (*R*)-saturated alcohol 79, was minimized to < or =5% by controlling the reaction pH, temperature, substrate concentration, and biomass level. Despite the toxicity effects limiting the volumetric productivity, a reproducible and saleable process was demonstrated at pilot scale with high enantioselectivity (e.e. > 95%) and overall yield greater than 80% [104]. The whole cell approach proved to be a valuable alternative to chemical reduction routes.

Figure 19. Integrin receptor agonist: Enzymatic preparation of (*R*)-allylic alcohol.

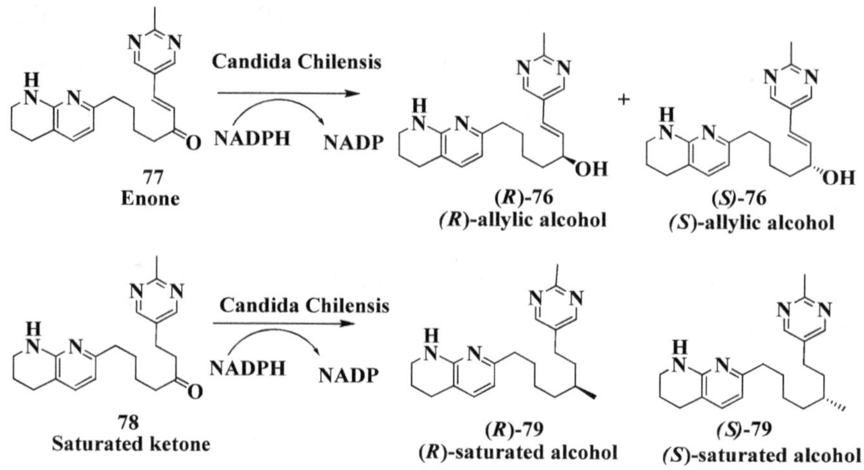

2.20. NK1 Receptor Antagonists: Enzymatic Synthesis of (S)-3,5-Bistrifluoromethylphenyl Ethanol

The synthesis of (*S*)-3,5-bistrifluoromethylphenyl ethanol, (*S*)-**80**, (Figure 20), an intermediate for the synthesis of NK-1 receptor antagonists **81** [104] was demonstrated from a ketone **82** via asymmetric enzymatic reduction process [105]. The isolated enzyme alcohol dehydrogenase from *Rhodococcus erythropolis* reduced the poorly water soluble substrate with an excellent enantiomeric excess (>99.9%) and good conversion (>98%). The optimized process was demonstrated up to pilot scale using concentration (390 mM) using a easy isolation process achieving overall isolation yields (>90%). Process improvements at preparative scale, demonstrated increase in the substrate input to 580 mM achieving a space time yield of 260 g/L/day [105].

Figure 20. NK1 receptor antagonists: Enzymatic synthesis of (*S*)-3,5-bistrifluoromethylphenyl ethanol.

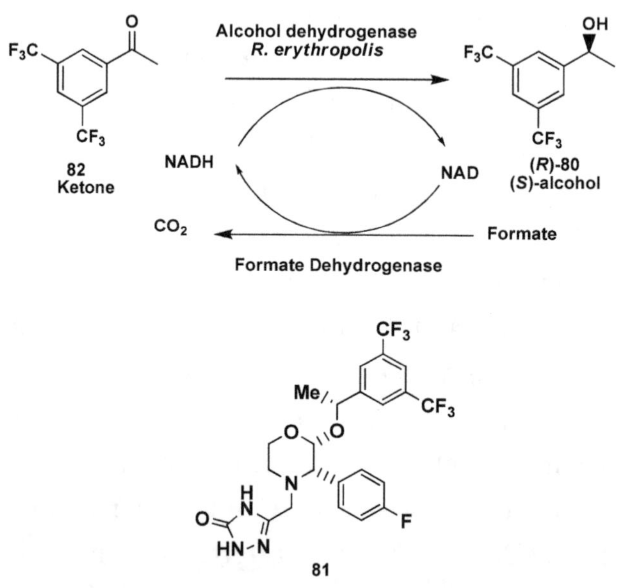

3. Enzymatic Preparation of Chiral Amino Acids

The reductive amination of α-keto acids using amino acid dehydrogenases to be one of the most useful methods because the enzymes have good stability, broad substrate specificity and very high enantioselectivity and can be used at high substrate concentrations as keto acids are soluble in aqueous system. The reductive aminations process coupled to an enzymatic cofactor regeneration system are most prominent method for preparation of chiral amino acids. For most enzymes, the required cofactor is NADH but NADPH is required in some cases. Yeast formate dehydrogenase is commonly used for NADH regeneration and glucose dehydrogenase usually from *Bacillus* species may be used for either NADH or NADPH regeneration. There are excellent reviews on the amino acid dehydrogenases and examples of their synthetic utilities [106–109].

3.1. Tigemonam: Enzymatic Synthesis of (S)-β-Hydroxyvaline

(*S*)-β-hydroxyvaline **83** (Figure 21), is a key chiral intermediate required for the total synthesis of orally active monobactam [110], Tigemonam **84**. Chiral amino acids have been made from corresponding keto acids by reductive amination process [111]. The synthesis of (*S*)-β-hydroxyvaline **83** from α-keto-β-hydroxyisovalerate **85** by reductive amination using leucine dehydrogenase from *Bacillus sphaericus* ATCC 4525 has been demonstrated [112]. The NADH required for this reaction was regenerated by either formate dehydrogenase from *Candida boidinii* or glucose dehydrogenase from *Bacillus megaterium*. The required substrate **85** was generated either from α-keto-β-bromoisovalerate or its ethyl esters by hydrolysis with sodium hydroxide *in situ.* In this process, an overall reaction yield of 98% and an enantiomeric excess of 99.8% were obtained for the L-β-hydroxyvaline **83**.

Figure 21. Tigemonam: Enzymatic synthesis of (*S*)-β-hydroxyvaline.

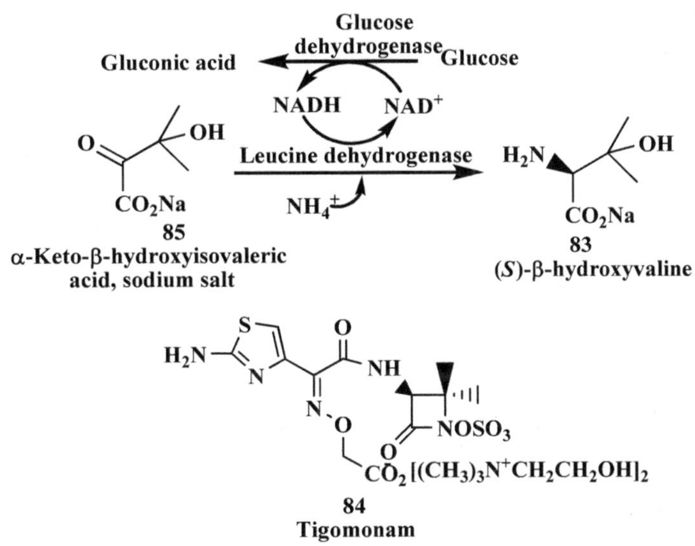

3.2. Atazanavir: Enzymatic Synthesis of (S)-Tertiary-Leucine

Atazanavir **86** is an acyclic aza-peptidomimetic, a potent HIV protease inhibitor [53,54]. Synthesis of atazanavir required (*S*)-tertiary leucine **87** (Figure 22). An enzymatic reductive amination of ketoacid **88** to amino acid **87** by recombinant *Escherichia coli* expressing leucine dehydrogenase from *Thermoactinimyces intermedius* has been demonstrated. The reaction required ammonia and NADH as a cofactor. NAD produced during the reaction was converted back to NADH using recombinant *Escherichia coli* expressing formate dehydrogenase from *Pichia pastoris*. A reaction yield of >95% with an e.e. of >99.5% was obtained for **87** at 100 g/L substrate [113]. Leucine dehydrogenase from *Bacillus* strain has also been cloned and expressed and used in reductive amination process [114,115].

Figure 22. Atazanavir (anti-viral agent): Enzymatic synthesis of (*S*)-tertiary-leucine.

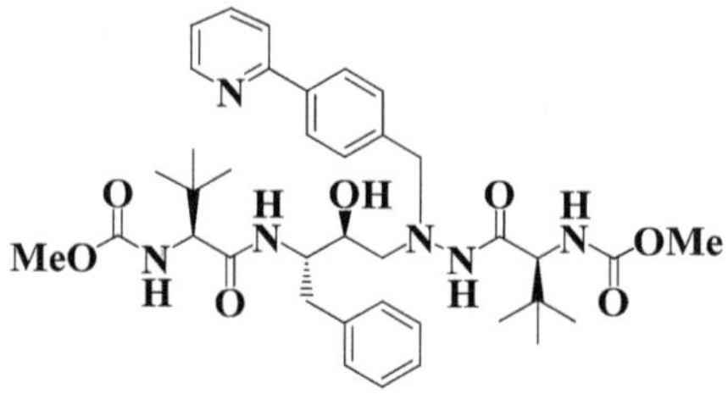

86
Atazanavir

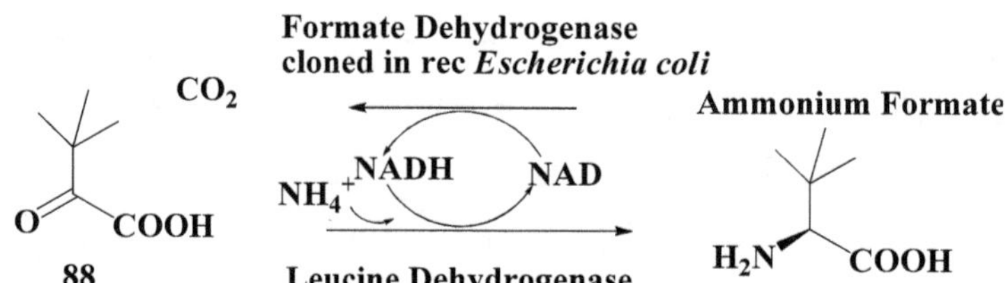

88
Keto acid

(*S*)-87
(*S*)-tert. Leucine

3.3. Vanlev: Enzymatic Synthesis of (S)-6-Hydroxynorleucine

Vanlev **89** (Figure 23) is an antihypertensive drug which acts by inhibiting angiotensin-converting enzyme (ACE) and neutral endopeptidase (NEP) [116]. (*S*)-6-Hydroxynorleucine **90** is a key intermediate in the synthesis of Vanlev. The synthesis and complete conversion of 2-keto-6-hydroxyhexanoic acid **91** to (*S*)-6-hydroxynorleucine **90** was demonstrated by reductive amination using beef liver glutamate dehydrogenase [117]. As depicted, compound **91**, in equilibrium with 2-hydroxytetrahydropyran-2-carboxylic acid sodium salt **92**, was converted to **90**. The reaction requires ammonia and NADH. NAD produced during the reaction was recycled to NADH by the oxidation of glucose to gluconic acid using glucose dehydrogenase from *Bacillus megaterium*. The reaction was complete in about 3 h at 100 g/L substrate input with a reaction yields of 92% and e.e. of 99.8% for (*S*)-6-hydroxynorleucine. The synthesis and isolation of keto acid **91** required several steps. In a second, more convenient process the ketoacid was prepared by treatment of racemic 6-hydroxy norleucine **90** [produced by hydrolysis of 5-(4-hydroxybutyl) hydantoin **93**] with (*R*)-amino acid oxidase (Figure 24) After the e.e. of the unreacted (*S*)-6-hydroxynorleucine had risen to 99.8%, the reductive amination procedure was used to convert the mixture containing the 2-keto-6-hydroxyhexanoic acid entirely to (*S*)-6-hydroxynorleucine in 97% yield with 99.8% e.e. from racemic 6-hydroxynorleucine at 100 g/L substrate input [117]. The (*S*)-6-hydroxynorleucine prepared by the enzymatic process was converted chemically to Valev **89** [118].

Figure 23. Vanlev: Enzymatic synthesis of (*S*)-6-hydroxynorleucine by reductive amination.

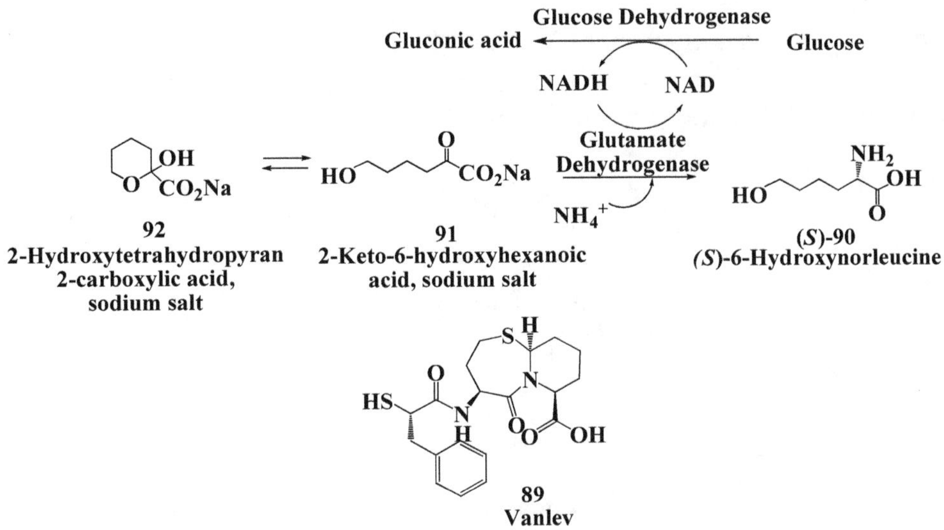

Figure 24. Vanlev: Enzymatic conversion of racemic 6-hydroxy norleucine to (*S*)-6-hydroxymorleucine.

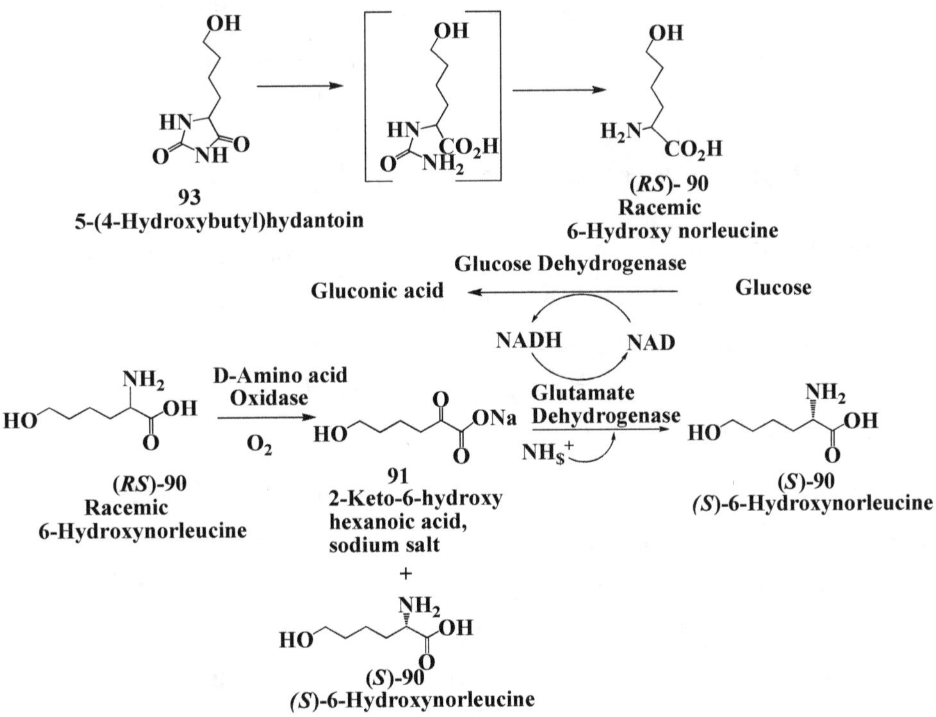

3.4. Vanlev: Enzymatic Synthesis of Allysine Ethylene Acetal

(*S*)-2-Amino-5-(1,3-dioxolan-2-yl)-pentanoic acid [(*S*)-allysine ethylene acetal] **94** (Figure 25) is one of three building blocks used in an alternative synthesis of Vanlev **89**. Synthesis of **94** was demonstrated by reductive amination of ketoacid acetal **95** using phenylalanine dehydrogenase [PDH] from *Thermoactinomyces intermedius* [119]. The reaction required ammonia and NADH; NAD produced during the reaction was recyled to NADH by the oxidation of formate to CO_2 using formate dehydrogenase [FDH]. *T. intermedius* PDH was cloned and expressed in *Escherichia coli* and recombinant culture was used as a source of PDH. Expression of *T. intermedius* PDH in *P. pastoris*, inducible by methanol, allowed generation of both enzymes in a single fermentation as methanol grown cells of

P. pastoris also contained formate dehydrogease. A total of 197 kg of **94** was produced in three 1,600-L batches using a 5% concentration of substrate **95** with an average yield of 91 M % and e.e. >98% [119]. (*S*)-allysine ethylene acetal was converted to Vanlev **89** [118].

Figure 25. Vanlev: Enzymatic synthesis of allysine ethylene acetal.

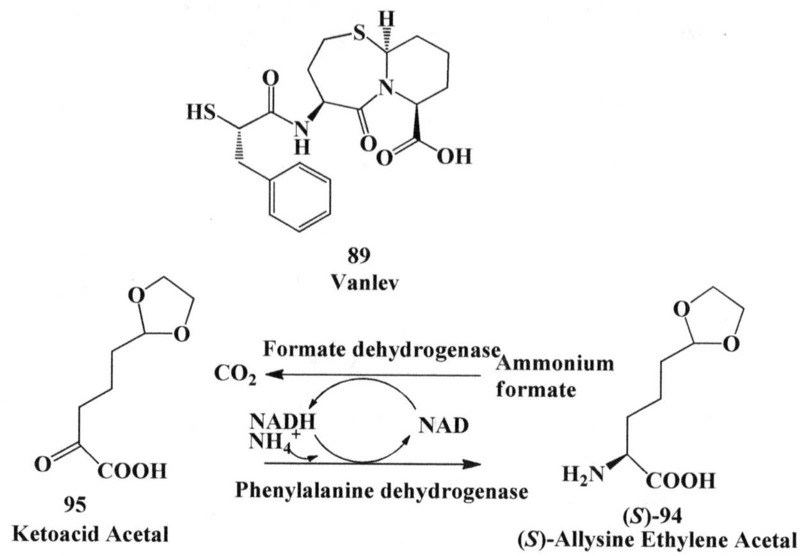

3.5. Saxagliptin: Enzymatic Reductive Amination of 2-(3-Hydroxy-1-Adamantyl)-2-Oxoethanoic Acid

Dipeptidyl peptidase 4 (DPP-4) is a ubiquitous proline-specific serine protease responsible for the rapid inactivation of incretins, including glucagon-like peptide 1 (GLP-1) and glucose-dependent insulinotropic peptide. To alleviate the inactivation of GLP-1, inhibitors of DPP-IV are being evaluated for their ability to provide improved control of blood glucose for diabetics [120–123]. Januvia developed by Merck is a marketed DPP4 Inhibitor [122].

Saxagliptin **96** [121,122] (Figure 26), a DPP-IV inhibitor developed by Bristol-Myers Squibb and now approved for type 2 diabetic treatment by Food and Drug administration, requires (*S*)-*N*-boc-3-hydroxyadamantylglycine **97** as an intermediate. A process for conversion of the keto acids **98** to the corresponding amino acid **99** using (*S*)-amino acid dehydrogenases was developed. A modified form of a recombinant phenylalanine dehydrogenase cloned from *Thermoactinomyces intermedius* and expressed in *Pichia pastoris* or *Escherichia coli* was used for this process. NAD produced during the reaction was recycled to NADH using formate dehydrogenase. The modified phenylalanine dehydrogenase contains two amino acid changes at the C-terminus and a 12 amino acid extension of the C-terminus [124].

Production of multi-kg batches was originally carried out with extracts of *Pichia pastoris* expressing the modified phenylalanine dehydrogenase from *Thermoactinomyces intermedius* and endogenous formate dehydrogenase. The reductive amination process was further scaled up using a preparation of the two enzymes expressed in single recombinant *E. coli*. The amino acid **99** was directly protected as its boc derivative without isolation to afford intermediate. Yields before isolation were close to 98% with 100% e.e. [124].

Figure 26. Saxagliptin: Enzymatic reductive amination of 2-(3-hydroxy-1-adamantyl)-2-oxoethanoic acid.

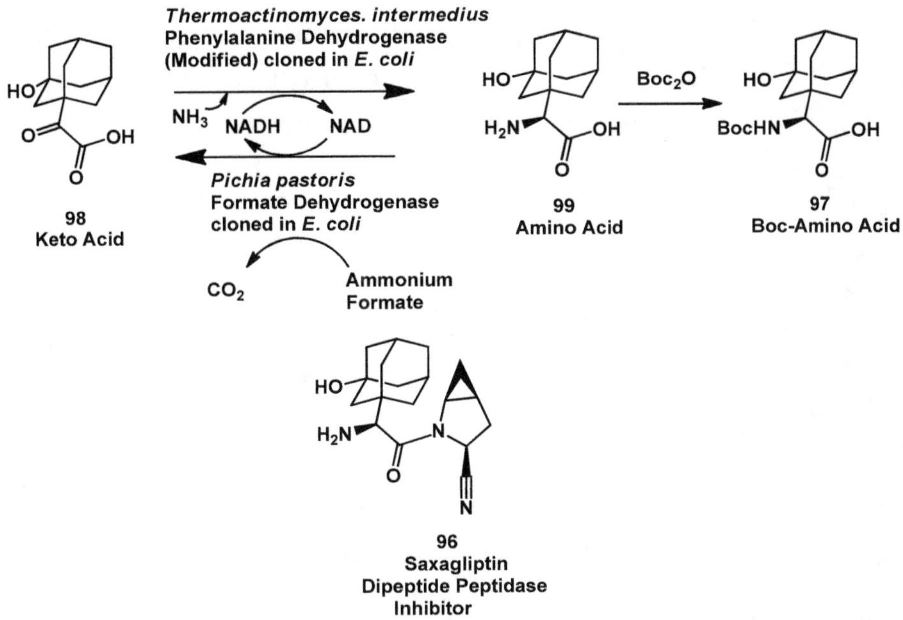

Reductive amination was also conducted using cell extracts from *E.coli* strain SC16496 expressing PDHmod and cloned FDH from *Pichia pastoris*. Cell extracts after polyethyleneamine treatment, clarification and concentration were used to complete the reaction in 30 h with >96% yield and >99.9% e.e. of product **99**. This process has now been used to prepare several hundred kg of boc-protected amino acid **97** to support the development of Saxagliptin [124].

3.6. Enzymatic Synthesis of (S)-Neopentylglycine

The enantioselective synthesis of (*S*)-neopentylglycine **100** (Figure 27) has been developed by Groeger *et al.* [125]. Recombinant whole cell containing leucine dehydrogenase and formate dehydrogenase was used in the reductive amination of the corresponding α-keto acid **101**. The desired (*S*)-neopentylglycine was obtained with >95% conversion and a high enantioselectivity of >99% e.e. at substrate concentrations of up to 88 g/L. Spiroheterocyclic compounds [morpholine-4-carboxylic acid amides of heterocyclic cyclohexylalanine and neopentylglycine derivatives and their analogs] are useful as reversible inhibitors of cysteine proteases such as cathepsin S useful in the treatment of variety of autoimmune diseases [126].

Figure 27. Enzymatic synthesis of (*S*)-neopentylglycine.

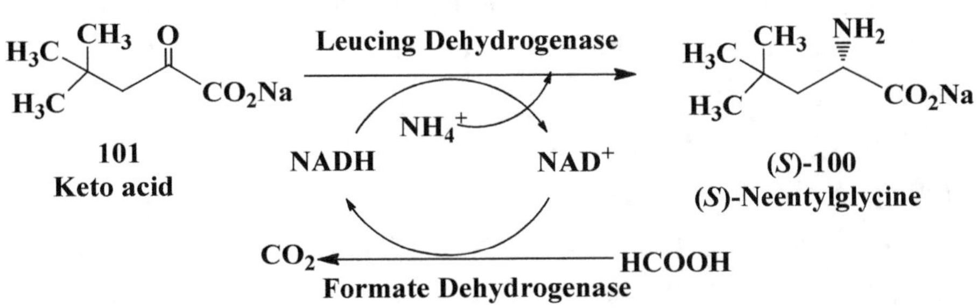

3.7. Glucogen like Peptide: Enzymatic Deracemization Racemic Amino Acid to (S)-Amino Acid

The (*S*)-amino-3-[3-{6-(2-methylphenyl)}pyridyl]-propionic acid 102 (Figure 28) is a key intermediate required for synthesis of GLP-1 mimics or GLP-1 receptor modulators. Such receptor modulators are potentially useful for the treatment of type II diabetes treatment [127,128].

Figure 28. Glucogen like peptide: The (*S*)-amino-3-[3-{6-(2-methylphenyl)}pyridyl]-propionic acid.

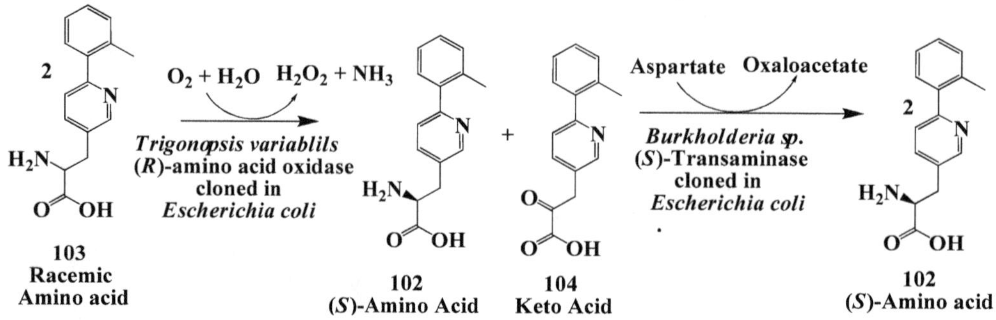

(*S*)-Amino-3-[3-{6-(2-methylphenyl)}pyridyl]-propionic acid was prepared by enzymatic deracemization process [129] in 72% isolated yield with >99.4% e.e. from racemic amino acid **103** using combination of two enzymes (*R*)-amino acid oxidase from *Trigonopsis variabilis* expressed in *Escherichia coli* and (*S*)-aminotransferase from *Sporosarcina ureae* cloned and expressed in *Escherichia coli*. (*S*)-aspartate was used as amino donor. A (*S*)-aminotransferase was also purified from a soil organism identified as *Burkholderia sp.* and cloned and expressed in *Escherichia coli* and used in this process [131]. In enzymatic process racemic amino acid was first treated with (*R*)-amino acid oxidase for 4 h to convert racemic amino acid to mixture of (*S*)-amino acid and keto acid **104**. Subsequently in the same reaction mixture (*S*)-aminotransferase was charged to convert keto acid **104** to (*S*)-amino acid **102** to get 85% yield at the end of the biotransformation process. This process was scaled up to 100 L scale at a substrate input of 1.5 kg.

In an alternate process, the enzymatic dynamic resolution of racemic amino acid **103** was also demonstrated. (*R*)-selective oxidation with celite-immobilized (*R*)-amino acid oxidase from *Trigonopsis variabilis* expressed in *Escherichia coli* in combination with chemical imine reduction with borane-ammonia gave a 75% in process yield and 100 e.e. of (*S*)-amino acid **102** [129].

3.8. Preparation of (R)-Amino Acid

(*R*)-Amino acids are increasingly becoming important building blocks in the production of pharmaceuticals and fine chemicals, and as chiral directing auxiliaries and chiral synthons in organic synthesis [130,131]. Using both rational and random mutagenesis, Rozzell and Novick [132] have created the broad substrate range, nicotinamide cofactor dependent, and highly stereoselective (*R*)-amino acid dehydrogenase. This new enzyme is capable of producing (*R*)-amino acids via the reductive amination of the corresponding 2-keto acid with ammonia. This biocatalyst was the result of three rounds of mutagenesis and screening performed on the enzyme *meso*-diaminopimelate (*R*)-dehydrogenase from *Corynebacterium glutamicum*. The first round targeted the active site of the wild-type enzyme and produced mutants that were no longer strictly dependent on the native substrate.

The second and third rounds produced mutants that had an increased substrate range including straight- and branched-aliphatic amino acids and aromatic amino acids. The very high selectivity toward the (*R*)-enantiomer (95% to >99% e.e.) was shown to be preserved three rounds of mutagenesis and screening [132]. This new enzyme was active against variety of amino acids could complement and improve upon current methods for (*R*)-amino acid synthesis. The synthesis of (*R*)-cyclohexylalanine 105 (Figure 29) was developed by reductive amination of cyclohexylpyruvate **106** to yield (*R*)-**105** in 98% yield and >99% e.e. (*R*)-105 is a potential chiral intermediate for the synthesis of thrombin inhibitor Inogatran **107** [133].

Figure 29. Thrombin inhibitor inogatran: Enzymatic synthesis of (*R*)-cyclohexylalanine.

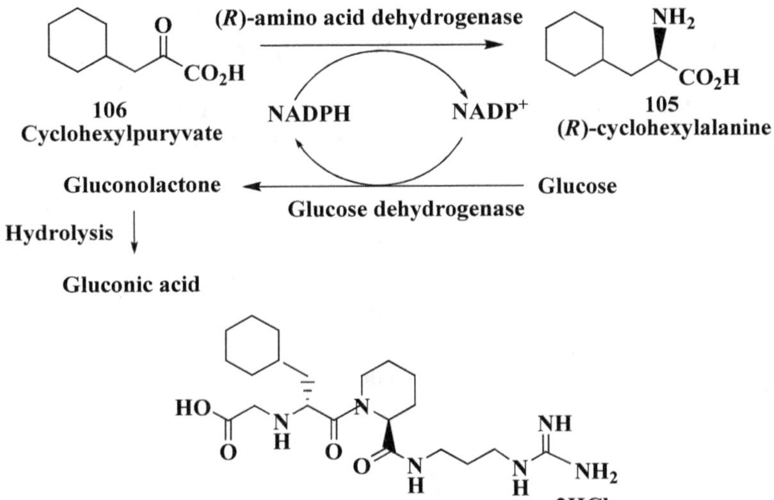

107
Inogatran

The deracemisation of DL-amino acids using L-amino acid oxidase from *Proteus myxofaciens* and amine-boranes as chemical reducing agents has been investigated. Amine-boranes were found to be of particular interest in terms of reactivity and chemoselectivity compared to sodium borohydride and cyanoborohydride. Starting from the racemic amino acids, a range of D-amino acids were prepared in yields of up to 90% and e.e. >99% [134].

3.9. Calcitonin Gene-Related Peptide Receptors (Antimigraine Drugs): Enzymatic Deracemization Process

The (*R*)-amino acid (*R*)-2-amino-3-(7-methyl-1 H-indazol-5-yl)propanoic acid (*R*)-**108**, (Figure 30) is a key intermediate needed for synthesis of antagonists of calcitonin gene-related peptide receptors **109** [135] Such antagonists are potentially useful for the treatment of migraine and other maladies [135,136].

(*R*)-Amino acid 108 was prepared in 68% isolated yield with >99% e.e. from racemic amino acid **110** using (*S*)-amino acid oxidase from *Proteus mirabilis* expressed in *Escherichia coli* in combination with a commercially available (*R*)-transaminase using (*R*)-alanine as amino donor [137]. The (*R*)-enantiomer was also prepared in 79% isolated yield with >99% e.e. from the corresponding keto acid **111** using the (*R*)-transaminase with racemic alanine as the amino donor. The rate and yield of this reaction could be accelerated by addition of lactate dehydrogenase (with NAD^+, formate and formate dehydrogenase to regenerate NADH) to remove the inhibitory pyruvate produced during the reaction. A (*R*)-transaminase was identified and purified from a soil organism identified as *Bacillus thuringiensis* and cloned and expressed in *Escherichia coli*. The recombinant (*R*)-transaminase was very effective for the preparation of (*R*)-**108** and gave a nearly complete conversion of **111** to (*R*)-**108** without the need for additional enzymes for pyruvate removal [137].

Figure 30. Calcitonin gene-related peptide receptors (antimigraine drugs): Enzymatic preparation of (*R*)-2-amino-3-(7-methyl-1*H*-indazol-5-yl)propanoic acid.

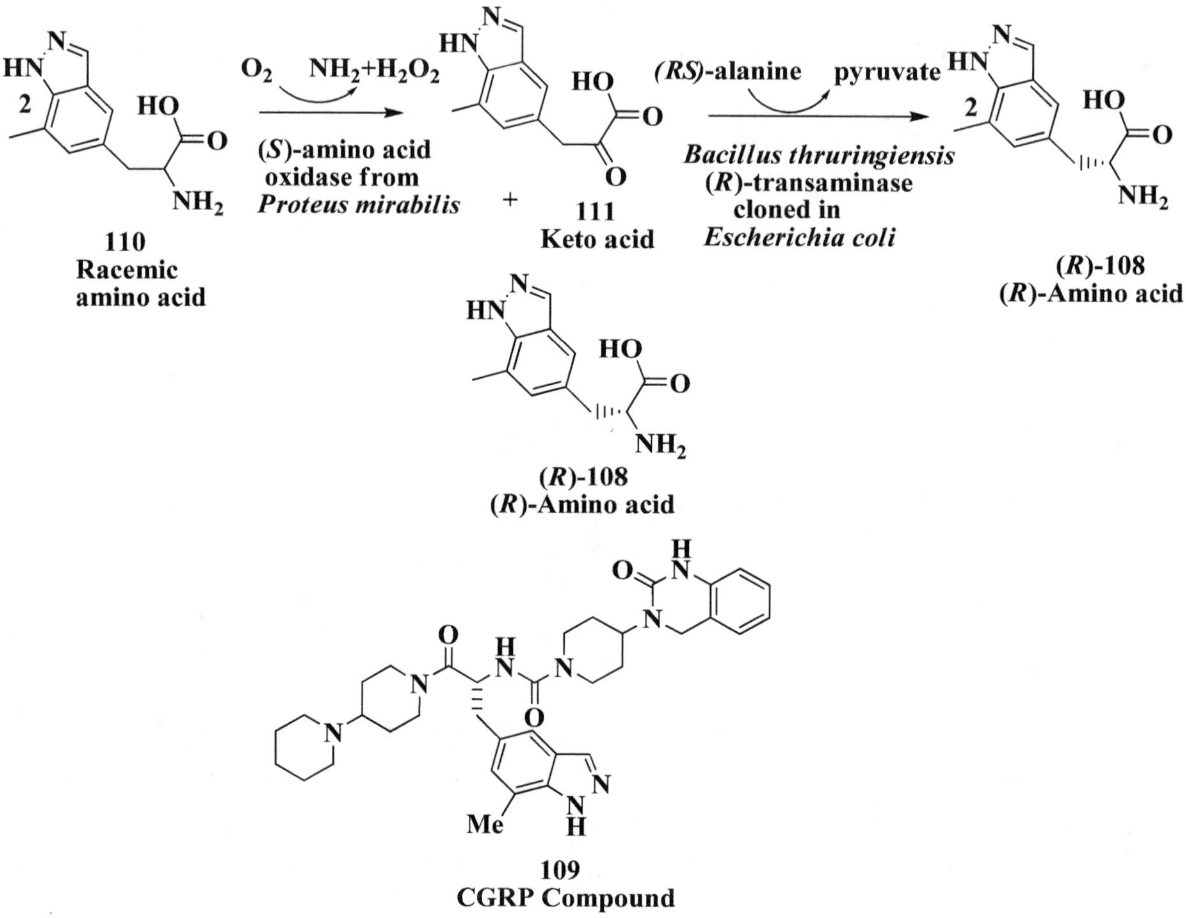

3.10. Corticotropin Releasing Factor (CRF)-1 Receptor Antagonist: Enzymatic Resolution by Transaminase

(*R*)-amines synthesis for Anxiety and depression are psychiatric disorders that constitute a major health concern worldwide. While numerous marketed treatments exits for both disorders, there continue to be need agents which may have increased efficacy and/or reduced side-effect profiles [138–140]. CRF-1 receptor antagonists have been proposed as novel pharmacological treatments for depression, anxiety and stress disorders [138–141]. (*R*)-sec-butylamine **112** and (*R*)-1-cyclopropylethylamine **113** (Figure 31) are key chiral intermediates for the synthesis of CRF-1 receptor antagonists such as **114** [141,142].

Figure 31. Corticotropin releasing factor (CRF)-1 receptor antagonist: Enzymatic synthesis of (*R*)-1-cyclopropylethylamine and (*R*)-*sec*-butylamine.

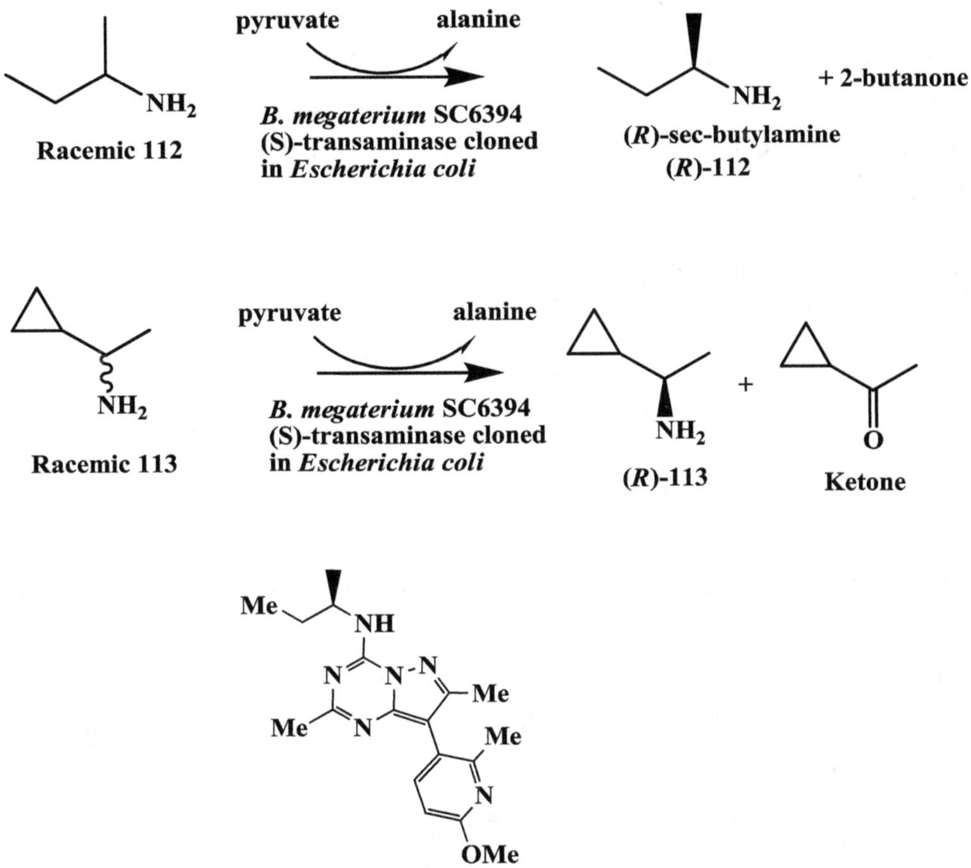

114
CRF-1 antagonist receptor

We have developed enzymatic resolution process for the preparation of (*R*)-sec-butylamine and (*R*)-1-cyclopropylethylamine [143]. Screening was carried out to identify strains useful for the preparation of (*R*)-1-cyclopropylethylamine and (*R*)-*sec*-butylamine from the racemic amines with an (*S*)-specific transaminase. Several *Bacillus megaterium* strains as well as several soil isolates were found to have the desired activity for the resolution of the racemic amines to give the (*R*)-enantiomers. Using an extract of the best strain, *Bacillus megaterium* SC6394, the reaction was shown to be a transamination requiring pyruvate as amino acceptor and pyridoxal phosphate as a cofactor. Initial batches of both amines were produced using whole cells of *Bacillus megaterium* SC6394. The transaminase was purified to homogeneity to obtain N-terminal as well as internal amino acid sequences. The sequences were used to design polymerase chain reaction (PCR) primers to enable cloning and expression of the transaminase in *Escherichia coli* SC16578. In contrast to using *Bacillus megaterium* process, pH control and aeration were not required for the resolution of *sec*-butylamine and an excess of pyruvate was not consumed by the recombinant cells. The resolution of *sec*-butylamine (0.68 M) using whole cells of *Escherichia coli* SC16578 was scaled up to give (*R*)-*sec*-butylamine·1/2 H_2SO_4 in 46.6% isolated yield with 99.2% *e.e.* An alternative isolation procedure was also used to isolate (*R*)-*sec*-butylamine as the free base. Using the same recombinant (*S*)-tansaminase, (*R*)-1-cyclopropylethylamine was obtained in 42% isolated yield (theoretical max. 50%) and 99% e.e. [143].

4. Conclusions

The production of single enantiomers of drug intermediates is increasingly important in the pharmaceutical industry. Biocatalysis provides organic chemists an alternate opportunity to prepare pharmaceutically important chiral compounds. The examples presented in this review are only from a few selected articles for synthresis of chiral alcohols and unnatural amino acids. Different types of biocatalytic reactions are capable of generating a wide variety of chiral compounds useful in the development of drugs.

The use of hydrolytic enzymes such as lipases, esterases, proteases, dehalogenases, acylases, amidases, nitrilases, epoxide hydrolases, and decarboxylases for the resolution of variety of racemic compounds and in the asymmetric synthesis of enantiomerically enriched chiral compounds. Dehydrogenases and aminotransferases has been successfully used along with cofactors and cofactor regenerating enzymes for the synthesis of chiral alcohols, aminoalcohols, amino acids and amines. Aldolases and decarboxylases have been effectively used in asymmetric synthesis by aldol condensation and acyloin condensation reactions. Monoxygenases have been used in enantioselective and regioselective hydroxylation, epoxidation, sulfoxidation and Baeyer-Villiger reactions. Dioxygenases have been used in the chemo-enzymatic synthesis of chiral diols. Enzymatic deracemization, dynamic resolution and stereoinversion, to achieve >50% yield and high e.e. by combination of chemo- and/or biocatalysts in sequential reactions or by a single biocatalyst. In the course of the last decade, progress in biochemistry, protein chemistry, molecular cloning, random and site-directed mutagenesis, directed evolution of biocatalysts under desired process conditions has opened up unlimited access to a variety of enzymes and microbial cultures as tools in organic synthesis. Future of bicatalysis for synthesis of chiral compounds looks very promising.

Acknowledgments

The author would like to acknowledge Ronald Hanson, Animesh Goswami, Amit Banerjee, Venkata Nanduri, Jeffrey Howell, Steven Goldeberg, Robert Johnston, Mary-Jo Donovan, Dana Cazzulino, Thomas Tully, Thomas LaPorte, Lawrence Parker, John Wasylyk, Michael Montana, Ronald Eiring, Rapheal Ko, Linda Chu, Clyde McNamee, Michael Montana for research collaboration.

References

1. Food and Drug Administration. FDA's statement for the development of new stereoisomeric drugs. *Chirality* **1992**, *4*, 338–340.
2. Oliver, M.; Voigt, C.A.; Arnold, F.H. Enzyme engineering by directed evolution. In *Enzyme Catalysis in Organic Synthesis*, 2nd ed.; VCH, New York, USA, 2002; Volume 1, pp. 95–138.
3. Kazlauskas, R.J. Enhancing catalytic promiscuity for biocatalysis. *Curr. Opin. Chem. Biol.* **2005**, *9*, 195–201.
4. Schmidt, M.; Bauman, M.; Henke, E.; Konarzycka-Bessler, M.; Bornscheuer, U.T. Directed evolution of lipases and esterases. *Meth. Enzymol.* **2004**, *388*, 199–207.

5. Reetz, M.T.; Torre, C.; Eipper, A.; Lohmer, R.; Hermes, M.; Brunner, B.; Maichele, A., Bocola, M.; Arand, M.; Cronin, A.; *et al.* Enhancing the enantioselectivity of an epoxide hydrolase by directed evolution. *Org. Lett.* **2004**, *6*, 177–180.
6. Rubin-Pitel, S.B.; Zhao, H. Recent advances in biocatalysis by directed enzyme evolution. *Comb. Chem. High T. Scr.* **2006**, *9*, 247–257.
7. Pollard, D.J.; Woodley, J.M. Biocatalysis for pharmaceutical intermediates: The future is now. *Trends Biotechnol.* **2007**, *25*, 66–73.
8. Otey, C.R.; Bandara, G.; Lalonde, J.; Takahashi, K.; Arnold, F.H. Preparation of human metabolites of propranolol using laboratory-evolved bacterial cytochromes P450. *Biotechnol. Bioeng.* **2006**, *93*, 494–499.
9. Huisman, G.W.; Lalonde, J.J. Enzyme evolution for chemical process applications. In *Biocatalysis in the Pharmaceutical and Biotechnology Industries*; Patel, R.N., Ed.; CRC Press:. Boca Raton, FL, USA, 2007; pp. 717–742.
10. Arnold, F.; Volkov, A. Directed evolution of biocatalysts. *Curr. Opin. Chem. Biol.* **1999**, *3*, 54–59.
11. Patten, P.A.; Howard, R.J.; Stemmer, W.P. Applications of DNA shuffling to pharmaceuticals and vaccines. *Curr. Opin. Biotechnol.* **1997**, *8*, 724–733.
12. Hibbert, E.G.; Baganz, F.; Hailes, H.C.; Ward, J.M.; Lye, G.J.; Woodley, J.M.; Dalby, P.A. Directed evolution of biocatalytic processes. *Biomol. Eng.* **2005**, *22*, 11–19.
13. Sylvestre, J.; Chautard, H.; Cedrone, F.M. Directed evolution of biocatalysts. *Org. Process Res. Dev.* **2006**, *10*, 562–571
14. Turner, N.J. Directed evolution drives the next generation of biocatalysts. *Nat. Chem. Biol.* **2009**, *5*, 567–573.
15. Bornscheuer, U.T.; Huisman, G.W.; Kazlauskas, R.J.; Lutz, S.; Moore, J.C.; Robins, K. Engineering the third wave of biocatalysis. *Nature* **2012**, *485*, 185–194.
16. Wells, A.S.; Finch, G.L.; Michels, P.C.; Wong, J.W. Use of enzymes in the manufacture of active pharmaceutical ingredients—A science and safety-based approach to ensure patient safety and drug quality. *Org. Process Res. Dev.* **2012**, *16*, 1986–1993.
17. Reetz, M.T. Biocatalysis in organic chemistry and biotechnology: Past, present, and future. *J. Am. Chem. Soc.* **2013**, *135*, 12480–12496.
18. Huisman, G.W.; Collier, S.J. On the development of new biocatalytic processes for practical pharmaceutical synthesis. *Curr. Opin. Chem. Biol.* **2013**, *17*, 284–292.
19. Bryan, M.C.; Dillon, B.; Hamann, L.G.; Hughes, G.J.; Kopach, M.E.; Peterson, E.A.; Pourashraf, M.; Raheem, I.; Richardson, P.; Richter, D.; *et al.* Sustainable practices in medicinal chemistry: Current state and future directions. *J. Med. Chem.* **2013**, *56*, 6007–6021.
20. DiCosimo, R. Nitrilases and nitrile hydratases. In *Biocatalysis in the Pharmaceutical and Biotechnology Industries*; Patel, R.N., Ed.; CRC Press: Boca Raton, FL, USA, 2007; pp. 1–26.
21. Patel, R.N. Biocatalysis: Synthesis of chiral intermediates for pharmaceuticals. *Curr. Org. Chem.* **2006**, *10*, 1289–1321.
22. Simeo, Y.; Kroutil, W.; Faber, K. Biocatalytic deracemization: Dynamic resolution, stereoinversion, enantioconvergent processes, and cyclic deracemization. In *Biocatalysis in the Pharmaceutical and Biotechnology Industries*; Patel, R.N., Ed.; CRC Press: Boca Raton, FL, USA, 2007; pp. 27–51.

23. Patel, R.N. Biocatalysis: Synthesis of key intermediates for development of pharmaceuticals. *ACS Catal.* **2011**, *1*, 1056–1074.
24. Turner, N.J. Enzyme catalyzed deracemization and dynamic kinetic resolution reactions. *Curr. Opin. Chem. Biol.* **2004**, *8*, 114–119.
25. Ishige, T.; Honda, K.; Shimizu, S. Whole organism biocatalysis. *Curr. Opin. Chem. Biol.* **2005**, *9*, 174–180.
26. Zhao, H.; Chockalingom, K.; Chen, Z. Directed evolution of enzymes and pathways for industrial biocatalysis. *Curr. Opin. Biotechnol.* **2002**, *13*, 104–110.
27. Schulze, B.; Wubbolts, M. Biocatalysis for industrial production of fine chemicals. *Curr. Opin. Biotechnol.* **1999**, *10*, 609–611.
28. Steinreiber, A.; Faber, K. Microbial epoxide hydrolases for preparative biotransformations. *Curr. Opin. Biotechnol.* **2001**, *12*, 552–558.
29. Tao, J.; Xu, J.-H. Biocatalysis in development of green pharmaceutical processes. *Curr. Opin. Chem. Biol.* **2009**, *13*, 43–50.
30. Patel, R.N. Biocatalytic hydrolysis (esters, amides, epoxides, nitriles) and biocatalytic dynamic kinetic resolution. In *Comprehensive Chirality*, **2012** *10*, 288-317.
31. Patel, R.N. Biocatalytic routes to chiral intermediates for development of drugs. In *Biocatalysis for Green Chemistry and Chemical Process Development*; Tao, J., Kazlauskas, R., Eds.; John Wiley & Sons, Inc.: Hoboken, USA city, 2010.
32. Jajoo, H.; Mayol, R.; LaBudde, J.; Blair, I. Metabolism of the antianxiety drug buspirone in human subjects. *Drug Metab. Dispos.* **1989**, *17*, 634–640.
33. Mayol, R. Buspirone Metabolite for the Alleviation of Anxiety. US6150365A, June 6, 2000.
34. Yevich, J.; New, J.; Lobeck, W.; Dextraze, P.; Bernstein, E.; Taylor, D.; Yocca, F.; Eison, M.; Temple, D., Jr. Synthesis and biological characterization of α-(4-fluorophenyl)-4-(5-fluoro-2-pyrimidinyl)-1-piperazinebutanol and analogs as potential atypical antipsychotic agents. *J. Med. Chem.* **1992**, *35*, 4516–4525.
35. Yevich, J.; Mayol, R.; Li, J.; Yocca, F. (*S*)-6-Hydroxy-Buspirone for Treatment of Anxiety, Depression and Related Disorders. US2003022899, January 30, 2003.
36. Patel, R.; Chu, L.; Nanduri, V.; Jianqing, L.; Kotnis, A.; Parker, W.; Liu, M.; Mueller, R. Enantioselective microbial reduction of 6-oxo-8-[4-[4-(2-pyrimidinyl)-1-piperazinyl]butyl]-8-azaspiro[4.5]decane-7,9-dione. *Tetrahedron Asymmetry* **2005**, *16*, 2778–2783.
37. Goldberg, S.; Nanduri, V.; Chu, L.; Johnston, R.; Patel, R. Enantioselective microbial reduction of 6-oxo-8-[4-[4-(2-pyrimidinyl)-1-piperazinyl]butyl]-8-azaspiro[4.5]decane-7,9-dione: Cloning and expression of reductases. *Enzyme Microb. Technol.* **2006**, *39*, 1441–1450.
38. Patel, R.N.; Banerjee, A.; McNamee, C.; Brzozowski, D.; Hanson, R.; Szarka, L. Enantioselective microbial reduction of 3,5-dioxo-6-(benzyloxy) hexanoic acid, ethyl ester. *Enzyme Microb. Technol.* **1993**, *15*, 1014-1021.
39. Sit, S.; Parker, R.; Motoe, I.; Balsubramanian, H.; Cott, C.; Brown, P.; Harte, W.; Thompson, M.; Wright, J. Synthesis, biological profile, and quantitative structure activity relationship of a series of novel 3-hydroxy-3-methylglutaryl coenzyme A reductase inhibitors. *J. Med. Chem.* **1990**, *33*, 2982–2999.

40. Roth, B.D. The discovery and development of atorvastatin, a potent novel hypolipidemic agent. *Prog. Med. Chem.* **2002**, *40*, 1–22.
41. Law, M.; Rudinka, A.R. Statin safety: A systematic review. *Am. J. Cardiol.*, 2006, *97(8A)* 52C-60C.
42. McTaggart, F.; Buckett, L.; Davidson, R.; Holdgate, G.; McCormick, A.; Schneck, D.; Smith, G.; Warwick, M. Preclinical and clinical pharmacology of Rosuvastatin, a new 3-hydroxy-3-methylglutaryl coenzyme A reductase inhibitor. *Am. J. Cardiol.* **2001**, *87*, 28B–32B.
43. Guo, Z.; Chen, Y.; Goswami, A.; Hanson, R.L.; Patel, R.N. Synthesis of ethyl and t-butyl (3*R*,5*S*)-dihydroxy-6-benzyloxyhexanoates via diastereo- and enantioselective microbial reduction. *Tetrahedron Asymmetry* **2006**, *17*, 1589–1602.
44. Goldberg, S.; Guo, Z.; Chen, S.; Goswami, A.; Patel, R.N. Synthesis of ethyl-(3*R*,5*S*)-dihydroxy-6-benzyloxyhexanoates via diastereo- and enantioselective microbial reduction: Cloning and expression of ketoreductase III from *Acinetobacter* sp. SC 13874. *Enzyme Microb. Technol.* **2008**, *43*, 544–549.
45. Davis, S.; Christopher, G.; John, H.; Gray, D.; Gruber, J.; Huisman, G.; Ma, S.; Newman, L.; Sheldon, R. Enzymatic Processes for the Production of 4-Substituted 3-Hydroxybutyric Acid Derivatives. US 7807423B2, October 5, 2010.
46. Jagoda, E.; Stouffer, B.; Ogan, M.; Tsay, H.M.; Turabi, N.; Mantha, S.; Yost, F.; Tu, J.I. Radioimmunoassay for hydroxyphosphinyl-3-hydroxybutanoic acid (SQ 33,600), a hypocholesterolemia agent. *Ther. Drug Monit.* **1993**, *15*, 213–219.
47. Wang, D.I.; Arnold, M.E.; Jemal, M.; Cohen. A.I. Determination of SQ 33,600, a phosphinic acid containing HMG CoA reductase inhibitor, in human serum by high-performance liquid chromatography combined with ionspray mass spectrometry. *Biol. Mass Spectrom.* **1992**, *21*, 189–194.
48. Patel, R.; McNamee, C.; Banerjee, A.; Howell, J.; Robison, R.; Szarka, L. Stereoselective reduction of β-keto ester by *Geotrich candidum. Enzyme Microb. Technol.* **1992**, *14*, 731–738.
49. Matsuyama, A.; Yamamoto, H.; Kobayashi, Y. Practical application of recombinant whole-cell biocatalysts for the manufacturing of pharmaceutical intermediates such as chiral alcohols. *Org. Process Res. Dev.* **2002**, *6*, 558–561.
50. Zalman, L.S.; Brothers, M.A.; Dragovich, P.S.; Zhou, R.; Prins, T.J.; Worland, S.T.; Patick, A.K. Inhibition of human rhinovirus-induced cytokine production by AG7088, a human rhinovirus 3C protease inhibitor. *Antimicrob. Agents Chemother.* **2000**, *44*, 1236–1241.
51. Dragovich, P.S.; Prins, T.J.; Zhou, R.; Webber, S.E.; Marakovits, J.T.; Fuhrman, S.A.; Patick, A.K.; Matthews, D.A.; Lee, C.A.; Ford, C.E.; *et al.* Structure-based design, synthesis, and biological evaluation of irreversible human rhinovirus 3C protease inhibitors. 4. incorporation of P_1 lactam moieties as l-glutamine replacements. *J. Med. Chem.* **1999**, *42*, 1213–1224.
52. Tao, J.; McGee, K. Development of a continuous enzymatic process for the preparation of (*R*)-3-(4-fluorophenyl)-2-hydroxy propionic acid. *Org. Process Res. Dev.* **2002**, *6*, 520–524.
53. Bold, G.; Faessler, A.; Capraro, H.-G.; Cozens, R.; Klimkait, T.; Lazdins, J.; JMestan, J.; Poncioni, B.; Roesel, J.; Stover, D.; *et al.* New aza-dipeptide analogs as potent and orally

absorbed HIV-1 protease inhibitors: Candidates for clinical development. *J. Med. Chem.* **1998**, *41*, 3387–3401.

54. Robinson, B.S.; Riccardi, K.A.; Gong, Y.F.; Guo, Q.; Stock, D.A.; Blair, W.S.; Terry, B.J.; Deminie, C.A.; Djang, F.; Colonno, R.J.; *et al.* BMS-232632, a highly potent human immunodeficiency virus protease inhibitor that can be used in combination with other available antiretroviral agents. *Antimicrob. Agents Chemother.* **2000**, *44*, 2093–2099.
55. Patel, R.N.; Chu, L.; Mueller, R.H. Diastereoselective microbial reduction of (S)-[3-chloro-2-oxo-1-(phenylmethyl)propyl]carbamic acid, 1,1-dimethylethyl ester. *Tetrahedron Asymmetry* **2003**, *14*, 3105–3109.
56. Bowers, N.I.; Skonezny, P.M.; Stein, G.L.; Franceschini, T.; Chiang, S.-J.; Anderson, W.L.; You, L.; Xing, Z. Pocess for preparing (2*R*,3*S*)-1,2-epoxy-3-(protected)amino-4-substituted butane and intermediates thereof. WO 2006127180 A1, November 30, 2006
57. Xu, Z.; Singh, J.; Schwinden, M.D.; Zheng, B.; Kissick, T.P.; Patel, B.; Humora, M.J.; Quiroz, F.; Dong, L.; Hsieh, D.-M.; *et al.* Process research and development for an efficient synthesis of the HIV protease inhibitor BMS-232632. *Org. Process Res. Dev.* **2002**, *6*, 323–328.
58. King, A.O.; Corley, E.G.; Anderson, R.K.; Larsen, R.D.; Verhoeven, T.R.; Reider, P.J.; Xiang, Y.B.; Belley, M.; Leblanc, Y.; Labelle, M.; *et al.* An efficient synthesis of LTD4 antagonist L-699,392. *J. Org. Chem.* **1993**, *58*, 3731–3735.
59. Shinkai, I.; King, A.O.; Larsen, R.D. Practical asymmetric synthesis of LTD4 antagonist. *Pure Appl. Chem.* **1994**, *66*, 1551–1556.
60. Zhao, M.; King, A.O.; Larsen, R.D.; Verhoeven, T.R.; Reider, P.J. A convenient and economical method for the preparation of DIP-Chloride™ and its application in the asymmetric reduction of aralkyl ketones. *Tetrahedron Lett.* **1997**, *36*, 2641–2644.
61. Shafiee, A.; Motamedi, H.; King, A. Purification, characterization and immobilization of an NADPH-dependent enzyme involved in the chiral specific reduction of the keto ester M, an intermediate in the synthesis of an anti-asthma drug, Montelukast, from *Microbacterium campoquemadoensis* (MB5614). *Appl. Microbiol. Biotechnol.* **1998**, *49*, 709–717.
62. Liang, J., Lalonde, J., Borup, B., Mitchell, V., Mundorff, E., Trinh, N., Kochrekar, D.A., Cherat, R.N., Pai, G.G. Development of a biocatalytic process as an alternative to the (−)-DIP-Cl-mediated asymmetric reduction of a key intermediate of Montelukast. *Org. Proc. Res. Dev.* **2010**, *14*, 193-198.
63. Holton, R.; Biediger, R.; Joatman, P. Semisynthesis of taxol and taxotere. In *Taxol: Science and Application*; Suffness, M., Ed.; CRC press: NewYork, NY, USA, 1995; pp. 97–123.
64. Kingston, D. Natural taxoids: Structure and chemistry. In *Taxol: Science and Application*; Suffness, M., Ed.; CRC press: New York, NY, USA, 1995; pp. 287–317.
65. Patel, R. Tour de paclitaxel: Biocatalysis for semisynthesis. *Annu. Rev. Microbiol.* **1995**, *98*, 361–395.
66. Patel, R.; Banerjee, A.; Howell, J.; McNamee, C.; Brozozowski, D.; Mirfakhrae, D.; Nanduri, V.; Thottathil, J.; Szarka, L. Microbial synthesis of (2*R*,3*S*)-*N*-benzoyl-3-phenyl isoserine ethyl ester-a taxol side-chain synthon. *Tetrahedron Asymmetry* **1993**, *4*, 2069–2084.
67. Junie, J.L.; Leonard, B.E. Drugs acting on sigma and phencyclidine receptors: A review of their nature, function, and possible therapeutic importance. *Clin. Neuropharmacol.* **1989**, *12*, 353–374.

68. Ferris, C.D.; Hirsch, D.J.; Brooks, B.P.; Snyder, S.H. σ Receptors: From molecule to man. *J. Neurochem.* **1991**, *57*, 729–737.
69. Patel, R.N.; Banerjee, A.; Liu, M.; Hanson, R.; Ko, R.; Howell, J.; Szarka, L. Microbial reduction of 1-(4-fluorophenyl)-4-[4-(5-fluoro-2-pyrimidinyl)-1-piperazinyl]butan-1-one. *Biotechnol. Appl. Biochem.* **1993**, *17*, 139–153.
70. Kagechika, H.; Kawachi, E.; Hashimoto, Y.; Shudo, K.; Himi, T. Retinobenzoic acids. 1. Structure-activity relationships of aromatic amides with retinoidal activity. *J. Med. Chem.* **1989**, *32*, 2583–2588.
71. Kagechika, H.; Shudo, K. Retinoids. Vitamin A for clinical applications. *Farumashia* **1990**, *26*, 35–40.
72. Moon, R.C.; Mehta, R.G. Anticarcinogenic effects of retinoids in animals. *Adv. Exp. Med. Biol.* **1986**, *206*, 399–411.
73. Patel, R.N.; Chu, L.; Chidambaram, R.; Zhu, J.; Kant, J. Enantioselective microbial reduction of 2-oxo-2-(1',2',3',4'-tetrahydro-1',1',4',4'-tetramethyl-6'-naphthalenyl)acetic acid and its ethyl ester. *Tetrahedron Asymmetry* **2002**, *13*, 349–355.
74. Prasad, C.V.C.; Wallace, O.B.; Noonan, J.W.; Sloan, C.P.; Lau, W.; Vig, S.; Parker, M.F.; Smith, D.W.; Hansel, S.B.; Polson, C.T.; *et al.* Hydroxytriamides as potent γ-secretase inhibitors. *Bioorg. Med. Chem. Lett.* **2004**, *14*, 3361–3371.
75. Nanduri, V.B.; Hanson, R.L.; Goswami, A.; Wasylyk, J.M.; LaPorte, T.L.; Katipally, K.; Chung, H.-J.; Patel, R.N. Biochemical approaches to the synthesis of ethyl 5-(*S*)-hydroxyhexanoate and 5-(*S*)-hydroxyhexanenitrile. *Enzyme Microb. Technol.* **2001**,*28*, 632–636.
76. Schenk, D.G.; Seubert, P. Potential treatment opportunities for Alzheimer's disease through inhibition of secretases and Aβ immunization. *J. Mol. Neurosci.* **2001**, *17*, 259–267.
77. Patel, R.N.; Goswami, A.; Chu, L.; Donovan, M.J.; Nanduri, V.; Goldberg, S.; Johnston, R.; Siva, P.J.; Nielsen, B.; Fan, J.; *et al.* Enantioselective microbial reduction of substituted acetophenones. *Tetrahedron Asymmetry* **2004**, *15*, 1247–1258.
78. Wittman, M.; Carboni, J.M.; Attar, R.; Balasubramanian, B.; Balimane, P.; Brassil, P.; Beaulieu, F.; Chang, C.; Clarke, W.; Dell, J.; *et al.* Discovery of a 1*H*-Benzoimidazol-2-yl)-1*H*-pyridin-2-one (BMS-536924) inhibitor of insulin-like growth factor I receptor kinase with *in vivo* antitumor activity. *J. Med. Chem.* **2005**, *48*, 5639–5643.
79. Carboni, J.M.; Hurlburt, W.W.; Gottardis, M.M. Synergistic methods and compositions for treating cancer. WO 2004030625 A2, April 15 2004.
80. Beaulieu, F.; Ouellet, C.; Zimmermann, K.; Velaparthi, U.; Wittman, M.D. Novel tyrosine kinase inhibitors WO 2004063151 A3 December 9, 2004
81. Hanson, R.L.; Goldberg, S.; Goswami, A.; Tully, T.P.; Patel, R.N. Purification and cloning of a ketoreductase used for the preparation of chiral alcohols. *Adv. Synth. Catal.* **2005**, *347*, 1073–1080.
82. Vacca, J.P. New advances in the discovery of thrombin and factor Xa inhibitors. *Curr. Opin. Chem. Biol.* **2000**, *4*, 394–400.
83. Gladwell, T.D. Bivalirudin: A direct thrombin inhibitor. *Clin. Ther.* **2002**, *24*, 38–58.
84. Fevig, J.M.; Wexler, R.R. Anticoagulants: Thrombin and factor Xa inhibitors. *Annu. Rep. Med. Chem.* **1999**, *34*, 81–100.

85. Williams, P.D.; Coburn, C.; Burgey, C.; Morrissette, M.M. Preparation of Triazolopyrimidines as Thrombin Inhibitors. WO 2002064211 A1, August 22, 2002.
86. Nelson, T.D.; LeBlond, C.R.; Frantz, D.E.; Matty, L.; Mitten, J.V.; Weaver, D.G.; Moore, J.C; Kim, J.M.; Boyd, R.; Kim, P.-Y.; *et al.* Stereoselective synthesis of a potent thrombin inhibitor by a novel P2-P3 lactone ring opening. *J. Org. Chem.* **2004**, *69*, 3620–3627.
87. Fukuroda, T.; Nishikibe, M. Enhancement of pulmonary artery contraction induced by endothelin-β receptor antagonism. *J. Cardiovasc. Pharmacol.* **1998**, *31*, S169–S171.
88. Sumner, M.J.; Cannon, T.R.; Mundin, J.W.; White, D.G.; Watts, I.S. Endothelin ETA and ETB receptors mediate vascular smooth muscle contraction. *Br. J. Pharmacol.* **1992**, *107*, 858–860.
89. Krulewicz, B.; Tschaen, D.; Devine, P.; Lee, S.S.; Roberge, C.; Greasham, R.; Chartrain, M. Asymmetric biosynthesis of key aromatic intermediates in the synthesis of an endothelin receptor antagonist. *Biocatal. Biotransformation* **2001**, *19*, 267–279.
90. Chaffman, M.; Brogden, R.N. Diltiazem. A review of its pharmacological properties and therapeutic efficacy. *Drugs* **1985**, *29*, 387–390.
91. Kawai, C.; Konishi, T.; Matsuyama, E.; Okazaki, H. Comparative effects of three calcium antagonists, diatiazem, verapamil and nifedipine, on the sinoatrial and atrioventricular nodes. Experimental and clinical studies. *Circulation* **1981**, *63*, 1035–1038.
92. Isshiki, T.; Pegram, B.; Frohlich, E. Immediate and prolonged hemodynamic effects of TA-3090 on spontaneously hypertensive (SHR) and normal Wistar-Kyoto (WKY) rats. *Cardiovasc. Drug Ther.* **1988**, *2*, 539–544.
93. Das, J.; Floyd, D.M.; Kimball, D.; Duff, K.J.; Lago, M.W.; Moquin, R.V.; Gougoutas, J.Z. Benzazepinone calcium channel blockers. 3. Synthesis and structure-activity studies of 3-alkylbenzazepinones. *J. Med. Chem.* **1992**, *35*, 773–780.
94. Patel, R.N.; Robison, R.S.; Szarka, L.J.; Kloss, J.; Thottathil, J.K.; Mueller, R.H. Stereospecific microbial reduction of 4,5-dihydro-4-(4-methoxyphenyl)-6-(trifluoromethyl-1*H*-1)-benzazepin-2-one. *Enzyme Microb. Technol.* **1991**, *13*, 906–912.
95. Arch, J.R.S. β3-adrenoceptors and other putative atypical β-adrenoceptors. *Pharmacol. Rev. Commun.* **1997**, *9*, 141–148.
96. Bloom, J.D.; Datta, M.D.; Johnson, B.D.; Wissner, A.; Bruns, M.G.; Largis, E.E.; Dolan, J.A.; Claus, T.H. Disodium (R,R)5-[2-(3-chlorophenyl)-2-hydroxyethyl]amino]propyl]-1,3-benzodioxole-2,2-dicarboxylate. A potent β-adrenergic agonist virtually specific for β3 receptors. *J. Med. Chem.* **1989**, *35*, 3081–3084.
97. Fisher, L.G.; Sher, P.M.; Skwish, S.; Michael, I.M.; Seiler, S.; Dickinson, K.E.J. BMS-187257, a potent, selective, and novel heterocyclic β-3 adrenergic receptor agonist. *Bioorg. Med. Chem. Lett.* **1994**, *6*, 2253–2258.
98. Patel, R.N.; Banerjee, A.; Chu, L.; Brzozowski, D.; Nanduri, V.; Szarka, L.J. Microbial synthesis of chiral intermediates for β-3-receptor agonists. *J. Am. Oil Chem. Soc.* **1998**, *75*, 1473–1482.
99. Iwata, H.; Tanaka, R.; Ishiguro, M. Structures of the alkaline hydrolysis products of penem antibiotic, SUN5555. *J. Antibiot.* **1990**, *43*, 901–903.
100. Jean-Paul Vandecasteele, J.-P. Enzymatic synthesis of L-carnitine by reduction of an achiral precursor: The problem of reduced nicotinamide adenine dinucleotide recycling. *Appl. Environ. Microbiol.* **1980**, *39*, 327–334.

101. Yamamoto, H; Matsuyama, A.; Kobayashi, Y. Synthesis of (*R*)-1,3-butanediol by enantioselective oxidation using whole recombinant *Escherichia coli* cells expressing (*S*)-specific secondary alcohol dehydrogenase. *Biosci Biotechnol Biochem.* **2002**, *66*, 925-927

102. Coleman, P.J.; Brashear, K.M.; Askew, B.C. Nonpeptide α vβ 3 antagonists. Part 11: Discovery and preclinical evaluation of potent α vβ3 antagonists for the prevention and treatment of osteoporosis. *J. Med. Chem.* **2004**, *47*, 4829–4837.

103. Pollard, D.J.; Telari, K.; Lane, J.; Humphrey, G.; McWilliams, C.; Nidositko, S.; Salmon, P.; Moore, J. Asymmetric reduction of alpha, beta-unsaturated ketone to (*R*) allylic alcohol by *Candida chilensis. Biotechnol. Bioeng.* **2006**, *93*, 674-684.

104. Brands, K.M.J.; Payack, J.F.; Rosen, J.D.; Nelson, T.D.; Candelario, A.; Huffman, M.A.; Zhao, M.M.; Bridgette, J.L.; Zhiguo, C.J.; Song, D.M.; *et al*. Efficient synthesis of NK1 receptor antagonist aprepitant using a crystallization-induced diastereoselective transformation. *J. Am. Chem. Soc.* **2003**, *125*, 2129–2135.

105. Pollard, D.; Truppo, M.; Pollard, J.; Chen, C.-Y.; Moore, J. Effective synthesis of (*S*)-3,5-bistrifluoro methylphenyl ethanol by asymmetric enzymatic reduction. *Tetrahedron Asymmetry* **2006**, *17*, 554–559.

106. Patel, R.N. Chemo-enzymatic synthesis of pharmaceutical intermediates. *Expert Opin. Drug Dis. Dev.* **2008**, *3*, 187–245.

107. Ohshima, T.; Soda, K. *Stereoselective Biocatalysis: Amino Acid Dehydrogenases and Their Applications in Stereoselective Biocatalysis*; Patel, R.N., Ed.; Marcel and Dekker Pub: New York, New York, USA 2000, pp. 877–903.

108. Wandrey, C.; Wichman, R.; Leuchtenberger, W. Continuous Enzymic Transformation of Water-Soluble α-Keto Carboxylic Acids into the Corresponding Amino Acids. EP 0023346 B1 July 13, 1983.

109. Brunhuber, N.M.; Blanchard, J.S. The biochemistry and enzymology of amino acid dehydrogenases. *Crit. Rev. Biochem. Mol. Biol.* **1994**, *29*, 415–467.

110. Gordon, E.M.; Ondetti, M.A.; Pluscec, J.; Cimarusti, C.M.; Bonner, D.P.; Sykes, R.B. O-Sulfated β-lactam hydroxamic acids (monosulfactams). Novel monocyclic β-lactam antibiotics of synthetic origin. *J. Am. Chem. Soc.* **1982**, *104*, 6053–6060.

111. Godfrey, J.D.; Mueller, R.H.; van Langen, D.J. β-Lactam synthesis: Cyclization *versus* 1,2-acyl migration-cyclization. The mechanism of the 1,2-acyl migration-cyclization. *Tetrahedron Lett.* **1986**, *27*, 2793–2796.

112. Hanson, R.L.; Singh, J.; Kissick, T.P.; Patel, R.N.; Szarka, L.; Mueller, R. Synthesis of L-β-hydroxyvaline from α-keto-β-hydroxyisovalerate using leucine dehydrogenase from *Bacillus* species. *Bioorg. Chem.* **1990**, *18*, 116–130.

113. Hanson, R., Goldberg, S., Patel, R. (unpublished results).

114. Stoyan, T.; Recktenwald, A.; Kula, M.R. Cloning, sequencing and overexpression of the leucine dehydrogenase gene from *Bacillus cereus*. *J. Biotechnol.* **1997**, *54*, 77–80.

115. Menzel, A.; Werner, H.; Altenbuchner, J.; Groger, H. From enzymes to "designer bugs" in reductive amination: A new process for the synthesis of L-tert-leucine using a whole cell. *Eng. Life Sci.* **2004**, *4*, 573–576.

116. Robl, J.; Sun, C.; Stevenson, J.; Ryono, D.; Simpkins, L.; Cimarusti, M.; Dejneka, T.; Slusarchyk, W.; Chao, S.; Stratton, L.; *et al.* Dual metalloprotease inhibitors: Mercaptoacetyl-based fused heterocyclic dipeptide mimetics as inhibitors of angiotensin-converting enzyme and neutral endopeptidase. *J. Med. Chem.* **1997**, *40*, 1570–1577.
117. Hanson, R.L.; Schwinden, M.D.; Banerjee, A.; Brzozowski, D.B.; Chen, B.-C.; Patel, B.P.; McNamee, C.G.; Kodersha, G.A.; Kronenthal, D.R.; Patel, R.N.; *et al.* Enzymatic synthesis of L-6-hydroxynorleucine. *Bioorg. Med. Chem.* **1999**, *7*, 2247–2252.
118. Patel, R. Enzymatic synthesis of chiral intermediates for Omapatrilat, an antihypertensive drug. *Biomol. Eng.* **2001**, *17*, 167–182.
119. Hanson, R.L.; Howell, J.; LaPorte, T.; Donovan, M.; Cazzulino, D.; Zannella, V.; Montana, M.; Nanduri, V.; Schwarz, S.; Eiring, R.; *et al.* Synthesis of allylsine ethylene acetal using phenylalanine dehydrogenase from *Thermoactinomyces intermedius. Enzyme Microb. Technol.* **2000**, *26*, 348–358.
120. Gallwitz, B. Glucagon-like peptide-1-based therapies for the treatment of type 2 diabetes mellitus. *Treat. Endocrinol.* **2005**, *4*, 361–370.
121. Sinclair, E.M.; Drucker, D.J. Glucagon-like peptide 1 receptor agonists and dipeptidyl peptidase IV inhibitors: New therapeutic agents for the treatment of type 2 diabetes. *Curr. Opin. Endocrinol. Diabet.* **2005**, *12*, 146–151.
122. Augeri, D.J.; Robl, J.A.; Betebenner, D.A.; Magnin, D.R.; Khanna, A.; Robertson, J.G.; Wang, A.; Simpkins, L.M.; Taunk, P.; Huang, Q.; *et al.* A highly potent, long-acting, orally active dipeptidyl peptidase IV inhibitor for the treatment of type 2 diabetes. *J. Med. Chem.* **2005**, *48*, 5025–5037.
123. David J Augeri, David A Betebenner, Lawrence G Hamann, David R Magnin, Jeffrey A Robl, Richard B Sulsky, Cyclopropyl-fused pyrrolidine-based inhibitors of dipeptidyl iv, processes for their preparation and their use. WO 2001068603 A2, September 20, 2001
124. Hanson, R.L.; Goldberg, S.L.; Brzozowski, D.B.; Tully, T.P.; Cazzulino, D.; Parker, W.L.; Lyngberg, O.K.; Vu, T.C.; Wong, M.K.; Patel, R.N. Preparation of an amino acid intermediate for the dipeptidyl peptidase IV inhibitor, saxagliptin, using a modified phenylalanine dehydro genase. *Adv. Synth. Catal.* **2007**, *349*, 1369–1378.
125. Groeger, H.; May, O.; Werner, H.; Menzel, A.; Altenbuchner, J. A "second-generation process" for the synthesis of L-neopentylglycine: Asymmetric reductive amination using a recombinant whole cell catalyst. *Org. Process Res. Dev.* **2006**, *10*, 666–669.
126. Emmanuel, M.J.; Frye, L.L.; Hickey, E.R.; Liu, W.; Morwick, T.M.; Spero, D.M.; Sun, S.; Thomson, D.S.; Ward, Y.D.; Young, E.R.R. Novel Spiroheterocyclic Compounds [Morpholine-4-Carboxylic Acid Amides of Heterocyclic Cyclohexylalanine and Neopentylglycine Derivatives and Their Analogs], Useful as Reversible Inhibitors of Cysteine Proteases such as Cathepsin S. WO 2001019816 A1, May 22, 2001.
127. Haque, T.S.; Ewing, W.R.; Mapelli, C.; Lee, V.G.; Sulsky, R.B.; Riexinger, D.J.; Martinez, R.L.; Zhu. Y.Z. Human Glucagon-like-Peptide-1 Modulators and Their Use in the Treatment of Diabetes and Related Conditions. WO2007082264, A2, January 11, 2007.
128. Qian, F.; Ewing, W.R.; Mapelli, C.; Riexinger, D.J.; Lee, V.G.; Sulsky, R.B.; Zhu, Y.; Haque, T.S.; Martinez, R.L.; Naringrekar, V.; *et al.* Sustained release GLP-1 receptor modulators. US 20070099835 A1, May 3, 2007.

129. Chen, Y.; Goldberg, S.L.; Hanson, R.L.; Parker, W.L.; Gill, I.; Tully, T.P.; Montana, M.; Goswami, A.; Patel, R.N. Enzymatic preparation of an (*S*)-amino acid from a racemic amino acid. *Org. Process Res. Dev.* **2011**, *15*, 241–248.

130. Straathof, A.J.J.; Panke, S.A. The production of fine chemicals by biotransformations. *Curr. Opin. Biotechnol.* **2002**, *13*, 548–556.

131. Bommarius, A.S.; Schwarm, M.; Drauz, K. Biocatalysis to amino acid-based chiral pharmaceuticals-examples and perspectives. *J. Mol. Catal. B Enzym.* **1998**, *5*, 1–11.

132. Vedha-Peters, K.; Gunawardana, M.; Rozzell, J.D.; Novick, S.J. Creation of a broad-range and highly stereoselective D-amino acid dehydrogenase for the one-step synthesis of D-amino acids. *J. Am. Chem. Soc.* **2006**, *128*, 10923–10929.

133. Gustafsson, D.; Elg, M.; Lenfors, S.; Boerjesson, I.; Teger-Nilsso, A.-C. Effects of inogatran, a new low-molecular-weight thrombin inhibitor, in rat models of venous and arterial thrombosis, thrombolysis and bleeding time. *Blood Coagul. Fibrinolysis* **1996**, *7*, 69–79.

134. Alexandre, F.-R.; Pantaleone, D.P.; Taylor, P.P.; Fotheringham, I.G.; Ager, D.J.; Turner, N.J. Amine-boranes: Effective reducing agents for the deracemisation of DL-amino acids using L-amino acid oxidase from *Proteus myxofaciens. Tetrahedron Lett.* **2002**, *43*, 707–710.

135. Chaturvedula, P.V.; Chen, L.; Civiello, R.; Degnan, A.P.; Dubowchik, G.M.; Han, X.; Jiang, J.J.; Macor, J.E.; Poindexter, G.S.; Tora, G.O.; *et al.* Anti-Migraine Spirocycles. US 7842808 B2, January 5, 2007.

136. Han, X., Civiello, R.L., Conway, C.M., Cook, D.A., Davis, C.D., Macci. R., Pin, S.S., Ren, S.X., Schartman, R., Signor, L.J., Thalody, G., Widmann, K.A., Xu, C., Chaturvedula, P.V., Macor, J.E., Dubowchik, G.M. The synthesis and SAR of calcitonin gene-related peptide (CGRP) receptor antagonists derived from tyrosine surrogates. Part 1. *Bioorg Med Chem Lett.* **2012,** *22*, 4723-4727.

137. Hanson, R.L.; Davis, B.L.; Goldberg, S.L.; Johnston, R.M.; Parker, W.L.; Tully, T.P.; Montana, M.A.; Patel, R.N. Enzymatic preparation of a D-amino acid from a racemic amino acid or keto acid. *Org. Process Res. Dev.* **2008**, *12*, 1119–1129.

138. Ising, M.; Zimmermann, U.S.; Künzel, H.E.; Uhr, M.; Foster, A.C.; Learned-Coughlin, S.M.; Holsboer, F.; Grigoriadis, D.E. High-affinity CRF1 receptor antagonist NBI-34041: Preclinical and clinical data suggest safety and efficacy in attenuating elevated stress response. *Neuropsychopharmacology* **2007**, *32*, 1941–1949.

139. Overstreet, D.H.; Griebel, G. Antidepressant-like effects of CRF1 receptor antagonist SSR125543 in an animal model of depression. *Eur. J. Pharmacol.* **2004**, *497*, 49–53.

140. Tache, Y.; Martinez, V.; Wang, L.; Million, M. CRF1 receptor signaling pathways are involved in stress-related alterations of colonic function and viscerosensitivity: Implications for irritable bowel syndrome. *Br. J. Phamacol.* **2004**, *141*, 1321–1330.

141. Yu, W.-L.; Lawrence, F.; Wong, H.; Lelas, S.; Zhang, G.; Lindner, M.D.; Wallace, T.; McElroy, J.; Lodge, N.J.; Gilligan, P.; *et al.* The pharmacology of DMP696 and DMP904, non-peptidergic CRF1 receptor antagonists. *CNS Drug Rev.* **2006**, *11*, 21–52.

142. Gilligan, P.J.; Clarke, T.; He, L.; Lelas, S.; Li, Y.-W.; Heman, K.; Fitzgerald, L.; Miller, K.; Zhang, G.; Marshall, A.; *et al.* 8-(4-Methoxyphenyl)pyrazolo[1,5-*a*]-1,3,5-triazines: Selective

and centrally active corticotropin-releasing factor receptor-1 (CRF1) antagonists. *J. Med. Chem.* **2009**, *52*, 3084–3092.

143. Hanson, R.L.; Davis, B.L.; Chen, Y.; Goldberg, S.L.; Parker, W.L.; Tully, T.P.; Montana, M.A.; Patel, R.N. Preparation of (*R*)-amines from racemic amines with an (*S*)-amine transaminase from *Bacillus megaterium*. *Adv. Synth. Catal.* **2008**, *350*, 1367–1375.

11

Lipases Immobilization for Effective Synthesis of Biodiesel Starting from Coffee Waste Oils

Valerio Ferrario [1], Harumi Veny [2], Elisabetta De Angelis [3], Luciano Navarini [3], Cynthia Ebert [1], and Lucia Gardossi [1,*]

[1] Dipartimento di Scienze Chimiche e Farmaceutiche, Università degli Studi di Trieste, Piazzale Europa 1, Trieste 34127, Italy; E-Mails: vferrario@units.it (V.F.); ebert@units.it (C.E.)
[2] Department of Chemical Engineering, Faculty of Engineering, University of Malaya, Malaysia; E-Mail: my_harumi@yahoo.com (H.V.)
[3] illycaffè S.p.A., via Flavia 110, Trieste 34147, Italy; E-Mails: elisabetta.deangelis@illy.com (E.D.A.); luciano.navarini@illy.com (L.N.)

* Author to whom correspondence should be addressed; E-Mail: gardossi@units.it

---

Abstract: Immobilized lipases were applied to the enzymatic conversion of oils from spent coffee ground into biodiesel. Two lipases were selected for the study because of their conformational behavior analysed by Molecular Dynamics (MD) simulations taking into account that immobilization conditions affect conformational behavior of the lipases and ultimately, their efficiency upon immobilization. The enzymatic synthesis of biodiesel was initially carried out on a model substrate (triolein) in order to select the most promising immobilized biocatalysts. The results indicate that oils can be converted quantitatively within hours. The role of the nature of the immobilization support emerged as a key factor affecting reaction rate, most probably because of partition and mass transfer barriers occurring with hydrophilic solid supports. Finally, oil from spent coffee ground was transformed into biodiesel with yields ranging from 55% to 72%. The synthesis is of particular interest in the perspective of developing sustainable processes for the production of bio-fuels from food wastes and renewable materials. The enzymatic synthesis of biodiesel is carried out under mild conditions, with stoichiometric amounts of substrates (oil and methanol) and the removal of free fatty acids is not required.

Key words: lipases; immobilization; biodiesel; oil from spent coffee ground

---

1. Introduction

Next generation biofuels is about utilization of non-food based feedstock and more sustainable process technology. The biodiesel industrial production worldwide is predominantly by chemical transesterification from food based feedstock. This is because with chemical transesterification the highest yield can be achieved in a relatively short reaction time. However, the current major cost of biodiesel production is from the feedstock [1]. Therefore, various non food feedstocks, such as waste oil and non edible oil, have been considered and studied in view of their potential use as raw materials [2–6].

In most cases the chemical process is performed by an alkaline catalyst, so that acid impurities and especially free fatty acids (FFA) contained in feedstock must be removed before transesterification to avoid the formation of saponification products. For instance, a basic solution is mixed with the extracted oil so that additional pre-treatments steps are introduced in the productive cycle [7]. In this regard, lipase-catalyzed transesterification of non food feedstock is becoming more attractive for the biodiesel industry, not only because it is sustainable and environmental friendly, but also because the free fatty acids can be esterified by lipases. Recently, attention has been given to oil extracted from coffee waste and examples of chemical conversion of oil from spent coffee ground into biodiesel have been reported [8,9]. The coffee grounds are mainly composed of proteins, carbohydrates, and lipids. It must be underlined that coffee is one of the largest agricultural products and, according to the U.S. Department of Agriculture, the world's coffee production is 16.34 billion pounds per year [8].

The spent coffee grounds contain, on an average, 15% (w/w) oil, which can be extracted by solvents such as trichloroethylene [10], *n*-hexane, ether and dichloromethane [10,11] with yields ranging from 6–28% (w/w) [9,10]. This is quite significant as compared to other major biodiesel feedstock such as rapeseed oil (37%–50%), palm oil (20%), and soybean oil (20%) [12]. Furthermore, the biodiesel from coffee possesses better stability than biodiesel from other sources due to its high antioxidant content, which hinders the rancimat process [13,14].

The work of Kondamundi reports the extraction of 15% (w/w) of oil from spent ground, which was dried prior to extraction in order to reduce the moisture content. A small amount of FFA, monoglycerides (MG), and diglycerides (DG) was also observed in the oil but a conversion in biodiesel of 100% was obtained by chemical transesterification. The oil and biodiesel formed in that process were found to be stable over one month without any observable physical changes and analysis demonstrated that biodiesel obtained from spent coffee grounds is a strong candidate as an alternative to diesel [8].

One additional issue in the development of an economically sustainable production of biodiesel by enzymatic transesterification is represented by the cost of the biocatalyst. The economic impact of biocatalysts can be reduced by immobilizing the lipases on solid supports and then recycling them. Therefore, it is important to select immobilized lipases that not only express high activity but also allow for repeated use thanks to improved stability.

Lipases are one of the classes of enzymes most largely employed in industry also because of their potential to work in non-aqueous environments [15] and they are applied at industrial scale for the transesterification of fats and oils in the food sector [16]. However, to the best of our knowledge, there is still a lack of immobilized lipases commercially available and suitable for application in biodiesel synthesis [17].

Indeed, in most cases, the lipase catalyzed synthesis of biodiesel has been studied by employing biocatalysts originally developed for interesterification of food oils [7]. However, the latter process implies the application of lipases in highly hydrophobic environments whereas biodiesel synthesis involves the use of relatively high percentages of hydrophilic short chain alcohols that have inactivating effects on enzymes [18,19]. Moreover, as the reaction proceeds, glycerol is produced. Consequently, the biocatalyst will behave quite differently in the two processes. Phenomena, such as partition of hydrophilic components, will deserve explicit attention and investigation when planning a methanolysis process.

By analyzing the immobilized lipases available on the market, Lipozyme TL IM (lipase from *Thermomyces lanuginosa*) results probably the most widely applied at industrial scale in triglyceride transformations in the food sector. It is immobilized by spraying the liquid lipase concentrate onto silica particles together with food grade granulation additives. After subsequent drying in fluid beds, the granules are ready for use in interesterification of triglycerides. However, Lipozyme TL IM will disintegrate when dispersed into water or hydrophilic media [20].

In the present work, the enzymatic synthesis of biodiesel was approached by studying immobilized biocatalysts specifically developed for this application. A particular attention was devoted to the nature of the immobilization carrier and the prevention of aggregation and methanol aspecific adsorption, phenomena that might cause enzyme inactivation. Finally, the selected biocatalysts were applied in the transesterification of oil extracted from spent coffee ground. Factors affecting the efficiency of lipase immobilization were also analyzed.

2. Results and Discussion

2.1. Selection of Lipases and Immobilization Supports

The present work follows our previous detailed investigation of structural and conformational properties of a series of lipases [21]. Lipases have been evolved for transforming insoluble hydrophobic substrates so that their surface present unusual features that make these proteins adapt for approaching lipophilic surfaces. Accommodation and transformation of bulky triglycerides is strictly related to the accessibility of the active site. The latter, as also the surrounding superficial area, is hydrophobic. A flexible protein domain, called lid, shields the opening of the active site when the protein is exposed to a hydrophilic environment, thus occluding the substrate access. Therefore, in principle, the immobilized lipase should maintain an open active conformation throughout the alcoholysis process despite the presence of polar hydrophilic components, such as the alcohol and the glycerol, in the reaction mixture.

The present study focuses attention on two lipases that demonstrated different conformational behavior when exposed to a hydrophilic environment, namely lipase B from *Candida antarctica* (CaLB) and *Burkholderia* (*Pseudomonas*) *cepacia* (PcL) (Figure 1). P revious MD simulations indicated that, upon exposure to water, the accessibility of the active site of these lipases is affected only at a minor extent when exposed to polar (e.g., aqueous) media [21].

Figure 1. Tridimensional models of CaLB (**a**) (PDB code 1TCA) and PcL (**b**) (PDB code 1YS1). The structures are colored according to their secondary structures; lids are highlighted in red. PcL (**b**) has the second "putative" lid highlighted in green and the Ca^{2+} ion represented as orange sphere.

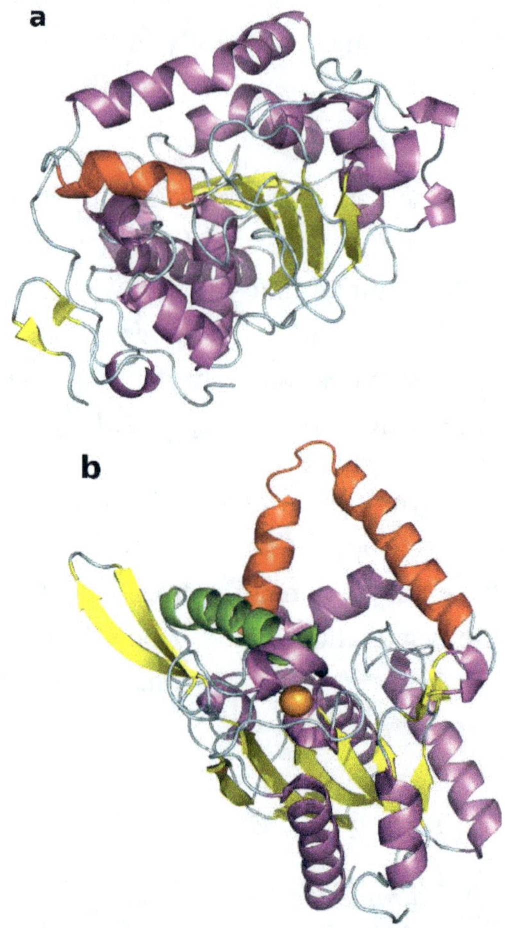

CaLB is characterized by a small lid [22] and we have previously demonstrated that after 10 ns of MD simulations in explicit water the small lid domain undergoes only some modest conformational changes. Moreover, because of the small size of the lid, there is no closing of the active site. Indeed, the final conformation presents no significant difference in the hydrophobic surface exposed to the bulky aqueous medium [21].

Regarding PcL, MD simulations demonstrated that a β-hairpin domain contributes to the stabilization of a "putative" second lid of PcL in its open conformation by forming two hydrogen bonds between Asn257-Thr224 and between Gln262-Gln215. Only the first one is lost during 20 ns MD in water, so that the upper part of the hairpin moves away from the α-helix but this second putative lid does not change its position considerably and does not occlude the active site [21].

Here, we show (Figure 2) the starting open conformation and the result after 20 ns MD simulations. No significant variation of the superficial domains occurs when the protein is embedded in water and, most importantly, the conformation achieved leads to a partial coverage of the active site. This can be deduced from the mapping of the hydrophobic/hydrophilic areas performed by GRID analysis [23], which reveals how a considerable part of the hydrophobic active site remains exposed to the solvent (Figure 2b).

Figure 2. GRID analysis of hydrophobic (yellow) and hydrophilic (blue) surface of PcL. On the left: open conformation corresponding to the crystal structure obtained in the presence of an inhibitor (PDB code 1YS1). On the right: partially closed conformation computed after 20 ns MD simulation in explicit water.

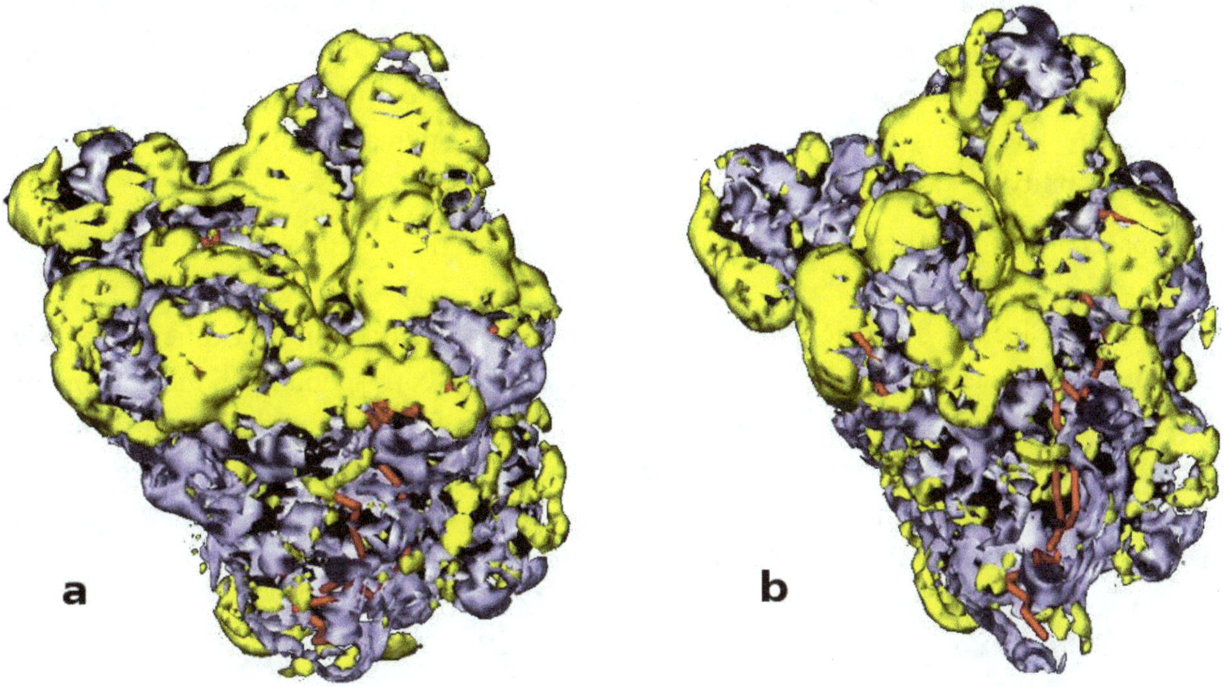

From the point of view of substrate specificity, lipase B from *Candida antarctica* (CaLB) shows a preference towards short and medium size fatty acids, although it has been investigated extensively also in biodiesel applications [24,25]. It is recognized that the active site of CaLB is situated in the core of the protein, the binding site has a funnel-like shape and highly hydrophobic amino acid residues envelop the cavity inner walls. Consequently, CaLB has, compared to other lipases, a very limited available space in the active site pocket that explains its substrate specificity [26].

Lipase from *Burkholderia cepacia* (PcL), instead, displays a marked specificity for long chain fatty acids and therefore, it appears to be a promising lipase for applications in biodiesel synthesis, as also reported in the literature [27].

In the present investigation, the selection and study of immobilized lipases suitable for biodiesel synthesis was guided by several considerations. Firstly, it must be taken into account that the enzymatic synthesis of biodiesel, unlike oil interesterification, involves the use of relatively high percentages of hydrophilic short chain alcohols that have inactivating effects on enzymes [19]. Moreover, as the reaction proceeds, glycerol is produced. Therefore, the immobilization of lipases for biodiesel synthesis should aim at preventing any detrimental interaction between the biocatalysts and these hydrophilic components, such as aspecific adsorption that might accentuate enzyme inactivation. Moreover, hydrophilic particles generally aggregate in hydrophobic media. On that basis, hydrophobic supports should be preferred.

However, it must be underlined that lipases recognize and accept very hydrophobic substrates that interact with the lipophilic area corresponding to the opening of the active site. As a continuation of our previous work, we have simulated the behavior of PcL at a water-octane interface (Figure 3) by carrying out a molecular dynamic (MD) simulation using a *coarse grained force field* (MARTINI) [28]. At the beginning of the MD simulation the opening of the active site (yellow dots) is oriented towards the aqueous phase; at the end, the lipase orients the opening of the active site towards the hydrophobic phase. This simulation suggests that the hydrophobic region of the enzyme will be also responsible for the establishment of hydrophobic interactions with hydrophobic supports. Indeed, this feature enables the exploitation of polymeric resins for achieving lipase purification and immobilization in one single step starting from crude enzymatic solutions, where lipases represent the only hydrophobic component [29]. Therefore, such interactions would promote an unfavorable orientation of the enzyme and scarce substrate accessibility.

On that basis, it seems difficult to define an ideal support for lipase immobilization *a priori*.

Figure 3. Molecular dynamic (MD) simulation of PcL embedded in an explicit bi-phasic system octane-water. The hydrophobic side of the enzyme, corresponding to the opening of the active site, is highlighted in yellow beads. The octane phase is in grey points in the figure. At the starting point of the simulation **(a)** the water-octane interface is not completely defined; during the simulation the interface was formed **(b)** and the enzyme is oriented with its active site towards the octane part.

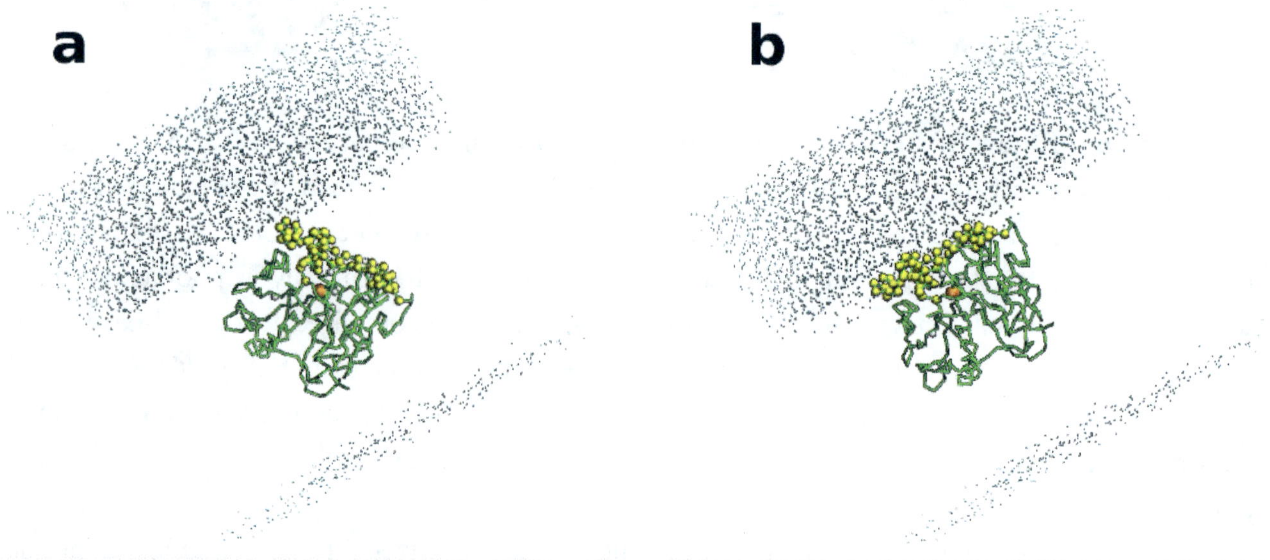

In order to elucidate the effect of the hydrophilic/hydrophobic nature of the carriers, the present study evaluated three different biocatalysts immobilized on highly and medium hydrophobic organic resins specifically developed by Sprin S.p.A. (Trieste, Italy) for biodiesel synthesis (Table 1). Then, the biocatalysts were compared with two preparations immobilized on hydrophilic siliceous carriers, namely diatomaceous earth (Celite). Table 1 reports the characterization of all the biocatalysts using standard assays.

A commercial formulation of PcL, Lipase PS-IM, was initially considered. It is commercialized by Amano and it consists in lipase PcL adsorbed on diatomaceous earth powder. However, despite the high hydrolytic activity expressed by the enzyme, the preparation was unsuitable for the methanolysis because the hydrophilic powder formed aggregates upon addition of the methanol to the oil.

Table 1. Activity of immobilized lipases tested in the study. CaL = lipase B from *Candida antarctica*. PcL = lipase from *Pseudomonas cepacia*.

Immobilized formulation	Carrier	Hydrophylicity	Particle size	Residual water [a] (%)	Synthetic activity (U/g dry)	Hydrolitic activity [d] (U/g dry)
PcL-S	Styrenic Porous	---	Beads, 300–500 μm	<5	159 [b]	463
PcL PS-IM	Siliceous Non porous	++	Powder <20 μm	<5	820 [b]	15,400
CaL-S	Styrenic Porous	--	Beads, 300–500 μm	<5	3450 [c]	978
CaL-M	Methacrylic Porous	-	Beads, 300–500 μm	<5	3020 [c]	490
CaLB on Celite® R-640	Siliceous Porous	+++	Rods 5 × 3 mm	15	2050 [c]	274

[a]The residual water content in the final immobilized preparation was determined on aluminum plates. A known amount of biocatalyst is dried at 110 °C for 6 h. Solvent content is defined as the % of weight loss after drying. [b]Synthetic activity: transesterification of vinyl acetate with 1-phenyl-1-ethanol. [c]Synthetic activity: esterification of lauric acid with 1-propanol. [d]Hydrolysis of tributyrin.

In light of the aggregation observed using the fine Celite particles, the immobilization of CaLB on hydrophilic carrier was carried out by selecting a commercial porous product, Celite® R-640 (Fluka), which consists in calcined Celite extruded in the form of big porous rods (5 mm high with a diameter of 3 mm). Porosity confers to the immobilization matrix a remarkable ability of adsorbing more than 100% of water (w/w) as well as a cumulative pore volume of 0.8 cm^3/g [30]. Immobilization was carried out by following a protocol previously developed in our lab for application of biocatalysts in low water media [31].

CaL-S and PcL-S are lipases adsorbed on highly hydrophobic styrenic resin whereas CaL-M corresponds to CaLB covalently immobilized on a methacrylic polymer [32]. It appears clear that there is no linear correlation between hydrolytic and synthetic activity. It must be also underlined that no direct comparison is possible between the hydrolytic activity of CaL-S and CaL-M since all enzymes immobilized through physical adsorption undergo partial leaching when suspended in aqueous media and this interfere with kinetic evaluation.

Moreover, the different substrate specificity of CaLB and PcL makes impracticable any direct comparison between the two enzymes.

2.2. Methanolysis of Triolein and Effect of Methanol Concentration

Methanolysis was conducted at 30 °C by using PcL-S (10% w/w of triolein) and by adding three equivalents of methanol obtaining 13% of conversion after 6 h and only 35% after 26 h. The data confirm the severe inactivating effect of methanol. Therefore, in the following experiments, multistep additions of methanol were used, by adding one equivalent molar of methanol at 0, 100 and 200 min. Proton NMR of the crude mixture (see Electronic Supplementary Information, Figure S1) showed the disappearance of the triglyceride peak already after 6 h of reaction. However, upon recycling, the biocatalyst displayed less than 40% of the initial activity. The recycles were carried out by introducing fresh triolein in each cycle and by determining the residual methyl oleate content prior to addition of methanol. Therefore, the PcL-S preparation, although very active, appeared unsuitable for multiple applications in biodiesel synthesis.

This unexpected behavior might be ascribable to different factors deserving specific investigations, which, however, are out of the scope of the present work. Since PcL displayed such poor efficiency when immobilised on the same support used for CaLB, the cause might reside either in the inherent properties of the protein or in the composition of the crude enzymatic formulation used for the immobilization. PcL is commercialized as a powder containing roughly 1% of protein diluted with cyclodextrins. The effect of this large amount of additives on the final immobilized formulation deserves further attention, although, it must be underlined that purification steps would cause an unacceptable increase of the cost of the biocatalyst.

One further point of attention is represented by the presence of a calcium ion in the structure of PcL (Figure 1b), which strongly affects the protein stability and this might be lost upon recycling [33].

Finally, a specific effect of some micro-components of triglyceride mixture on PcL stability cannot be excluded. Actually, several compounds generated by atmospheric oxidation of triglycerides, such as hydroperoxides and secondary products, may cause significant lipase inactivation in low water medium. Activity decay is usually time-dependent and shows a non-linear dependence on inactivating compound concentration [34].

When methanolysis was catalyzed by CaLB immobilized on Celite® R-640 (10% w/w of triolein) conversions lower than 5% were observed after 48 h. By adding 2.5% w/w of enzyme at the end of the reaction, there was no improvement in the conversion. It must be underlined that the biocatalyst was able to catalyze tributirin hydrolysis and propil laurate synthesis as shown in Table 1. The data indicate that the stop of reaction is not due to inactivation of enzyme but rather to some other factor connected to the reaction system. Most probably, the hydrophilic and hydrated pores (15% water by weight) of Celite® R-640 do not favor the partition of the oil towards the lipase.

Finally, the two preparations CaL-S and CaL-M immobilized on polymeric organic resins were monitored at different methanol to triolein molar ratios. Figure 4 reports the percentage of activity of the two CaLB formulations, defined as percentage of methyl ester yield evaluated after 25 h of reaction (theoretical maximum conversion in methyl ester is 33%, 66% and 100% for methanol to triolein molar ratio of 1:1, 2:1 and 3:1, respectively). Results indicate that the lipase activity of both CaL-S and CaL-M decreased with the increase of methanol concentration. By using a molar ratio of 3:1 the enzyme inactivation is so fast and severe that the reaction stops after 29% and 5% of conversion, respectively. CaL-S appears more resistant to the inactivating effect of methanol, since a conversion of 54% is achieved at a molar ratio of 2:1.

Figure 4. Effect of methanol concentration on CaL-S (star) and CaL-M (square) expressed as the conversion in methyl ester achieved after 25 h. The transesterification of triolein was carried out at 30 °C with molar ratios of 1:1, 2:1 and 3:1. Therefore, theoretical maximum conversions achievable were 33%, 66% and 100%, respectively.

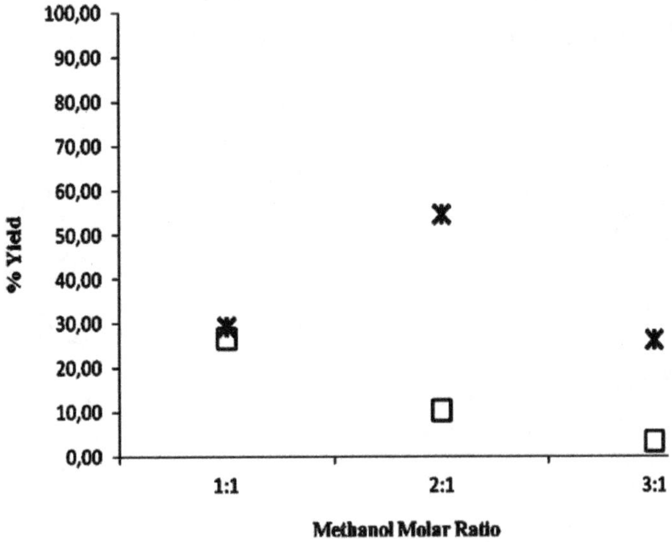

The different behavior of the two preparations is visible also from the kinetic profile of the reactions carried out at 1:1 molar ratio of triolein and methanol and 10% of biocatalyst (w/w % of triolein) (Figure 5). Apparently, the CaL-M preparation is less efficient even at such low concentration of methanol. In fact, as expected, initial rates are similar since comparable enzymatic units were employed. However the reaction profile of CaL-M indicates that the reaction slows down and, surprisingly, a lower final conversion is achieved, although the thermodynamic equilibrium of the reaction does not depend on the biocatalyst by definition. No further increase of the yield could be achieved upon addition of fresh enzyme and this would suggest that the poorer performance of CaL-M is not due to enzyme inactivation.

Figure 5. Reaction profiles (24 h) of transesterification of triolein using a 1:1 molar ratio of oil and methanol (30 °C). CaL-S (star) and CaL-M (square). The maximum theoretical yield is 33%.

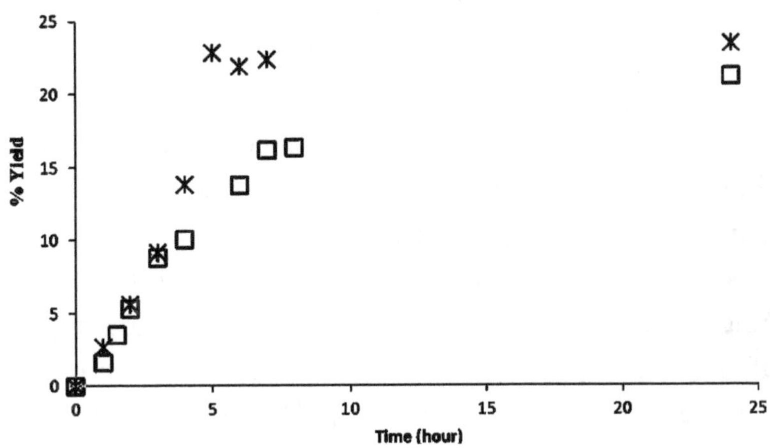

CaL-S and CaL-M show a good dispersion of the polymeric supports throughout the reaction with no formation of aggregates. Figure 6 illustrates the appearance of the reaction system at the end of transesterification with stepwise additions of methanol (molar ratio 1:1).

Figure 6. The appearance of the reaction system at the end of the methanolysis using the three preparations of CaLB considered in the study.

CaL-S **CaL-M** **CaLB on Celite® R-640**

Methanolysis reaction (molar ratio 1:1) using CaL-S gives 24% of conversion after 6h. It must be underlined that we also carried out methanolysis under the same conditions by using *Novozyme 435* obtaining about 10 % of formation of methyl oleate.

As reported before, the transesterification catalyzed by CaL-S and CaL-M did not achieve a quantitative conversion of the triglyceride. In order to verify whether some methanol was lost during the reaction and the sampling procedure, the transesterification was conducted for 48 h under similar operating conditions but without sampling. The final yield of methyl oleate was 29% for CaL-S but the NMR analysis proved no presence of residual methanol in the reaction medium (see Electronic Supplementary Information, Figure S2). The data suggest that during the transesterification process, some methanol is adsorbed on the immobilization support and this phenomenon is more pronounced in the case of CaL-M, as indicated by the lower conversion achieved. Therefore, the nature of the support appears to play a major role, since an increased local concentration of methanol is expected to be detrimental for the stability of the lipase. This is also in accordance with the lower efficiency of CaL-M observed at higher methanol concentration, as illustrated in Figure 5 above.

The effect of the nature of the immobilization carrier on the partition of hydrophilic components was studied by measuring the water activity of the multiphase reaction systems. Water activity was measured before and after the addition of the immobilized enzymes by means of a hygrometer and after 24 h equilibration as described previously [30,35]. The initial reaction system was a mixture of triolein and methanol with molar ratio 1:1. The values of water activity of the different reaction systems are shown in Table 2.

Table 2. Water activity of reaction systems in the presence of different preparations of immobilized CaLB.

Reaction System	Water activity (a_w)
Reactants mixture	0.64
CaL-S	0.48
CaL-M	0.27
CaLB on Celite® R-640	0.27

All the supports here considered are porous and have the ability to bind a significant amount of water (from 50–100% of their weight) although the residual water (evaluated by weight variation upon drying) resulted to be <5% (w/w) in CaL-S and CaL-M (Table 1). Results show that no immobilized enzyme brings free water to the reaction system that could cause competing hydrolytic reactions [36].

On the contrary, the hydrophilic carriers adsorb some water from the reactants, as indicated by the decrease of a_W values, even in the case of Celite R-640, which contains a residual 15% of water in its pores. Significantly, CaL-S, which is the most hydrophobic support, can bind a lower amount of water and, most probably, less methanol as well.

In order to investigate in more detail the effect of methanol on CaL-S, the kinetic profile of the methanolysis at different concentrations of methanol was studied. Figure 7 shows how, by working with three equivalents of methanol, the inactivating effect is visible already after the first hour of reaction. The residual activity of CaL-S after 1h of reaction at a methanol/triglyceride molar ratio of 3:1 was assayed also by using the method of propil laurate synthesis (see materials and methods). Data indicates that the activity of CaL-S decreases from the initial 3450–1600 (U/g dry), thus confirming the partial inactivation of the enzyme. In the other two cases, the reaction proceeded until the equilibrium was reached and no appreciable variation of conversion was observed by adding fresh enzyme at the end of the reaction (2.5% w/w). This also confirms that the stopping of the reaction was not due to lipase inactivation.

Figure 7. Effect of methanol concentration on CaL-S evaluated at 30 °C by monitoring the transesterification of triolein at different methanol/oil molar ratio.

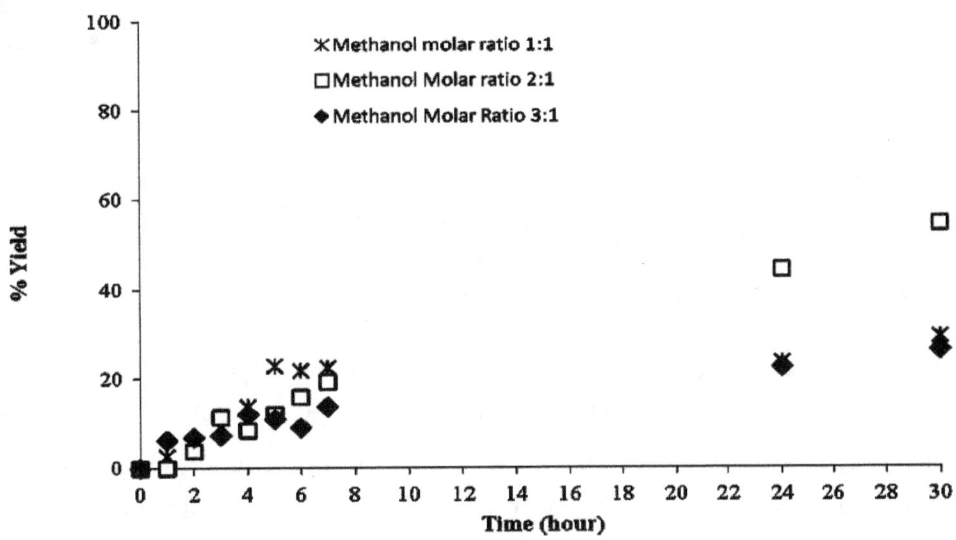

2.3. Recyclability of Cal-S

The stability study was performed using a methanol to oil molar ratio of 1:1 and by stopping the reaction after 4 h in each cycle and measuring the conversion in terms of formation of methyl oleate. Then, the biocatalyst was recovered and reused without further treatment/rinsing by simply introducing fresh triolein and methanol. It must be noted that for each cycle, after introducing fresh triolein, the residual methyl oleate content deriving from the previous cycle was determined prior to the addition of methanol and the starting of the methanolysis. The results indicate that CaLB on styrene is stable after 10 cycles, without any rinsing treatment (Figure 8). Moreover, the biocatalyst has shown its stability even after 22 days of exposure to the reaction conditions. Therefore, no appreciable inhibition due to oil, glycerol or product was detected. Importantly, no aggregation of the particle was observed. The behavior of the preparation indicates that the styrenic resin prevents the undesirable hydrophilic interactions (glycerol and methanol adsorption, protein aggregation, particle aggregation) which are likely to occur in the hydrophobic oils.

Figure 8. Recyclability of CalB immobilized on styrenic support expressed as the percentage of methyl oleate formed after 4 h at 30 °C in the transesterification of triolein with methanol (1:1 molar ratio).

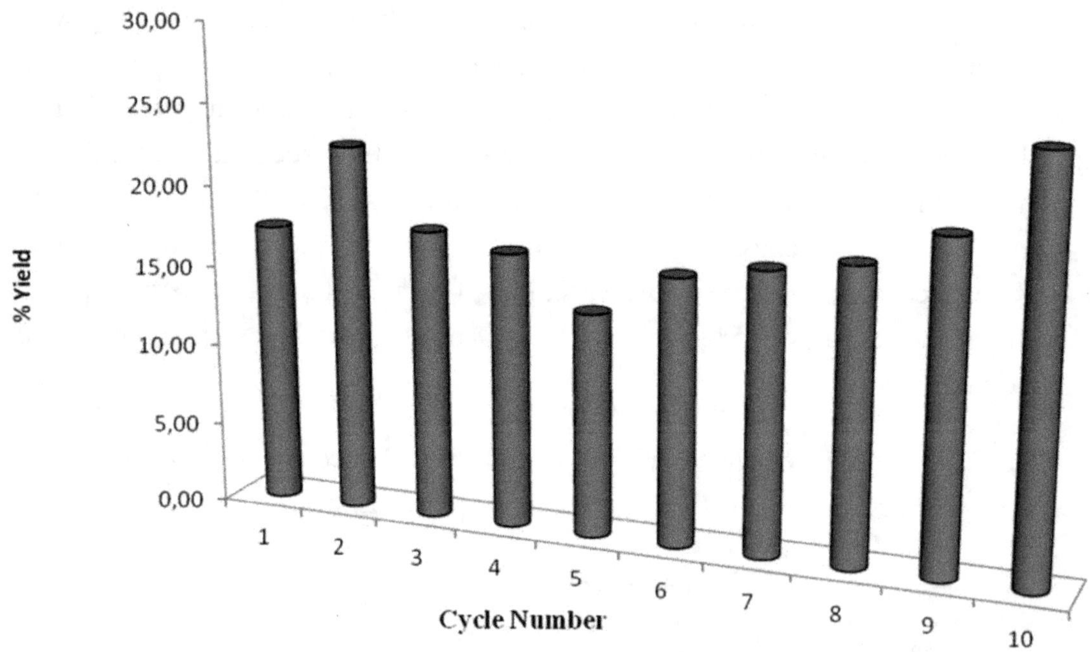

Notably, Liù and co-workers reported how CaLB Novozymes 435 undergoes a severe inactivation under comparable conditions, with a loss of roughly 10% of activity after each synthetic cycle [24].

2.4. Enzymatic Methanolysis of Oil Extracted from Espresso Spent Coffee Ground

As a final step in the investigation of lipase catalyzed methanolysis, the immobilized CaLB was applied to the transformation of a non-edible oil from food waste. The oil was extracted from espresso spent coffee ground properly dried [8] according to the procedure already reported [37]. The oil from spent coffee ground appears as a transparent homogeneous amber colored liquid (see Electronic

Supplementary Information, Figure S3). Table 3 illustrates the composition in terms of fatty acids determined by GC-MS and compared to the oil used in the work of Kondamundi. Of course, fatty acids can vary not only as a function of the type of coffee used but also as consequence of the method used for coffee brewing.

Table 3. Composition of the oil from espresso spent coffee ground in terms of fatty acids and compared to the oil used in the work of Kondamundi *et al.* [8].

Fatty acids	Oil from spent coffee ground as in [8]	Oil from espresso spent coffee ground
C14	n.d.	n.d.
C16	51.4	58.29
C18	8.3	8.79
C18:1	n.d.	3.41
C18:2	40.3	27.80
C18:3	n.d.	0.09
C20	n.d.	1.61
C22	n.d.	n.d.

The transesterification was carried out using CaL-S on oil from espresso spent coffee ground without pretreatment of free fatty acid and at 30 °C.

The reaction was monitored via HPLC by evaluating the formation of the methyl esters of the main fatty acids (palmitic, oleic, linoleic, linolenic, stearic, arachidonic). Because of the complexity of the reaction mixture, a semi-quantitative evaluation of the conversion was obtained by summing the areas of these products of the conversion. About 55% of biodiesel yield was achieved by performing one single addition of three equivalents of methanol at the beginning of the reaction whereas multistep additions of one equivalent of methanol (time = 0, 100 min, 200 min) led to 72% conversion after 30 h of reaction. No optimization of the process (e.g., range of time for addition of methanol) was carried out.

The decrement of the H^1-NMR signals relative to the triglycerides was used as a further indicator of the proceeding of the reaction. The area of the protons at 2.5 ppm (signal related to two protons in α-position to carbonyl group of fatty acid) and the area of the protons at 3.6 ppm (signal related to the three protons of methyl esters) indicate that the conversion is nearly quantitative within 24 h (Figure 9).

The GC-MS analysis of the products obtained from the methanolysis (see Electronic Supplementary Information, Figure S4) confirms the formation of the methyl esters of all the fatty acids present in the oil from spent coffee ground, as reported in Table 3. Therefore, CaLB demonstrates to be a versatile biocatalyst, accepting all the different fatty acids present in the oil from spent coffee ground.

It must be underlined that the chemical (KOH) methanolysis of oil from spent coffee ground reported by Kondamudi required the use of an excess of 40 volumes of methanol and was carried out under reflux at 70 °C. The process described in the present manuscript is carried out at 30 °C and requires no excess of methanol, so that the only co-product of the reaction is the glycerol. Moreover, at the end of the reaction the biocatalyst can be simply filtrated and no acid must be used to neutralize the reaction mixture.

Figure 9. H^1-NMR spectra of the oil from espresso spent coffee ground (**a**) and of the reaction mixture after 5 h (**b**) and 24 h (**c**).

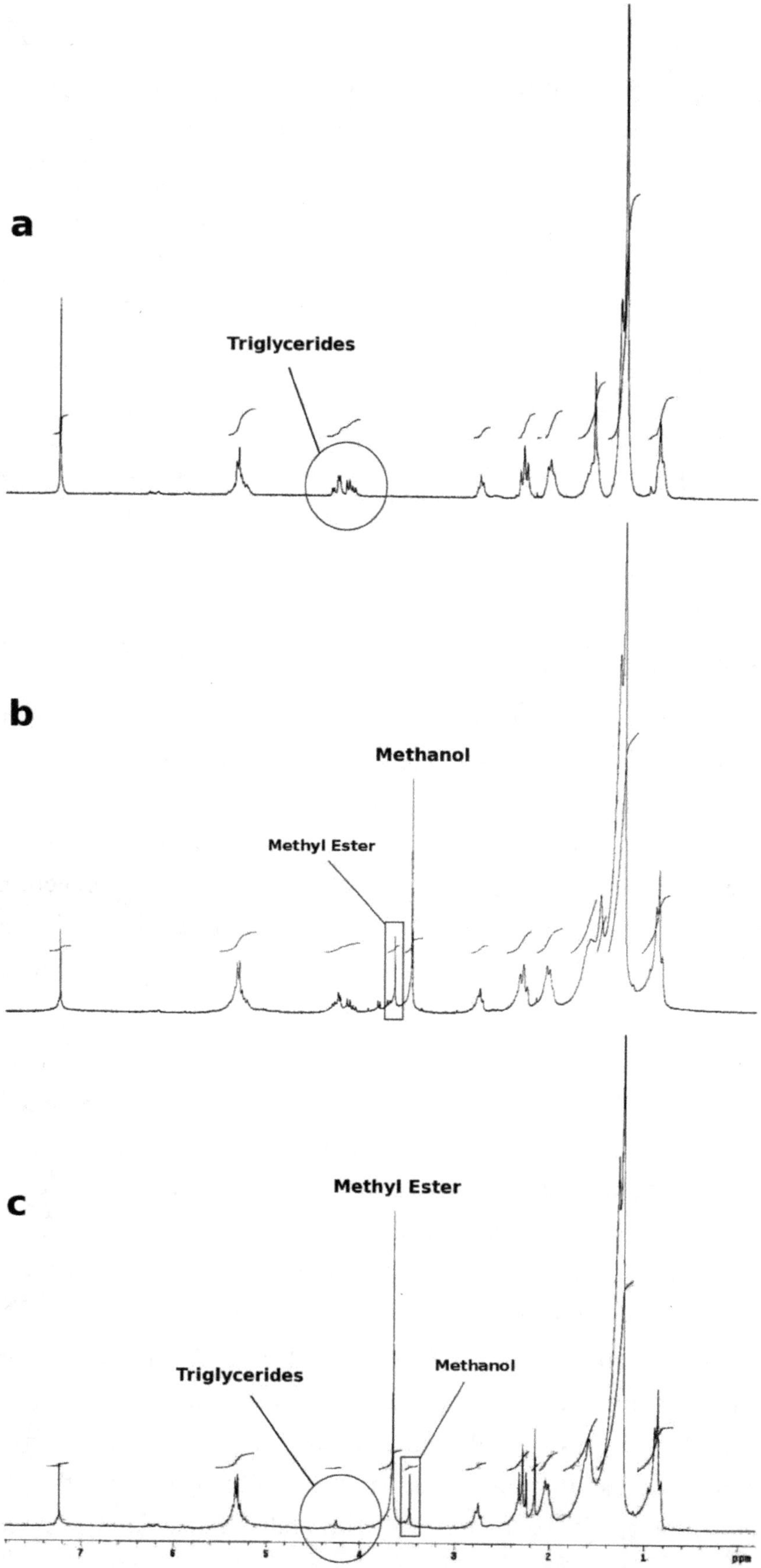

3. Experimental Sections

3.1. Enzymes

Native CaLB employed during this study was Lipozyme CalB L from Novozyme (DK). The commercial preparation is constituted by CaLB (11 mg/mL) diluted in a solution of glycerol, sorbitol and salts to stabilize the protein. The pH value of the enzymatic solution is about 4.2.

Native PcL was Lipase PS "AMANO" SD from AMANO (Japan). The commercial preparation is a yellowish lyophilized powder constituted by PcL (about 1%) diluted with dextrins.

Immobilized CaL-M (methacrylic) [32] and CaL-S (styrenic) were kindly donated by Sprin S.p.A. (Trieste, Italy). Organic polymeric resins are in the form of beads with particle size of 300–500 μm. Characterization is reported in Table 1.

The residual water content in the final immobilized preparation was determined on aluminum plates. A known amount of biocatalyst was dried at 110 °C for 6 h. Water content is defined as the % of weight loss after drying.

3.2. Immobilization of CalB on Celite®R-640

The immobilization was done according to the procedure already reported [31]. The Lipozyme solution was adjusted to pH 8 prior to immobilization. 2.5 mL of enzymatic solution, 1 g of dry Celite® R640 and 12 mL of dry toluene were mixed and kept at 25 °C for one day. The organic solvent was removed and the enzymatic preparation was washed with acetone on a büchner filter under reduced pressure for three times (support/solvent ratio 1/2 w/v). The residual water content was then evaluated as reported above.

3.3. Chemicals

Triolein (99%), methyl esters standards and all other chemical reagents are from Sigma Aldrich.

3.4. Monitoring the Formation of Fatty acid Methyl Ester (Biodiesel)

Methyl oleate formed from transesterification of triolein was measured by Gilson HPLC equipped with an auto injector. The column was ODS Hypersil SUPERCHROM set at isothermal oven temperature of 54 °C. The detector was UV/VIS-155 with dual wavelength of 200 nm and 210 nm. The mobile phase was 100% Acetonitrile and 0.05% of TFA with flow of 1 mL/min. Biodiesel (methyl esters of fatty acids-FAME) was analyzed by following methods from ref. [10]. Biodiesel yield was calculated by summing the areas of methyl esters present in the reaction mixture and on the basis of calibration curves built up with standards FAME.

3.5. Synthetic Activity of PcL

Activity was determined by transesterification reaction of vinyl acetate with 1-phenyl-1-ethanol, which is a standard synthetic method reported by AMANO Enzymes company. Transesterification reaction was carried out at 25 °C and mixing at 200 rpm. 1-phenylethyl acetate formed during the reaction was measured by Gilson HPLC equipped with an auto injector. The column was ODS

Hypersil SUPERCHROM set at isothermal oven temperature of 30 °C. The detector was a Gilson UV/VIS-155 with dual wavelength of 210 nm and 260 nm. The mobile phase was 40% acetonitrile and 0.05% of TFA and 60% H_2O with a flow of 1 mL/min.

3.6. Synthetic Activity of CaLB:

The assay allows the determination of enzymatic activity through the synthesis of propyl laurate formed by enzymatic esterification of lauric acid and 1-propanol. One unit corresponds to the amount of enzyme that produces 1 μmol of propyl laurate in one minute at 55 °C without solvent. In a vial of 20 mL, 1.2 g of lauric acid and 0.36 g of 1-propanol were added. The reaction mixture was thermostated at 55 °C and mixed at 250 rpm in order to allow the melting of lauric acid. A withdrawal of 0.1 mL was diluted in 1.3 mL of hexane. 0.2 ml of this solution was added to 1.6 mL of hexane and analyzed by HPLC (t = 0). The reaction was started by adding 40–50 mg of biocatalyst to the main solution and maintained under continuous stirring at 250 rpm. Withdrawals were performed at different times and analysed by HPLC.

3.7. Hydrolytics Activity of Lipases

The activity of enzymatic preparations was assayed by following the tributyrin hydrolysis and by titrating, with 0.1 M sodium hydroxide, the butyric acid that is released during the hydrolysis. An emulsion composed by 1.5 mL tributyrin, 5.1 mL gum arabic emulsifier (0.6% w/v) and 23.4 mL water was prepared in order to obtain a final molarity of tributyrin of 0.17 M. Successively, 2 mL of K-phosphate buffer (0.1M, pH 7.0) were added to 30 mL of tributyrin emulsion and the mixture was incubated in a thermostated vessel at 30 °C, equipped with a mechanical stirrer. After pH stabilization, about 50 mg of immobilized protein were added. The consumption of 0.1 M sodium hydroxide was monitored for 15–20 min. One unit of activity was defined as the amount of immobilized enzyme required to produce 1 μmol of butyric acid per min at 30 °C.

3.8. Water Activity

Water activity of the reaction system was determined at 30 °C by using a hygrometer (DARAI-Trieste, Italy) as previously described [30]. Measurements were carried out by sealing the sensor into the open end of 5 mL glasses vials, thermostated at 30 °C, until constant reading. All samples were previously equilibrated for 24 h. The hygrometer was firstly calibrated for 24 h with saturated salt solution of $MgCl_2$ (a_w = 0.3273), NaCl (a_w = 0.7532) and NH_4NH_2 (a_w = 0.927) using ultrapure water [30,35]. The values of standard salt solution from calibrated hygrometer were 0.39, 0.74 and 0.87.

3.9. Monitoring the Methanolysis by HPLC

Methyl esters formed from the transesterification were evaluated by means of a Gilson HPLC equipped with auto injector. The column was ODS Hypersil SUPERCHROM set at isothermal oven temperature of 54 °C. The detector was UV/VIS-155 with dual wavelength of 200 nm and 210 nm. The mobile phase was 100% acetonitrile and 0.05% of TFA with flow of 1 mL/min. Biodiesel

(FAME) content and its standards were analyzed by following methods from [38]. Biodiesel yield was calculated based on total area of the methyl ester mixture in the final product.

3.10. Lipase Catalyzed Methanolysis

Transesterification was carried out in 10mL screw capped vials in orbital shaker with temperature control. The temperature was set at 30 °C with shaking rate of 250 rpm. Prior to addition of methanol, the immobilized lipase (10% w/w of oil) was mixed with the feedstock (2 g) for a few minutes. 10 μL samples were withdrawn and analyzed by HPLC.

3.11. Recycling of CaL-S

The stability study was performed using a methanol to oil molar ratio of 1:1, stopping the reaction after 4 hours in each cycle and by measuring the conversion in terms of formation of methyl oleate. Then, the biocatalyst was recovered and reused without further treatment/rinsing by simply introducing fresh triolein and methanol. For each cycle, in order to avoid overestimation of the product concentration, after introducing fresh triolein, the residual methyl oleate content deriving from the previous cycle was determined prior to the addition of methanol and the starting of the new methanolysis cycle. The reactions were performed at 30 °C.

3.12. ^{1}H-NMR

Spectra were recorded on a NMR Varian 200 Gemini, operating at 200 MHz. Samples were prepared by diluting 10 μL of the reaction mixture in 700 μL of $CDCl_3$.

3.13. Extraction of Oil and GC-MS Analysis of Fatty Acids

The spent coffee ground was dried [8] and extracted according to the procedure already reported [37], obtaining yields around 20% (w/w). Characterization was performed according to a method previously described [39] by mean of GC MS Agilent technologies 6890 N and Agilent 5973 MSD as detector. The column was a Zebron ZB-FFAP Capillary Column 60 m × 0.25 mm × 0.25 μm. For the analysis, the temperature raised from 170–235 °C by increasing 2 °C/min. Gas phase was helium at constant flow of 1.3 mL/min.

3.14. Computational Study: Molecular Dynamic Simulations

The molecular dynamic (MD) simulations to analyse the behaviour of lipases upon water exposure were performed using the software GROMACS 4 with the GROMOS-96 53a6 force field. The crystal structure of *Pseudomonas cepacia* lipase (PDB 1YS1) and the crystal structure of *Candida antarctica* lipase B (PDB 1TCA) were implemented in the force field in *gro* file format by using the tool pdb2gmx of the GROMACS software package, which also adds the necessary hydrogen atoms. The proteins were solvated separately with explicit SPC water in a virtual box of 343 nm^3 each. The molecular dynamic simulations were performed in an NPT environment simulating the temperature of 300 K and keeping the pressure constant (Berendsen thermostat and pressure); the cut-off for electrostatic interactions was set to 1.4 nm and the limit for the Van der Waals interactions set to 1.4

nm. For the minimization procedures, the PME (Particle Mesh Ewald) algorithm was used for the calculation of the electrostatic interactions, setting the limit at 1.0 nm. Before the MD simulation, the systems were minimized at least for 10,000 steps.

The MD simulation of *Pseudomonas cepacia* lipase at water-octane interface was performed using the software GROMACS 4 with the MARTINI force field. Crystal structure of *Pseudomonas cepacia* lipase (PDB 1YS1) was implemented in the MARTINI force field in *gro* file format by using the necessary scripts available on the MARTINI web site [40] and the DSSP program [41] for the necessary secondary structure definition. The protein was solvated with explicit water-octane 1:1 mixture in virtual cubic boxes of 512 nm^3. The MD simulation was performed in a NPT environment and in conditions specified above. The time step for integration was set to 4 fs (usually this value is set to 2 fs). Minimization procedures were performed before the MD simulation using a cut-off for electrostatic interactions and by setting the threshold at 1.4 nm. The steepest descendent algorithm was used. Minimization was performed for at least 10,000 steps.

3.15. Computational Study: GRID Mapping of Surface of Proteins

The GRID analysis [32] was performed on the output of the dinamised crystal structure of *Pseudomonas cepacia* lipase (PDB code 1YS1) by choosing a cage big enough to include the whole protein. The grid nodes were set every 0.5 Å. The probes used for the calculation of the molecular interaction fields were DRY (hydrophobic probe) and WATER (H2O probe). Isopotential surfaces were visualized by setting energy values of -0.3 $kcalmol^{-1}$ for the DRY probe and -2.5 $kcalmol^{-1}$ for the WATER probe.

4. Conclusions

The present study shows that immobilization plays a major role in determining the efficiency of lipases in the methanolysis of oils. Information on the structure and conformation of lipases provides a rational basis for the selection of biocatalysts and their immobilization. *Ad hoc* immobilization methodologies are necessary for fully exploiting the catalytic potential of lipases while favoring effective mass transfer and partition of substrates/products.

Lipase B from *Candida antartica* (CaLB) displayed the highest efficiency when immobilized on the most hydrophobic supports (CaL-S). The biocatalyst showed good stability after 10 cycles of reaction in presence of equimolar concentration of methanol. Hydrophilic immobilization supports demonstrated to be less suitable for reactions involving the bulky and hydrophobic triglycerides, most probably because of unfavorable partition phenomena. Moreover, data suggest that these porous carriers are more prone to absorb water or other hydrophilic components such as methanol, which affect negatively the stability of lipases.

The study of methanolysis of oil from spent coffee ground indicates that new types of non-edible oils can be considered for developing competitive and sustainable enzymatic biodiesel synthesis. Immobilized CaLB was able to catalyze the conversion of all different fatty acids present in the raw material. Optimization studies will be required for implementing appropriate feed-batch processes to decrease the reaction time and for the evaluation of various types of oils from different coffee wastes.

Acknowledgements

Harumy Veny is grateful to ICS-UNIDO (Trieste, Italy) for a fellowship. Thanks are due to Sara Cantone and Diana Fattor (SPRIN S.p.A., Technology for Sustainable Chemistry, Trieste, Italy) for enzyme samples and useful discussions.

References

1. Medina, A.R.; González-Moreno, P.A.; Esteban-Cerdán, L.; Molina-Grima, E. Biocatalysis: Towards ever greener biodiesel production, *Biotechnol. Adv.* **2009**, *27*, 398–408.
2. Mittelbach, M.; Remschmidt, C. *Biodiesel - The Comprehensive Handbook*, 1st ed.; Börsedruk Ges. m.b.H., Vienna, 2004.
3. Uma, B.H.; Kim, Y.S. Review: A chance for Korea to advance algal-biodiesel technology. *J. Ind. Eng. Chem.* **2009**, *15*, 1–7.
4. Chisti, Y. Biodiesel from Microalgae. *Biotechnol. Adv.* **2007**, *25*, 294–306.
5. Shah, S.; Sharma, S.; Gupta, M.N. Biodiesel preparation by lipase-catalyzed transesterification of jatropha oil. *Energy Fuels* **2004**, *18*, 154–159.
6. Oliveira, L.S.; Franca, A.S.; Camargos, R.R.S.; Ferraz, V.P. Coffee oil as a potential feedstock for biodiesel production. *Bioresour. Technol.* **2008**, *99*, 3244–3250.
7. Calabrò, V.; Ricca, E.; de Paola, M.G.; Curcio, S.; Iorio, G. Kinetics of enzymatic trans-esterification of glycerides for biodiesel production. *Bioprocess Biosyst. Eng.* **2010**, *33*, 701–710.
8. Kondamundi, N.; Mohapatra, S.K.; Misra, M. Spent coffee grounds as a versatile source of green energy. *J. Agric. Food Chem.* **2008**, *56*, 11757–11760.
9. Caetano, N.S.; Silva, V.F.M.; Mata, T.M. Valorization of coffee grounds for biodiesel production. *Chem. Eng. Trans.* **2012**, *26*, 267–272.
10. Khan, N.A.; Brown, J.B. The composition of coffee oil and its component fatty acids. *J. Am. Oil Chem. Soc.* **1953**, *30*, 606–609.
11. Nunes, A.A.; Franca, A.S.; Oliveira, L.S. Activated carbons from waste biomass: An alternative use for biodiesel production solid residues. *Bioresour. Technol.* **2009**, *100*, 1786–1792.
12. Gui, M.M.; Lee, K.T.; Bhata, S. Feasibility of edible oil *vs.* non-edible oil *vs.* waste edible oil as biodiesel feedstock. *Energy* **2008**, *33*, 1646–1653.
13. Yanagimoto, K.; Ochi, H.; Lee, K.G.; Takayuki, S. Antioxidative activities of fractions obtained from brewed coffee. *J. Agricolture Food Chem.* **2004**, *52*, 592–596.
14. Campo, P.; Zhao, Y.; Suidan, M.T.; Venosa, A.D.; Sorial, G.A. Biodegradation kitetics and toxicity of vegetable oils triacylglycelols under aerobic conditions. *Chemosphere* **2007**, *68*, 2054–2062.
15. Schmid, R.D.; Verger, R. Lipases: Interfacial enzyme with attractive application. *Angew. Chem. Int. Ed.* **1998**, *37*, 1608–1633.
16. Mittelbach, M. Lipase-catalyzed alcoholysis of sunflower oil. *J. Am. Chem. Soc.* **1990**, *67*, 168–170.
17. Nielsen, P.M.; Brask, J.; Fjerbaek, L. Enzymatic biodiesel production: Technical and economical considerations. *Eur. J. Lipid Sci. Technol.* **2008**, *110*, 692–700.

18. Du, W.; Xu, Y.Y.; Liu, H.; Li, Z.B. Study on acyl migration in immobilized lipozyme TL-catalyzed transesterification of soybean oil for biodiesel production. *J. Mol. Catal. B Enzym.* **2005**, *37*, 68–71.
19. Du, W.; Xu, Y.; Liu, D.; Zeng, J. Comparative study on lipase-catalyzed transformation of soybean oil for biodiesel production with different acyl acceptors. *J. Mol. Catal. B Enzym.* **2004**, *30*, 125–129.
20. Christensen, M.W.; Andersen, L.; Husum, T.L.; Kirk, O. Industrial lipase immobilization. *Eur. J. Lipid Sci. Technol.* **2003**, *105*, 318–321.
21. Ferrario, V.; Ebert, C.; Knapic, L.; Fattor, D.; Basso, A.; Spizzo, P.; Gardossi, L. Conformational changes of lipases in aqueous media: A comparative computational study and experimental implications. *Adv. Synth. Catal.* **2011**, *353*, 2466–2480.
22. Skjot, M.; de Maria, L.; Chatterjee, R.; Svendsen, A.; Patkar, S.A.; Ostergraad, P.R.; Brask, J. Understanding the plasticity of the alpha/beta hydrolase fold: Lid swapping on the Candida antarctica lipase B results in chimeras with interesting biocatalytic properties. *ChemBioChem* **2009**, *10*, 520–527.
23. Goodford, P.J. A computational procedure for determining energetically favorable binding sites on biologically important macromolecules. *J. Med. Chem.* **1985**, *28*, 849–857
24. Xu, Y.; Du, W.; Liu, D. Study on the kinetics of enzymatic interesterification of triglycerides for biodiesel production with methyl acetate as the acyl acceptor. *J. Mol. Catal. B Enzym.* **2005**, *32*, 241–245.
25. Watanabe, Y.; Shimada, Y.; Sugihara, A.; Tominaga, Y. Conversion of degummed soybean oil to biodiesel fuel with immobilized *Candida antartica* lipase. *J. Mol. Catal. B Enzym.* **2002**, *17*, 151–155.
26. Anderson, E.; Larsson, K.; Kirk, O. One biocatalyst—many applications: The use of Candida antarctica. *Biocatal. Biotransform.* **1998**, *16*, 181–204.
27. Salis, A.; Pinna, M.; Monduzzi, M.; Solinas, V. Biodiesel production from triolein and short chain alcohols through biocatalysis. *J. Biotechnol.* **2005**, *119*, 291–299.
28. Marrink, S.J.; Tieleman, D.P. Perspective on the Martini model. *Chem. Soc. Rev.* **2013**, doi:10.1039/C3CS60093A.
29. Friedrich, T.; Stuermer, R. Production of immobilized lipase from Pseudomonas and application for enantioselective reactions. U.S. Patent 6 596 520, 2003.
30. Gardossi, L. Immobilization of Enzymes and Control of Water Activity in Low-Water Media: Properties and Applications of Celite R-640 (Celite Rods). In *Methods in Biotechnology: Enzyme in Non-Aqueous Solvents: Methods and Protocols*; Vulfson, E.N., Halling, P.J., Holland, H., Eds.; Humana Press, Inc.: Totowa, NJ, USA, 2001; pp. 151–172.
31. Basso, A.; de Martin, L.; Ebert, C.; Gardossi, L.; Linda, P. High isolated yields in thermodynamically controlled peptide synthesis in toluene catalysed by thermolysin adsorbed on Celite R-640. *Chem. Commun.* **2000**, *6*, 467–468.
32. Basso, A.; Braiuca, P.; Cantone, S.; Ebert, C.; Linda, P.; Spizzo, P.; Caimi, P.; Hanefeld, U.; Degrassi, G.; Gardossi, L. In silico analysis of enzyme surface and glycosylation effect as a tool for efficient covalent immobilization of CalB and PGA on Sepabeads®. *Adv. Synth. Catal.* **2007**, *349*, 877–886.

33. Tanaka, A.; Sugimoto, H.; Muta, Y.; Mizuno, T.; Senoo, K.; Obata, H.; Inouye, K. Differential scanning calorimetry of the effects of Ca2+ on the thermal unfolding of Pseudomonas cepacia lipase. *Biosci. Biotechnol. Biochem.* **2003**, *67*, 207–210.
34. Pirozzi, D. Improvement of lipase stability in the presence of commercial triglycerides. *Eur. J. Lipid Sci. Technol.* **2003**, *105*, 608–613.
35. Ulijn, R.V.; de Martin, L.; Halling, P.J.; Janssen, A.E.M.; Gardossi, L.; Moore, B.D. Solvent selection for solid-to-solid synthesis. *Biotechnol. Bioeng.* **2002**, *80*, 509–515.
36. Kaeida, M.; Samukawa, T.; Kondo, A.; Fukuda, H. Effect of methanol and water contents on production of biodiesel fuel from plant oil catalyzed by various lipases in a solvent-free system. *J. Biosci. Bioeng.* **2001**, *91*, 12–15.
37. Ferrari, M.; Ravera, F.; De Angelis, E.; Suggi Liverani, F.; Navarini, L. Interfacial properties of coffee oils. *Colloids Surf. A Physicochem. Eng. Aspects* **2010**, *365*, 79–82.
38. Holcapek, M.; Jandera, P.; Fischer, J.; Prokes, B. Analytical monitoring of the production of biodiesel by high-performance liquid chromatography with various detection methods. *J. Chromatogr. A* **1999**, *858*, 13–31.
39. D'Amelio, N.; de Angelis, E.; Navarini, L.; Schievano, E.; Mammi, S. Green coffee oil analysis by high-resolution nuclear magnetic resonance spectroscopy. *Talanta* **2013**, *110*, 118–127.
40. Available online: http://md.chem.rug.nl/cgmartini/index.php/home (accessed on 12/05/2010).
41. Wolfgang, K.; Sander, C. Dictionary of protein secondary structure: Pattern recognition of hydrogen-bonded and geometrical features. *Biopolymers* **1983**, *22*, 2577–2637.

12

Carbonic Anhydrases and their Biotechnological Applications

Christopher D. Boone, Andrew Habibzadegan, Sonika Gill and Robert McKenna *

Biochemistry & Molecular Biology, University of Florida, P.O. Box 100245, Gainesville, FL 32610, USA; E-Mails: cdboone@ufl.edu (C.D.B.); ahabibzadegan@ufl.edu (A.H.); sonika22@ufl.edu (S.G.)

* Author to whom correspondence should be addressed; E-Mail: rmckenna@ufl.edu

---

Abstract: The carbonic anhydrases (CAs) are mostly zinc-containing metalloenzymes which catalyze the reversible hydration/dehydration of carbon dioxide/bicarbonate. The CAs have been extensively studied because of their broad physiological importance in all kingdoms of life and clinical relevance as drug targets. In particular, human CA isoform II (HCA II) has a catalytic efficiency of 10^8 M^{-1} s^{-1}, approaching the diffusion limit. The high catalytic rate, relatively simple procedure of expression and purification, relative stability and extensive biophysical studies of HCA II has made it an exciting candidate to be incorporated into various biomedical applications such as artificial lungs, biosensors and CO_2 sequestration systems, among others. This review highlights the current state of these applications, lists their advantages and limitations, and discusses their future development.

Keywords: carbonic anhydrase; artificial lungs; biosensor; carbon sequestration

---

1. Introduction

Three analogous families of carbonic anhydrases (CA) exist within nature: α-CAs (predominant within animals), β-CAs (predominant within plants), and the γ-CAs (predominant within Archaea) [1–5]. In total, there are 15 human α-CA isoforms, all of which differ in their catalytic rates, inhibitor sensitivity and selectivity, cellular localization and tissue distribution [1,6,7]. The 12 catalytically active human isoforms (HCAI–VA, VB–VII, IX, XII–XIV) exhibit a wide range of catalytic efficiencies ($k_{cat}/K_M = 10^3–10^8$ M^{-1} s^{-1}). The acatalytic human CA-related proteins (HCA-RPs VIII, X and XI) are inactive due to the evolutionarily loss of one or more of the zinc-coordinating histidine residues, leading to loss of the zinc metal from the active site [1,8].

CAs are involved in various physiological roles fluid secretion, acid/base balance and thus pH regulation, gluconeogenesis, ureagenesis, gastric acid production, and transport of CO_2 from tissues to the lungs (in the form of bicarbonate) through blood [4,9,10]. CO_2 released as a part of respiration by tissues is not very soluble in blood and thus, in order to be transported, is converted to HCO_3^- by HCA II. Furthermore, the role of CA in diseases such as glaucoma has long been known. Over secretion of aqueous humor in the eye causes increased intra-ocular pressures consequently leading to a condition called glaucoma. Reduction in CA activity decreases the secretion of HCO_3^- and aqueous humor, thereby reducing the pressure [1,7,11,12].

The best characterized of these enzymes is HCA II, found within the cytosol of many cells and organs [9,13–15]. Known to possess a remarkably high catalytic efficiency, with a k_{cat} of $1.4 \times 10^6\ s^{-1}$ and a k_{cat}/K_M of $1.5 \times 10^8\ M^{-1}s^{-1}$ [6,16,17], HCA II aids in the conversion of water and carbon dioxide into bicarbonate and a proton through a two-step ping pong mechanism:

$$E{:}Zn\text{-}OH^- + CO_2 \overset{H_2O}{\rightleftarrows} E{:}Zn\text{-}H_2O + HCO_3^- \qquad \text{(reaction 1)}$$

$$E{:}Zn\text{-}H_2O + B \rightleftarrows E{:}ZnOH^- + BH^+ \qquad \text{(reaction 2)}$$

In the hydration direction shown, in the first step the zinc-bound hydroxide acts as a nucleophile, attacking the carbon dioxide and ultimately forming bicarbonate. This leads to a water molecule bound to the zinc (reaction 1). The second step (reaction 2) regenerates the zinc-bound hydroxide through a proton transfer mechanism via His64 in HCA II [18] to solvent, B [19–22].

This review will discuss the current state of utilizing HCA II in the biomedical field to aid in the development of an artificial lung system, as a biosensor for trace elements in complex media and in CO_2 sequestration among confined spaces, followed by a short discussion on other systems. HCA II is a particularly attractive candidate for these applications because of its relatively high stability [23], ability to be expressed in large quantities from *E. coli* [24], and the relatively easy purification from either affinity or conventional chromatography [25,26]. Additionally, the available high-resolution X-ray and medium neutron crystallographic structures of HCA II [27–30] allow for the rational engineering via site-directed mutagenesis for enhanced catalytic activity [31,32] and stability [33–35] for various industrial [36,37] and medical [1,38–40] applications.

2. Artificial Lungs

One of the current prevailing health problems within the United States is respiratory failure [41]. A common treatment for this condition is the utilization of mechanical ventilators [42]. Unfortunately, these ventilators can create many problems for patients who are treated with them, including decreased lung efficacy because of over-pressurized or over-distended lung tissue [43]. An artificial lung is a device capable of assisting with respiration without input from the lungs. This technology could supersede ventilators in treating respiratory failure. There are still many challenges facing this new technology, however, that must be dealt with before artificial lungs replace mechanical ventilators [44].

As of now, the main issue preventing effective artificial lungs concerns inadequate transfer of CO_2 per square inch across the polymetric hollow fiber membranes (HFM) present within this technology. Currently, a 1–2 m^2 surface area is required to sufficiently transfer CO_2 through the membrane [45–48]. A surface area of this size lacks the practicalities of functioning effectively within the human body [49,50]. One way that has been effective at increasing the transfer of CO_2 lies in immobilizing CA onto the HFM. The enzyme, dissolved in a phosphate buffer, is added to the surface of the HFM. Cyanogen bromide within acetonitrile activates the HFM, allowing covalent bonds to form between CA and the HFM. Transfer rates of CO_2 measured with CA treated compared to untreated HFM show a 75% higher rate of CO_2 removal rate present in with the treated HFM. This finding indicates the possibilities for smaller artificial lungs to be engineered with CA incorporation, which could function effectively within the human body [51].

Another method that has been shown to also increase CO_2 transfer across the HFM of artificial lungs relies on impeller devices that increase the rate of blood mixing. Unfortunately, this method cannot be combined with the CA method at this time. When these two methods were combined, the shear forces of the impeller device denatured CA, leading to a loss of enzyme function. This creates the need for a more stable form of CA that will not be denatured by the shear forces. If such a stable CA variant can be engineered, these methods might be combined, which could lead to smaller more efficient artificial lungs [52].

3. Biosensors

Quantification of trace analytes in complex media containing chemically similar molecules is lacking in many traditional chemical systems. As such, the development of sensors based on biological molecules, termed biosensors, can achieve such specificity and sensitivity [53]. The high affinity of HCA II for zinc (4 pM) [54] has been used to quantify trace amounts of zinc in sea and waste water [55] for concerns over toxicity to certain plants, invertebrates and fish [56,57]. Optimally, this biosensor would operate along the sea bed and relay a fluorescence signal up to the ocean surface that is released upon binding of a strong inhibitor, dansylamide, upon binding of zinc in the active site of apo-CA [58]. However, the slow dissociation rate of zinc from the CA active site ($t_{1/2} \approx 90$ days [54]) limits the reusability and efficiency of the system. The relative abundance of natural zinc in the environment compared to that of the binding affinity also limits the production of apo-HCA II [23]. As such, an HCA II variant (E117Q) that contained both a lowered binding affinity (nM) and a much faster dissociation time ($t_{1/2} \approx 3$ sec) for zinc was developed to circumvent these limitations [59]. Other studies have aimed to improve the fluorescence signaling upon zinc binding in the active site of HCA II via incorpororation of the H36C variant, which then selectively labeled with a thiol-reactive fluorophore [60] that would interact upon ligation of an inhibitor, azosulfonamide, acting as a fluorescence acceptor [61].

Other metals that bind to HCA II consist primarily of transition metals in the +2 oxidation state, which include: Cd^{2+}, Co^{2+}, Cu^{2+}, Hg^{2+}, Fe^{2+}, Mn^{2+}, Ni^{2+}, Pb^{2+} and In^{3+} [53,54,60,62]. However, only the binding affinity of HCA II for Cu^{2+} and Hg^{2+} is greater than that of Zn^{2+} [54], so variants with a lowered binding affinity for these metals would have to be developed before detection of other metals is feasible. Sulfonamide inhibitors, however, do not bind tightly to CA with metals other than Zn^{2+} or Co^{2+} bound in the active site [63], promoting the need for a novel development of metal ligation. This limitation can been superseded because several of these divalent ions (Cu^{2+}, Co^{2+} and Ni^{2+}) exhibit weak d-d absorbance bands in the visible regions that can be directly measured via fluorescence energy transfer lifetimes [60]. This can be extended to the other metals that do not exhibit d-d absorbance bands, (Hg^{2+} and Cd^{2+}) since the binding of these metals to apo-CA causes a quenching of the fluorescence of an active site fluorophore [64]. Since biologically prevalent divalent metals such as Mg^{2+} and Ca^{2+} do not bind to CA and interfere with the assay, biosensors employed in biomedical applications are especially useful [65–70].

4. CO_2 Sequestration

Elevated CO_2 levels in the human body have detrimental effects ranging from impaired judgment to death. CO_2 control is important in confined spaces where there is little buffering ability to absorb this gas, such as spacecraft or submarines. These life support systems employ a small amount of CA dissolved in thin aqueous buffered films and compressed between porous polypropylene membranes [71]. The concentrations of CO_2 commonly experienced in these systems (~0.1% v/v) are ideal for selective capture by CA. Analysis of the produced (scrubbed) gas shows that the CA-containing setup selectively lets N_2 and O_2 through, with ratios of 1400:1 and 900:1, respectively [71,72]. The relatively low concentration of CO_2 readily dissolves in the thin layer of enzyme containing buffer and across the membrane, where it is removed via vacuum or carrier gas. Engineered CA-based bioreactors outperformed chemical methods using diethylamine solutions, with much higher selectivity, 400:1 and 300:1 for N_2 and O_2, respectively [72]. In addition, the presence of CA increased CO_2 transport across the polypropylene membrane by ~70% [71]. Another benefit of these CA-bioreactor systems is that they are very efficient at ambient pressures and temperatures [72], improving overall cost-efficiency. However, the longevity of the systems has raised concerns as the need to keep the membranes wet, or at least humid, will add cost and operational difficulties to their practical use.

5. Pharmalogical Considerations

CAs have been employed in CO_2-responsive cationic hydrogels in antidote delivery to treat analgesic overdose without losing therapeutic levels of drug [73]. Alternate medicines, such as opioids, have very potent analgesic effects but overdoses can cause respiratory hypoventilation, which leads to increased CO_2 and decreased O_2 levels in the body ultimately leading to an acidosis-induced death. The CA treatment involves a feedback-regulated antidote delivery system that responds to high CO_2 levels or decrease in pH [73]. The cationic hydrogel is based on N,N-dimethylaminoethyl methacrylate (DMAEMA) polymers that have modified to have a $pK_a \sim 7.5$, which makes it an adequate blood pH monitor. Incorporation of CA as a CO_2 sensor improved the efficiency of these antidote-delivery systems [73]. Other hydrogels have been designed with a switchable co-block polymer can undergo a transition change from gel to sol upon exposure to CO_2 [74]. This study demonstrates that stimuli-triggered drug delivery could be incorporated with CAs utilizing CO_2, bicarbonate or pH changes as signaling molecules.

High-resolution X-ray [75,76] and neutron structures [29] of HCA II in complex with acetazolamide (Diamox), a tight binding inhibitor used in the treatment of glaucoma [12,77], has accelerated research into structural-based rational design of an isozyme specific inhibitor. Current research interest includes specific inhibition of HCA IX, which has been shown to be overexpressed in a wide array of cancer cell lines [11,39,78–80]. In short, tumor cells proliferate in acidic environments which could be presumably due to the catalytic activity of HCA IX on the cell surface. Selective inhibition of HCA IX could provide a means for targeted tumor eradication. A suitable drug candidate has not been discovered as of yet, but an impressive library of inhibitors designed from different functional groups with varying binding affinities and specificities for various CA isoforms has been steadily growing with entries and has been summarized elsewhere [2,10,38,40,81,82].

6. Blood Substitutes

A continual source of blood is required for use in trauma injuries or major surgeries, and, as natural blood is often in limited supply, there has been progress in the development of blood substitutes which primarily consist of 4–5 cross-linked stroma-free hemoglobin (polySFHb) molecules [83]. The major drawback of these substitutes, however, was the inadequate CO_2 removal rates. Increased CO_2 levels in the body leads to acidosis, and if left untreated will end up in coma and death [9]. Incorporation of catalase (CAT), superoxide dismutase (SOD) and CA to the PolySFHb substitute (PolySFHb-SOD-CAT-CA) was introduced to overcome this limitation with encouraging activity [84]. Blood substitutes have also been shown to be advantageous over transfused whole blood in that they can be sterilized, stored for long periods and contains no blood antigens [83].

7. Conclusions

The various aforementioned biotechnological aspects of the different CA-associated systems emphasize the usefulness of this enzyme. The advancement of fast and cost-effective genome sequencing, molecular biology techniques that boost overexpression of protein and direct evolution techniques that can select for highly active and stable CAs provide an optimistic view as to the advancement in the efficiency and selectivity in current systems. It is also likely that further developments in these fields will lead to novel biomedical applications of CAs.

References

1. Aggarwal, M.; Boone, C.D.; Kondeti, B.; McKenna, R. Structural annotation of human carbonic anhydrases. *J. Enzyme Inhib. Med. Chem.* **2013**, *28*, 267–277.
2. Krishnamurthy, V.M.; Kaufman, G.K.; Urbach, A.R.; Gitlin, I.; Gudiksen, K.L.; Weibel, D.B.; Whitesides, G.M. Carbonic anhydrase as a model for biophysical and physical-organic studies of proteins and protein-ligand binding. *Chem. Rev.* **2008**, *108*, 946–1051.
3. Rowlett, R.S. Structure and catalytic mechanism of the beta-carbonic anhydrases. *Biochim. et Biophy. Acta* **2010**, *1804*, 362–373.
4. Supuran, C.T. Carbonic anhydrases—an overview. *Curr. Pharm. Des.* **2008**, *14*, 603–614.
5. Hewett-Emmett, D.; Tashian, R.E. Functional diversity, conservation, and convergence in the

evolution of the alpha-, beta-, and gamma-carbonic anhydrase gene families. *Mol. Phylogenetics Evolut.* **1996**, *5*, 50–77.

6. Lindskog, S. Structure and mechanism of carbonic anhydrase. *Pharmacol. Ther.* **1997**, *74*, 1–20.
7. Alterio, V.; Di Fiore, A.; D'Ambrosio, K.; Supuran, C.T.; de Simone, G. Multiple binding modes of inhibitors to carbonic anhydrases: How to design specific drugs targeting 15 different isoforms? *Chem. Rev.* **2012**, *112*, 4421–4468.
8. Bergenhem, N.C.; Hallberg, M.; Wisén, S. Molecular characterization of the human carbonic anhydrase-related protein (hca-rp viii). *Biochim. et Biophy. Acta* **1998**, *1384*, 294–298.
9. Sly, W.S.; Hu, P.Y. Human carbonic anhydrases and carbonic anhydrase deficiencies. *Annu. Rev. Biochem.* **1995**, *64*, 375–401.
10. Supuran, C.T.; Scozzafava, A. Carbonic anhydrases as targets for medicinal chemistry. *Bioorganic Med. Chem.* **2007**, *15*, 4336–4350.
11. Pastorekova, S.; Parkkila, S.; Pastorek, J.; Supuran, C.T. Carbonic anhydrases: Current state of the art, therapeutic applications and future prospects. *J. Enzyme Inhib. Med. Chem.* **2004**, *19*, 199–229.
12. Aggarwal, M.; McKenna, R. Update on carbonic anhydrase inhibitors: A patent review (2008–2011). *Expert Opin. Ther. Pat.* **2012**, *22*, 903–915.
13. Chegwidden, W.R.; Carter, N.D. Introduction to the Carbonic Anhydrases. In *The Carbonic Anhdyrases: New horizons*; Chegwidden, W.R., Carter, N.D., Edwards, Y.H., Eds.; Birkhäuser Verlag: Boston, MA, USA, 2000; pp. 13–29.
14. Christianson, D.W.; Fierke, C.A. Carbonic anhydrase: Evolution of the zinc binding site by nature and design. *Acc. Chem. Res.* **1996**, *29*, 331–339.
15. Duda, D.; McKenna, R. Carbonic Anhydrase, α-class. In *Handbook of Metalloproteins*; Messerschmidt, A., Ed.; John Wiley & Sons: New York, NY, USA, 2004; pp. 249–263.
16. Lindskog, S.; Coleman, J.E. Catalytic mechanism of carbonic-anhydrase. *Proc. Natl. Acad. Sci. USA* **1973**, *70*, 2505–2508.
17. Lindskog, S.; Silverman, D.N. The Catalytic Mechanism of Mammalian Carbonic Anhydrases. In *The Carbonic Anhdyrases: New Horizons*; Chegwidden, W.R., Carter, N.D., Edwards, Y.H., Eds.; Birkhäuser Verlag: Boston, MA, USA, 2000; pp. 175–195.
18. Tu, C.K.; Silverman, D.N.; Forsman, C.; Jonsson, B.H.; Lindskog, S. Role of histidine 64 in the catalytic mechanism of human carbonic anhydrase ii studied with a site-specific mutant. *Biochemistry* **1989**, *28*, 7913–7918.
19. Mikulski, R.L.; Silverman, D.N. Proton transfer in catalysis and the role of proton shuttles in carbonic anhydrase. *Biochim. et Biophy. Acta* **2010**, *1804*, 422–426.
20. Silverman, D.N. Carbonic anhydrase: Oxygen-18 exchange catalyzed by an enzyme with rate-contributing proton-transfer steps. *Methods Enzymol.* **1982**, *87*, 732–752.
21. Silverman, D.N.; Lindskog, S. The catalytic mechanism of carbonic anhydrase: Implications of a rate-limiting protolysis of water. *Acc. Chem. Res.***1988**, *21*, 30–36.
22. Silverman, D.N.; McKenna, R. Solvent-mediated proton transfer in catalysis by carbonic anhydrase. *Acc. Chem. Res.* **2007**, *40*, 669–675.
23. Avvaru, B.S.; Busby, S.A.; Chalmers, M.J.; Griffin, P.R.; Venkatakrishnan, B.; Agbandje-McKenna, M.; Silverman, D.N.; McKenna, R. Apo-human carbonic anhdrase ii revisited: Implications of the loss of a metal in protein structure, stability, and solvent network. *Biochemistry* **2009**, *48*, 7365–7372.

24. Murakami, H.; Marelich, G.P.; Grubb, J.H.; Kyle, J.W.; Sly, W.S. Cloning, expression, and sequence homologies of cdna for human carbonic anhydrase ii. *Genomics* **1987**, *1*, 159–166.
25. Krebs, J.F.; Fierke, C.A. Determinants of catalytic activity and stability of carbonic anhydrase ii as revealed by random mutagenesis. *J. Biol. Chem.* **1993**, *268*, 27458–27466.
26. Osborne, W.R.; Tashian, R.E. An improved method for the purification of carbonic anhydrase isozymes by affinity chromatography. *Anal. Biochem.* **1975**, *64*, 297–303.
27. Avvaru, B.S.; Kim, C.U.; Sippel, K.H.; Gruner, S.M.; Agbandje-McKenna, M.; Silverman, D.N.; McKenna, R. A short, strong hydrogen bond in the active site of human carbonic anhydrase ii. *Biochemistry* **2010**, *49*, 249–251.
28. Eriksson, A.E.; Jones, T.A.; Liljas, A. Refined structure of human carbonic anhydrase ii at 2.0 a resolution. *Proteins* **1988**, *4*, 274–282.
29. Fisher, S.Z.; Aggarwal, M.; Kovalesky, A.; Silverman, D.N.; McKenna, R. Neutron-diffraction of acetazolamide-bound human carbonic anhydrase ii reveals atomic details of drug binding. *J. Am. Chem. Soc.* **2012**, *134*, 14726–14729.
30. Fisher, S.Z.; Kovalevsky, A.Y.; Domsic, J.F.; Mustyakimov, M.; McKenna, R.; Silverman, D.N.; Langan, P.A. Neutron structure of human carbonic anhydrase ii: Implications for proton transfer. *Biochemistry* **2010**, *49*, 415–421.
31. Fisher, S.Z.; Tu, C.; Bhatt, D.; Govindasamy, L.; Agbandje-McKenna, M.; McKenna, R.; Silverman, D.N. Speeding up proton transfer in a fast enzyme: Kinetic and crystallographic studies on the effect of hydrophobic amino acid substitutions in the active site of human carbonic anhydrase ii. *Biochemistry* **2007**, *46*, 3803–3813.
32. Mikulski, R.; West, D.; Sippel, K.H.; Avvaru, B.S.; Aggarwal, M.; Tu, C.; McKenna, R.; Silverman, D.N. Water networks in fast proton transfer during catalysis by human carbonic anhydrase ii. *Biochemistry* **2013**, *52*, 125–131.
33. Fisher, Z.; Boone, C.D.; Biswas, S.M.; Venkatakrishnan, B.; Aggarwal, M.; Tu, C.; Agbandje-McKenna, M.; Silverman, D.; McKenna, R. Kinetic and structural characterization of thermostabilized mutants of human carbonic anhydrase ii. *Protein Eng. Des. Sel. PEDS* **2012**, *25*, 347–355.
34. Mårtensson, L.-G.; Karlsson, M.; Carlsson, U. Dramatic stabilization of the native state of human carbonic anhydrase ii by an engineered disulfide bond. *Biochemistry* **2002**, *41*, 15867–15875.
35. Boone, C.D.; Habibzadegan, A.; Tu, C.; Silverman, D.N.; McKenna, R. Structural and catalytic characterization of a thermally stable and acid-stable variant of human carbonic anhydrase ii containing an engineered disulfide bond. *Acta Crystallogr. Sect. D Biol. Crystallogr.* **2013**, *69*, 1414–1422.
36. Boone, C.D.; Gill, S.; Habibzadegan, A.; McKenna, R. Carbonic anhydrases and their industrial applications. *Curr. Top. Biochem. Res.* **2013**, *14*, 1–10.
37. Christianson, D.W.; Fierke, C.A. Carbonic anhydrase: Evolution of the zinc binding site by nature and by design. *Acc. Chem. Res.* **1996**, *29*, 331–339.
38. Supuran, C.T. Carbonic anhydrase inhibitors. *Bioorganic Med. Chem. Lett.* **2010**, *20*, 3467–3474.
39. Supuran, C.T. Inhibition of carbonic anhydrase ix as a novel anticancer mechanism. *World J. Clin. Oncol.* **2012**, *3*, 98–103.

40. Supuran, C.T. Carbonic anhydrases: Novel therapeutic applications for inhibitors and activators. *Nat. Rev. Drug Discov.* **2008**, *7*, 168–181.
41. Ware, L.B.; Matthay, M.A. The acute respiratory distress syndrome. *N. Engl. J. Med.* **2000**, *342*, 1334–1349.
42. Esteban, A.; Anzueto, A.; Frutos, F.; Alia, I.; Brochard, L.; Stewart, T.E.; Benito, S.; Epstein, S.K.; Apezteguia, C.; Nightingale, P.; *et al.* Characteristics and outcomes in adult patients receiving mechanical ventilation: A 28-day international study. *JAMA* **2002**, *287*, 345–355.
43. Maggiore, S.M.; Richard, J.C.; Brochard, L. What has been learnt from p/v curves in patients with acute lung injury/acute respiratory distress syndrome. *Eur. Respir. J.* **2003**, *22*, 22s–26s.
44. Haft, J.W.; Griffith, B.P.; Hirschl, R.B.; Bartlett, R.H. Results of an artificial-lung survey to lung transplant program directors. *J. Heart Lung Transplant.* **2002**, *21*, 467–473.
45. Hattler, B.G.; Federspiel, W.J. Gas Exchange in the Venous System: Support for the Failing Lung. In *The Artificial Lung*; Vaslef, S.N., Anderson, R.W., Eds.; Landes Bioscience: Georgetown, DC, USA, 2002; pp. 133–174.
46. Wegner, J.A. Oxygenator anatomy and function. *J. Cardiothorac. Vasc. Anesth.* **1997**, *11*, 275–281.
47. Federspiel, W.J.; Henchir, K.A. Artificial Lungs: Basic Principles and Current Applications. In *Encyclopedia of Biomaterials and Biomedical Engineering*; Wnek, G.E., Bowlin, G.L., Eds.; Marcel Dekker, Inc: New York, NY, USA, 2004; pp. 922–931.
48. Beckley, P.D.; Holt, D.W.; Tallman, R.D. Oxygenators for Extracorporeal Circulation. In *Cardiopulmonary Bypass: Principles and Techniques of Extracorporeal Circulation*; Mora, C.T., Ed.; Springer-Verlag: New York, NY, USA, 1995; pp. 199–219.
49. Okamoto, T.; Tashiro, M.; Sakanashi, Y.; Tanimoto, H.; Imaizumi, T.; Sugita, M.; Terasaki, H. A new heparin-bonded dense membrane lung combined with minimal systemic heparinization prolonged extracorporeal lung assist in goats. *Artif. Organs* **1998**, *22*, 864–872.
50. Watnabe, H.; Hayashi, J.; Ohzeki, H.; Moro, H.; Sugawara, M.; Eguchi, S. Biocompatibility of a silicone-coated polypropylene hollow fiber oxygenator in an in vitro model. *Ann. Thorac. Surg.* **1999**, *67*, 1315–1319.
51. Kaar, J.L.; Oh, H.-I.; Russell, A.J.; Federspiel, W.J. Towards improved artificial lungs through biocatalysis. *Biomaterials* **2007**, *28*, 3131–3139.
52. Arazawa, D.T.; Oh, H.-I.; Ye, S.-H.; Johnson, C.A., Jr.; Woolley, J.R.; Wagner, W.R.; Federspiel, W.J., Immobilized carbonic anhydrase on hollow fiber membranes accelerates CO_2 removal from blood. *J. Membr. Sci.* **2012**, *404*, 25–31.
53. Hunt, J.A.; Lesburg, C.A.; Christianson, D.W.; Thompson, R.B.; Fierke, C.A. Active-site Engineering of Carbonic Anhydrase and Its Applications to Biosensors. In *The Carbonic Anhydrases: New horizons*; Chegwidden, W.R., Carter, N.D., Edwards, Y.H., Eds.; Birkhäuser Verlag: Boston, MA, USA, 2000; pp. 221–240.
54. Lindskog, S.; Nyman, P.O. Metal-binding properties of human erythrocyte carbonic anhydrase. *Biochem. Biophys. Acta* **1964**, *85*, 462–474.
55. Thompson, R.B.; Jones, E.R. Enzyme-based fiber optic zinc biosensor. *Anal. Chem.* **1993**, *65*, 730–734.
56. Rout, G.R.; Das, P. Effect of metal toxicity on plant growth and metabolism: I. Zinc. *Agronomie* **2003**, *23*, 3–11.

57. Muyssen, B.T.; de Schamphelaere, K.A.; Janssen, C.R. Mechanisms of chronic waterborne zn toxicity in daphnia magna. *Aquat. Toxicol.* **2006**, *77*, 393–401.
58. Chen, R.F.; Kernohan, J.C. Combination of bovine carbonic anhydrase with a fluorescence sulfonamide. *J. Biol. Chem.* **1967**, *242*, 5813–5823.
59. Huang, C.-C.; Lesburg, C.A.; Kiefer, L.L.; Fierke, C.A.; Christianson, D.W. Reversal of the hydrogen bond to zinc ligand histidine-119 dramatically diminishes catalysis and enhances metal equilibriation kinetics in carbonic anhydrase. *Biochemistry* **1996**, *35*, 3439–3446.
60. Thompson, R.B.; Ge, Z.; Patchan, M.W.; Kiefer, L.L.; Fierke, C.A. Performance enhancement of fluorescence energy transfer-based biosensors by site-directed mutagenesis of the transducer. *SPIE* **1996**, *2508*, 136–144.
61. Thompson, R.B.; Patchan, M.W. Lifetime-based fluorescence energy transfer biosensing of zinc. *Anal. Biochem.* **1995**, *227*, 123–128.
62. Demille, G.R.; Larlee, K.; Livesey, D.L.; Mailer, K. Conformational change in carbonic anhydrase studied by perturbed directional correlations of gamma rays. *Chem. Phys. Lett.* **1979**, *64*, 534–539.
63. Harrington, P.C.; Wilkins, R.G. Interaction of acetazolamide and 4-nitrothiophenolate ion with bivalent metal ion derivatives of bovine carbonic anhydrase. *Biochemistry* **1977**, *16*, 448–454.
64. Thompson, R.B.; Gė, Z.; Patchan, M.W.; Huang, C.-C.; Fierke, C.A. Fiber optic biosensor for co(ii) and cu(ii) based on fluorescence energy transfer with an enzyme transducer. *Biosens. Bioelectron.* **1996**, *11*, 557–564.
65. Frederickson, C.J.; Giblin, L.J.; Krezel, A.; McAdoo, D.J.; Mueller, R.N.; Zeng, Y.; Balaji, R.V.; Masalha, R.; Thompson, R.B.; Fierke, C.A.; *et al.* Concentrations of extracellular free zinc (pzn)e in the central nervous system during simple anesthetization, ischemia and reperfusion. *Exp. Neurol.* **2006**, *198*, 285–293.
66. Thompson, R.B.; Peterson, D.; Mahoney, W.; Cramer, M.; Maliwal, B.P.; Suh, S.W.; Frederickson, C.; Fierke, C.; Herman, P. Fluorescent zinc indicators for neurobiology. *J. Neurosci. Methods* **2002**, *118*, 63–75.
67. Thompson, R.B.; Whetsell, W.O., Jr.; Maliwal, B.P.; Fierke, C.A.; Frederickson, C.J. Fluorescence microscopy of stimulated zn(ii) release from organotypic cultures of mammalian hippocampus using a carbonic anhydrase-based biosensor system. *J. Neurosci. Methods* **2000**, *96*, 35–45.
68. Bozym, R.; Hurst, T.K.; Westerberg, N.; Stoddard, A.; Fierke, C.A.; Frederickson, C.J.; Thompson, R.B. Chapter 14 determination of zinc using carbonic anhydrase-based fluorescence biosensors. *Methods Enzymol.* **2008**, *450*, 287–309.
69. Wang, D.; Hurst, T.K.; Thompson, R.B.; Fierke, C.A. Genetically encoded ratiometric biosensors to measure intracellular exchangeable zinc in escherichia coli. *J. Biomed. Opt.* **2011**, *16*, doi:10.1117/1.3613926.
70. McCranor, B.J.; Bozym, R.A.; Vitolo, M.I.; Fierke, C.A.; Bambrick, L.; Polster, B.M.; Fiskum, G.; Thompson, R.B. Quantitative imaging of mitochondrial and cytosolic free zinc levels in an in vitro model of ischemia/reperfusion. *J. Bioenerg. Biomembr.* **2012**, *44*, 253–263.
71. Simsek-Ege, F.A.; Bond, G.M.; Stringer, J. Matrix molecular weight cut-off for encapsulation of carbonic anhydrase in polyelectrolyte beads. *J. Biomater. Sci. Polym. Ed.* **2002**, *13*, 1175–1187.
72. Cowan, R.M.; Ge, J.; Qin, Y.J.; McGregor, M.L.; Trachtenberg, M.C. Co2 capture by means of an enzyme-based reactor. *Ann. N. Y. Acad. Sci.* **2003**, *984*, 453–469.

73. Satav, S.S.; Bhat, S.; Thayumanavan, S. Feedback regulated drug delivery vehicles: Carbon dioxide responsive cationic hydrogels for antidote release. *Biomacromol* **2010**, *11*, 1735–1740.
74. Han, D.; Boissiere, O.; Kumar, S.; Tong, X.; Tremblay, L.N.; Zhao, Y. Two-way co2-switchable triblock copolymer hydrogels. *Macromol* **2012**, *45*, 7440–7445.
75. Aggarwal, M.; Boone, C.D.; Kondeti, B.; Tu, C.; Silverman, D.N.; McKenna, R. Effects of cryoprotectants on the structure and thermostability of the human carbonic anhydrase ii-acetazolamide complex. *Acta Crystallogr. Sect. D Biol. Crystallogr.* **2013**, *69*, 860–865.
76. Sippel, K.H.; Robbins, A.H.; Domsic, J.; Genis, C.; Agbandje-McKenna, M.; McKenna, R. High-resolution structure of human carbonic anhydrase ii complexed with acetazolamide reveals insights into inhibitor drug design. *Acta Crystallogr. Sect. F Struct. Biol. Cryst. Commun.* **2009**, *65*, 992–995.
77. Aggarwal, M.; Kondeti, B.; McKenna, R. Insights towards sulfonamide drug specificity in alpha-carbonic anhydrases. *Bioorganic Med. Chem.* **2013**, *21*, 1526–1533.
78. Svastová, E.; Huliková, A.; Rafajová, M.; Zat'ovicová, M.; Gibadulinová, A.; Casini, A.; Cecchi, A.; Scozzafava, A.; Supuran, C.T.; Pastorek, J.; *et al.* Hypoxia activates the capacity of tumor-associated carbonic anhydrase ix to acidify extracellular ph. *FEBS lett.* **2004**, *577*, 439–445.
79. Vullo, D.; Franchi, M.; Gallori, E.; Pastorek, J.; Scozzafava, A.; Pastorekova, S.; Supuran, C.T. Carbonic anhydrase inhibitors: Inhibition of the tumor-associated isozyme ix with aromatic and heterocyclic sulfonamides. *Bioorganic Med. Chem. Lett.* **2003**, *13*, 1005–1009.
80. Winum, J.Y.; Rami, M.; Scozzafava, A.; Montero, J.L.; Supuran, C. Carbonic anhydrase ix: A new druggable target for the design of antitumor agents. *Med. Res. Rev.* **2008**, *28*, 445–463.
81. Supuran, C.T. Carbonic anhydrase inhibitors: An editorial. *Expert Opin. Ther. Pat.* **2013**, *23*, 677–679.
82. Supuran, C.T.; Scozzafava, A.; Casini, A. Carbonic anhydrase inhibitors. *Med. Res. Rev.* **2003**, *23*, 146–189.
83. Gould, S.A.; Moore, E.E.; Hoyt, D.B.; Ness, P.M.; Norris, E.J.; Carson, J.L.; Hides, G.A.; Freeman, I.H.; DeWoskin, R.; Moss, G.S. The life-sustaining capacity of human polymerized hemoglobin when red cells might be unavailable. *J. Am. Coll. Surg.* **2002**, *195*, 445–452.
84. Bian, Y.; Rong, Z.; Chang, T.M. Polyhemoglobin-superoxide dismutase-catalase-carbonic anhydrase: A novel biotechnology-based blood substitute that transports both oxygen and carbon dioxide and also acts as an antioxidant. *Artif. Cells Blood Substit. Immobil. Biotechnol.* **2012**, *40*, 28–37.

PERMISSIONS

All chapters in this book were first published in MDPI; hereby published with permission under the Creative Commons Attribution License or equivalent. Every chapter published in this book has been scrutinized by our experts. Their significance has been extensively debated. The topics covered herein carry significant findings which will fuel the growth of the discipline. They may even be implemented as practical applications or may be referred to as a beginning point for another development.

The contributors of this book come from diverse backgrounds, making this book a truly international effort. This book will bring forth new frontiers with its revolutionizing research information and detailed analysis of the nascent developments around the world.

We would like to thank all the contributing authors for lending their expertise to make the book truly unique. They have played a crucial role in the development of this book. Without their invaluable contributions this book wouldn't have been possible. They have made vital efforts to compile up to date information on the varied aspects of this subject to make this book a valuable addition to the collection of many professionals and students.

This book was conceptualized with the vision of imparting up-to-date information and advanced data in this field. To ensure the same, a matchless editorial board was set up. Every individual on the board went through rigorous rounds of assessment to prove their worth. After which they invested a large part of their time researching and compiling the most relevant data for our readers.

The editorial board has been involved in producing this book since its inception. They have spent rigorous hours researching and exploring the diverse topics which have resulted in the successful publishing of this book. They have passed on their knowledge of decades through this book. To expedite this challenging task, the publisher supported the team at every step. A small team of assistant editors was also appointed to further simplify the editing procedure and attain best results for the readers.

Apart from the editorial board, the designing team has also invested a significant amount of their time in understanding the subject and creating the most relevant covers. They scrutinized every image to scout for the most suitable representation of the subject and create an appropriate cover for the book.

The publishing team has been an ardent support to the editorial, designing and production team. Their endless efforts to recruit the best for this project, has resulted in the accomplishment of this book. They are a veteran in the field of academics and their pool of knowledge is as vast as their experience in printing. Their expertise and guidance has proved useful at every step. Their uncompromising quality standards have made this book an exceptional effort. Their encouragement from time to time has been an inspiration for everyone.

The publisher and the editorial board hope that this book will prove to be a valuable piece of knowledge for researchers, students, practitioners and scholars across the globe.

LIST OF CONTRIBUTORS

Jun-ichi Kadokawa
Graduate School of Science and Engineering, Kagoshima University, 1-21-40 Korimoto, Kagoshima 890-0065, Japan

Qingqing Meng
School of Pharmacy, Shanghai Jiao Tong University, 800 Dongchuan Road, Shanghai 200240, China

Linquan Bai
State Key Laboratory of Microbial Metabolism, Shanghai Jiao Tong University, 800 Dongchuan Road, Shanghai 200240, China

Yaxue Zhao and Huchen Zhou
School of Pharmacy, Shanghai Jiao Tong University, 800 Dongchuan Road, Shanghai 200240, China
State Key Laboratory of Microbial Metabolism, Shanghai Jiao Tong University, 800 Dongchuan Road, Shanghai 200240, China

Grant Schauer and Sanford Leuba
Program in Molecular Biophysics and Structural Biology, Hillman Cancer Center, University of Pittsburgh, 5117 Centre Ave., Pittsburgh, PA 15213, USA
Department of Cell Biology, Hillman Cancer Center, University of Pittsburgh, 5117 Centre Ave., Pittsburgh, PA 15213, USA

Nicolas Sluis-Cremer
Department of Medicine, Division of Infectious Diseases, 3550 Terrace St., Pittsburgh, PA 15261, USA

Mette Lübeck and Peter S. Lübeck
Section for Sustainable Biotechnology, Aalborg University Copenhagen, A C Meyers Vaenge 15, 2450 Copenhagen SV, Denmark

Annette Sørensen and Birgitte K. Ahring
Bioproducts, Sciences & Engineering Laboratory, Washington State University, 2710 Crimson Way, Richland, WA 99354, USA

Antonio Trincone
Institute of Biomolecular Chemistry, National Research Council, Via Campi Flegrei, 34, Pozzuoli 80078, Naples, Italy

Wan Chi Lam, Daniel Pleissner and Carol Sze Ki Lin
School of Energy and Environment, City University of Hong Kong, Tat Chee Avenue, Kowloon, Hong Kong, China

Albert J.J. Woortman and Gert O.R. Alberda van Ekenstein
Department of Polymer Chemistry, Zernike Institute for Advanced Materials, University of Groningen, Nijenborgh 4, 9747 AG Groningen, The Netherlands

Yi Jiang and Katja Loos
Dutch Polymer Institute (DPI), 5600 AX Eindhoven, The Netherlands

Rubén de Regil and Georgina Sandoval
Unidad de Biotecnología Industrial, CIATEJ, A.C. Av. Normalistas 800, Col. Colinas de la Normal, Guadalajara, Jal, C.P. 44270, Mexico

Mohammad S. Eram and Kesen Ma
Department of Biology, University of Waterloo, 200 University Avenue West, Waterloo, Ontario N2L 3G1, Canada

Ramesh N. Patel
SLRP Associates Consultation in Biotechnology, 572 Cabot Hill Road, Bridgewater, NJ 08807, USA

Valerio Ferrario, Cynthia Ebert and Lucia Gardossi
Dipartimento di Scienze Chimiche e Farmaceutiche, Università degli Studi di Trieste, Piazzale Europa 1, Trieste 34127, Italy

Harumi Veny
Department of Chemical Engineering, Faculty of Engineering, University of Malaya, Malaysia

Elisabetta De Angelis and Luciano Navarini
illycaffè S.p.A., via Flavia 110, Trieste 34147, Italy

Christopher D. Boone, Andrew Habibzadegan, Sonika Gill and Robert McKenna
Biochemistry & Molecular Biology, University of Florida, Gainesville, FL 32610, USA

Index

P

R

S

T

U

Y

www.ingramcontent.com/pod-product-compliance
Lightning Source LLC
LaVergne TN
LVHW081013210926
848485LV00004B/850